MEASURING THE UNIVERSE

Astronomy is an observational science, renewed and even revolutionized by new developments in instrumentation. With the resulting growth of multi-wavelength investigation as an engine of discovery, it is increasingly important for astronomers to understand the underlying physical principles and operational characteristics for a broad range of instruments.

This comprehensive text is ideal for graduate students, active researchers, and instrument developers. It is a thorough review of how astronomers obtain their data, covering current approaches to astronomical measurements from radio to gamma rays. The focus is on current technology rather than the history of the field, allowing each topic to be discussed in depth. Areas covered include telescopes, detectors, photometry, spectroscopy, adaptive optics and high-contrast imaging, millimeter-wave and radio receivers, radio and optical/infrared interferometry, and X-ray and gamma-ray astronomy, all at a level that bridges the gap between the basic principles of optics and the subject's abundant specialist literature.

Color versions of figures and solutions to selected problems are available online at www.cambridge.org/9780521762298.

GEORGE H. RIEKE is Regents Professor of Astronomy and Planetary Sciences at the University of Arizona, Deputy Director of Steward Observatory, and a member of the US National Academy of Sciences. Professor Rieke is Science Lead for the Mid-Infrared Instrument (MIRI) for NASA's James Webb Space Telescope and was Principal Investigator of the Multiband Imaging Photometer for Spitzer (MIPS). He has also led construction of a broad range of groundbased instruments and has taught core graduate courses on instrumentation throughout his career.

MEASURING THE UNIVERSE

A Multiwavelength Perspective

GEORGE H. RIEKE

University of Arizona

Illustrated by Shiras Manning

CAMBRIDGE
UNIVERSITY PRESS

CAMBRIDGE
UNIVERSITY PRESS

University Printing House, Cambridge CB2 8BS, United Kingdom

One Liberty Plaza, 20th Floor, New York, NY 10006, USA

477 Williamstown Road, Port Melbourne, VIC 3207, Australia

4843/24, 2nd Floor, Ansari Road, Daryaganj, Delhi - 110002, India

79 Anson Road, #06-04/06, Singapore 079906

Cambridge University Press is part of the University of Cambridge.

It furthers the University's mission by disseminating knowledge in the pursuit of education, learning and research at the highest international levels of excellence.

www.cambridge.org
Information on this title: www.cambridge.org/9781108405232

© G. H. Rieke 2012

First published 2012
Reprinted 2015
First paperback edition 2017

A catalogue record for this publication is available from the British Library

Library of Congress Cataloging in Publication data
Rieke, G. H. (George Henry)
Measuring the universe : a multiwavelength perspective / George H. Rieke,
University of Arizona.
pages cm
Includes bibliographical references and index.
ISBN 978-0-521-76229-8
1. Radio astronomy.   2. Infrared astronomy.   3. Gamma ray astronomy.
4. X-ray astronomy.   I. Title.
QB476.5.R54 2012
522–dc23
2011046216

ISBN 978-0-521-76229-8 Hardback
ISBN 978-1-108-40523-2 Paperback

Additional resources for this publication at www.cambridge.org/9781108405232

To Foster Rieke (my father),
Trevor Weekes, and Frank Low,
who introduced me to the
challenge and fun of
scientific instrumentation

Contents

Preface

Progress in astronomy is fueled by new technical opportunities (Harwit, 1984). For a long time, steady and overall spectacular advances in the optical were made in telescopes and, more recently, in detectors. In the last 60 years, continued progress has been fueled by opening new spectral windows: radio, X-ray, infrared (IR), gamma ray. We haven't run out of possibilities: submillimeter, hard X-ray/ gamma ray, cold IR telescopes, multi-conjugate adaptive optics, neutrinos, and gravitational waves are some of the remaining frontiers. To stay at the forefront requires that you be knowledgeable about new technical possibilities.

You will also need to maintain a broad perspective, an increasingly difficult challenge with the ongoing explosion of information. Much of the future progress in astronomy will come from combining insights in different spectral regions. Astronomy has become panchromatic. This is behind much of the push for Virtual Observatories and the development of data archives of many kinds. To make optimum use of all this information requires you to understand the capabilities and limitations of a broad range of instruments so you know the strengths and limitations of the data you are working with.

As Harwit (1984) shows, before about 1950, discoveries were driven by other factors as much as by technology, but since then technology has had a discovery shelf life of only about five years! Most of the physics we use is more than 50 years old, and a lot of it is more than 100 years old. You can do good astronomy without being totally current in physics (but you need to know physics very well). The relevant technology, in comparison, changes rapidly and you absolutely must be up to date on it, or you will rapidly fall behind in the field.

There are many specialist texts and conferences on specific measurement techniques. However, treatments designed to bridge from simple principles of optics to these materials are rare. It is also uncommon to provide such a bridge across the entire electromagnetic spectrum so it can provide an overview at an appropriate level for anyone approaching astronomy at a professional level, be it as an observer, theoretician, or engineer. This book is intended to fill this gap. How to do so requires choices; there is a long history of instrumentation approaches as well as a diverse range of them in current application. My choice has been to concentrate solely on instrumentation that is currently in wide use. As a result, I can give a reasonably complete description of each; those interested in earlier approaches can find descriptions in other textbooks. I apologize for omitting areas, either inadvertently or from a conclusion that they were too specialized for a general treatment.

I have also avoided detailed descriptions of individual astronomical instruments or facilities. Although of current interest, such material is likely to become dated quickly. A better alternative than a book is to locate the website for the facility and instrument of interest.

Many have improved this text by critiquing earlier versions or by supplying material: Jill Bechtold, Olivier Guyon, Michael Hart, Philip Hinz, Eugene Lauria, David Lesser, Michael Lesser, Daniel Marone, Peter Michelson, Stephanie Moats, Jane Morrison, Richard Perley, Paul Smith, Christopher Walker, Martin Weisskopf, Michelle Wilson, and especially Megan Bagley, Nicholas Ballering, and Melissa Dykhuis. The book would have been a disaster without their help, and if it is still a disaster it is not their fault.

I thank Shiras for drawing the illustrations. Shiras and I thank Murray Stein very, very much for his contributions to the illustrations. Many of the figures have been redrawn to make the style consistent. I thank the original sources in all cases, specifically (in addition to references given in the text): 1.5 (adapted from Wikipedia, Robert H. Rohde); 2.11 (from Norman Koren, with permission); 3.11 (adapted from "The Last of the Great Observatories" by myself, 2006, copyright the Arizona Board of Regents; 4.3 (from PASP, with permission) ; 4.5 (from Tim Hardy and John Hutchings); 4.7 and 6.11 (Gemini Observatory/ AURA); 5.6 and 7.18 (from *ApJ*, with permission); 5.9 (Wikipedia, original by Bob Mellish); 6.23 and 6.24 (Russell Scaduto); 7.1 (Wikipedia, prepared by R. N. Tubbs); 7.7, 7.8, 7.11, 7.12, and 7.13 (Sebastian Egner); 7.15 and 10.8 (from *A&A*, with permission); 7.16 and 7.17 (Robert J. Vanderbei); 7.23 (Olivier Guyon); 8.2 (Patrick Agnese, CEA Electronics and Information Technology Laboratory (Leti)); 9.2 (Jim Condon). I also thank Mairi Sutherland for a careful editing of the final draft. Please address any comments about the book to me at grieke@as.arizona.edu. SI units are assumed by default, but alternatives are used where they are conventional. Solutions to selected problems, along with color versions of figures, are available at www.cambridge.org/9780521762298. Additional problems will be posted there, if they are submitted to me with solutions.

Further reading

Harwit, M. (1984). *Cosmic Discovery*. Cambridge, MA: The MIT Press.

1

Radiometry, optics, statistics

1.1 Basics of radiometry

This book is about measuring photons. Therefore, we begin with a short description of how their production is treated in radiation physics. To compute the emission of an object, consider a surface element of it, dA, projected onto a plane perpendicular to the direction of observation (Figure 1.1). The projected area is $dA \cos\theta$, where θ is the angle between the direction of observation and the outward normal to dA. The specific intensity, L_v, is the power (watts) leaving a unit projected area of the surface of the source (in m^2) into a unit solid angle (in ster-radians) and unit frequency interval (in Hz); the v subscript lets us know this is a frequency-based quantity. It has SI units of $W\,m^{-2}\,Hz^{-1}\,ster^{-1}$. In physics, the same quantity is usually called the spectral radiance (in frequency units); we will use the common astronomical terminology here and refer the reader to Rieke (2003) for a table of terminology conversions. Similarly, we can define a specific intensity per wavelength interval, L_λ, where λ indicates a wavelength-based quantity. The intensity, L, is the specific intensity integrated over all frequencies or wavelengths, and the radiant exitance, M, is the integral of the intensity over solid angle. It is a measure of the total power emitted per unit surface area in units of $W\,m^{-2}$. For a blackbody,

$$M = \sigma T^4 = 5.669 \times 10^{-8}\,W\,m^{-2}\,K^{-4}T^4 \qquad (1.1)$$

where σ is the Stefan–Boltzmann constant (physical and other constants are given in Appendix A) and T is the temperature in kelvin. The integral of the radiant exitance over the source gives the total output, that is, the *luminosity* of the source. The irradiance, E, is the power received from the source by a unit surface element a large distance away. It also has units of $W\,m^{-2}$, but is easily distinguished from radiant exitance if we think of a source so distant that it is entirely within the field of view of the detection system. In this case,

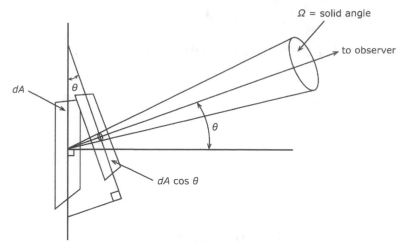

Figure 1.1. Emission geometry from a radiation source.

we might find it impossible to determine M, but E still has a straightforward meaning. The spectral flux density, that is, the irradiance per frequency or wavelength interval per unit of area, E_v or E_λ, is very commonly used in astronomy; it has units of $W\,m^{-2}\,Hz^{-1}$ (frequency units), or $W\,m^{-3}$ (wavelength units). In fact, astronomers have their own unit, the Jansky (Jy), which is $10^{-26}\,W\,m^{-2}\,Hz^{-1}$.

The energy of a photon is

$$\varepsilon_{ph} = hv = \frac{hc}{\lambda} = 6.626 \times 10^{-34}\,Js\,v$$

$$= 1.986 \times 10^{-25}\frac{Jm}{\lambda} \tag{1.2}$$

where $c = 2.998 \times 10^8\,m\,s^{-1}$ is the speed of light and $h = 6.626 \times 10^{-34}\,J\,s$ is Planck's constant. Conversion to photons/second can be achieved by dividing the various measures of power output by equation (1.2).

To make calculations simple, we generally deal only with *Lambertian* sources. For them, the intensity is constant regardless of the direction from which the source is viewed. In the astronomical context, a Lambertian source has no limb darkening (or brightening). In fact, we usually deal only with the special case, a black or grey body (a grey body has the same spectrum as a blackbody but with an emission efficiency, or emissivity ε, less than 1). We then have for the specific intensities in frequency and wavelength units:

$$L_v = \frac{\varepsilon\,[2hv^3n^2/c^2]}{e^{hv/kT} - 1} = \frac{1.474 \times 10^{-50}\,\varepsilon\,n^2v^3}{e^{4.800\times10^{-11}v/T} - 1}\,W\,m^{-2}\,ster^{-1}\,Hz^{-1}$$

$$L_\lambda = \frac{\varepsilon[2hc^2]}{\lambda^5 n^2(e^{\frac{hc}{\lambda kT}} - 1)} = \frac{1.191 \times 10^{-16}\varepsilon}{\lambda^5 n^2(e^{0.01438/\lambda T} - 1)} \, \mathrm{W\,m^{-3}\,ster^{-1}} \tag{1.3}$$

where $k = 1.381 \times 10^{-23} \, \mathrm{J\,K^{-1}}$ is the Boltzmann constant and n is the refractive index (virtually always $n \cong 1$ for astronomical sources). We also have

$$E_v = \frac{AL_v}{4\,r^2}$$
$$\tag{1.4}$$
$$E_\lambda = \frac{AL_\lambda}{4\,r^2}$$

where r is the distance to the source and A is its surface area. It is convenient to convert from frequency to wavelength units by means of

$$E_v = \frac{\lambda^2}{c} E_\lambda \tag{1.5}$$

which can be derived by differentiating

$$\lambda = \frac{c}{v} \tag{1.6}$$

to obtain the relation between $d\lambda$ and dv. An easy way to remember the conversion is that multiplying the flux density in $\mathrm{W\,m^{-2}\,Hz^{-1}}$ by c/λ^2 with $c = 2.998 \times 10^{10} \, \mathrm{cm/s}$ and λ in $\mu\mathrm{m}$ gives the flux density in $\mathrm{W\,cm^{-2}\,\mu m^{-1}}$.

1.2 Image formation and the wave nature of light

OK, some source has launched a stream of photons in our direction. How are we going to catch it and unlock its secrets? A photon is described in terms that are illustrated in Figure 1.2 (after Barbastathis 2004). In the figure, we imagine that time has been frozen, but the photon has been moving at the speed of light in the direction of the arrow. We often discuss the photon in terms of wavefronts, lines marking the surfaces of constant phase and hence separated by one wavelength.

As electromagnetic radiation, a photon has both electric and magnetic components, oscillating in phase perpendicular to each other and perpendicular to the direction of energy propagation. The amplitude of the electric field, its wavelength and phase, and the direction it is moving characterize the photon. The behavior of the electric field can be expressed as

$$E = E_0 \cos(\omega t + \varphi) \tag{1.7}$$

where E_0 is the amplitude, ω is the angular frequency, and φ is the phase.

Another formalism is to express the phase in complex notation:

$$E(t) = E_0 e^{-j\omega t} = E_0 \cos \omega t - jE_0 \sin \omega t \tag{1.8}$$

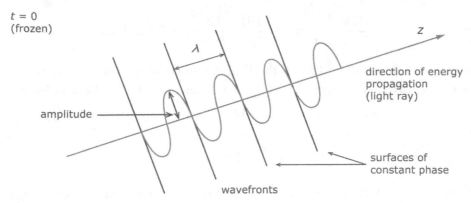

Figure 1.2. Terms describing the propagation of a photon, or a light ray.
Redrawn with modifications from Barbastathis (2004).

where j is the imaginary square root of –1. In this case, the quantity is envisioned as a vector on a two-dimensional diagram with the real component along the usual x-axis and the imaginary one along the usual y-axis. The angle of this vector represents the phase. The flux (energy flow per unit area per second) is given by the Poynting vector, or in simplified form:

$$S = \varepsilon_0 \, c \, E^2 = \varepsilon_0 \, c \, E_0^2 \cos^2(\omega t + \varphi) \tag{1.9}$$

where ε_0 is the permittivity of free space, $8.854 \times 10^{-12} \, \mathrm{F \, m^{-1}}$. The intensity of the light is the average of S over time.

For efficient detection of this light, we use optics to squeeze it into the smallest possible space. The definition of "the smallest possible space" is a bit complex. We want to map the range of directions toward the target onto a physical surface where we put an instrument or a detector. This goal implies that we form an image, and since light is a wave, images are formed as a product of interference. That is, the collector must bend the rays of light from a distant object to bring them to a focus at the image plane (see Figure 1.3). There, they must interfere constructively, requiring that they arrive at the image in phase, having traversed paths of identical length from the object. This requirement is summarized in Fermat's Principle, which is expressed in a number of ways including:

"The optical path from a point on the object through the optical system to the corresponding point on the image must be the same length for all neighboring rays."

The optical path is defined as the integral of the physical path times the index of refraction of the material that path traverses.

$$s = \int n(x) \, dx \tag{1.10}$$

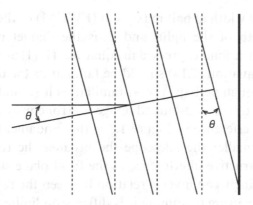

Figure 1.3. Approach of light wavefronts to an image plane.

What tolerance is there on the lengths of the optical paths? As shown in Figure 1.3, if the light waves impinge on the focal plane over too wide a range of angle, then the range of phases across the area where we detect the light can be large enough to result in a significant level of destructive interference. From the figure, this condition can be expressed as

$$l \tan(\theta_{max}) = \frac{\lambda}{2} \approx l \theta_{max} \qquad (1.11)$$

Here, l is some characteristic distance over which we collect the light – say the size of a detector at the focus of a telescope, and θ_{max} is the largest angle over which generally constructive interference occurs. Equation (1.11) seems to imply that the size of the detector enters into the fundamental performance of a telescope. However, there is another requirement on the optical system. The beam of light entering the telescope is characterized by an area at its base, A, normally the area of the telescope primary mirror, and a solid angle, $\Omega \approx \pi \theta^2$ where θ is the half-angle of the convergence of the beam onto the telescope (i.e., the angular diameter of the beam is 2θ). The product of A and Ω (and the refractive index n) can never be reduced as a beam passes through an optical system, by the laws of thermodynamics. This product

$$n \, A \, \Omega = C \qquad (1.12)$$

is called the etendue (French for extent or space). Since astronomical optics generally operate in air or vacuum with $n \cong 1$, we will drop the refractive index term in future discussions of the etendue. Assuming a round telescope and detector and a circular beam (for convenience – with more mathematics we could express the result generally), we find that an incoming beam with

$$\Theta = \frac{\lambda}{D} \qquad (1.13)$$

(where Θ is the full width at half maximum (FWHM) of the beam in radians, λ is the wavelength of the light, and D is the diameter of the telescope aperture) is just at the limit expressed in equations (1.11) and (1.12) regardless of detector size. Equation (1.13) should be familiar as the diffraction limit of the telescope (although perhaps not as familiar as it should be because often the FWHM of the diffraction-limited image is erroneously given as $1.22\lambda/D$). The destructive interference in Figure 1.3 is the fundamental image-forming mechanism. The smaller the telescope the broader the range of angles on the sky before destructive interference at the focal plane starts to reduce the signal, corresponding to an inverse relation between the resolution limit of a telescope and its aperture (assuming it is diffraction limited).

1.3 Power received by an optical system

1.3.1 Basic geometry

An optical system will receive a portion of the source power that is deter-mined by a number of geometric factors (plus efficiencies) – see Figure 1.4. If you refer to this diagram and follow it carefully, you will (almost) never get confused on how to do radiometry calculations in astronomy. As Figure 1.4 shows, the system accepts radiation only from a limited range of directions, called its field of view (FOV). The area of the source that is effective in

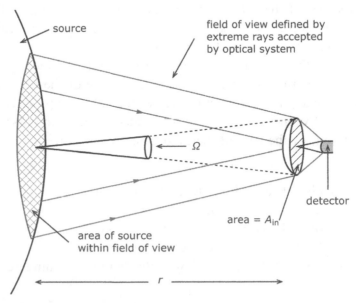

Figure 1.4. Geometry for detected signals.

producing a signal is determined by the FOV and the distance between the optical system and the source (or by the size of the source if it all lies within the FOV). This area emits with some angular dependence (e.g., for a Lambertian source, the dependence just compensates for the foreshortening and off-axis emission direction). The range of emission directions that is intercepted by the system is determined by the solid angle, Ω, that the entrance aperture of the system subtends as viewed from the source:

$$\Omega = \frac{A_{in}}{r^2} \tag{1.14},$$

where A_{in} is the area of the entrance aperture of the system and r is the distance from it to the source (we assume the entrance aperture is perpendicular to the line from its center to the center of the source). If the aperture is circular,

$$\Omega = 4 \pi \sin^2 \left(\frac{\theta}{2} \right) \tag{1.15}$$

where θ is the half-angle of the circular cone with its vertex on a point on the surface of the source and its base across the entrance aperture of the optical system.

If none of the emitted power is lost (by absorption or scattering) on the way to the optical system, then the power it receives is the intensity the source emits in its direction multiplied by (1) the source area within the system FOV times, (2) the solid angle subtended by the optical system as viewed from the source. We often get to make this situation even simpler, because for us many sources fall entirely within the FOV of the system. In this case, the spectral irradiance (that is, the flux density (equations (1.4)) is a convenient unit of measure.

1.3.2 How much gets to the detector?

Once the signal reaches the optical system, a number of things happen to it. Before reaching the detector, part of the signal is lost due to diffraction, optical imperfections, absorption, and scattering within the instrument. Once it reaches the detector, only a fraction of the signal is absorbed and utilized to produce the signal; this fraction is called the *quantum efficiency*. All of these efficiencies (i.e., quantum efficiency and transmittances < 1) must be multiplied together to get the net efficiency of conversion into the output signal.

For groundbased telescopes, there is another important component to the transmission term, the absorption by the atmosphere of the earth. The overall absorption is shown in Figure 1.5 – the "windows" through which

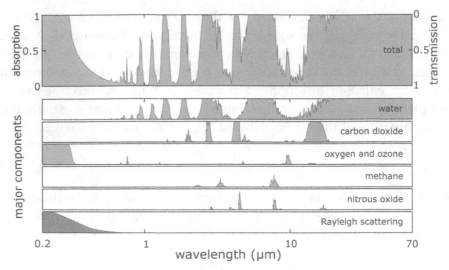

Figure 1.5. Atmospheric absorption and scattering. Redrawn with modifications from R. Rohde (n.d.).

astronomers can peer are obtained by turning the figure at the top upside down. The atmosphere is opaque at all wavelengths shorter than 0.3 microns and between 40 and 300 μm. Further details between 1 and 25 μm are given in Chapter 5. Figure 1.6 shows the transmission in the submillimeter and millimeter-wave. For still longer wavelengths (i.e., beyond 1 cm) the atmosphere is transparent until about 100 m, where absorption by the layer of free electrons in the ionosphere becomes strong.

Many of the long-lived atmospheric constituents are well-mixed; that is, the composition of the air is not a strong function of altitude. The pressure can be characterized by an exponential with a scale height (the distance over which the pressure is reduced by a factor of $1/e$) of 8 km. However, water is in equilibrium with processes on the ground and has an exponential scale height of about 2 km. In addition, the amount of water in the air is reduced rapidly with decreasing temperature. Typical high-quality observing sites have overhead water vapor levels equivalent to about 2 mm; Figure 1.6 shows the improvement in the submillimeter for a reduction by a factor of 4 in water vapor. Therefore, there is a premium in placing telescopes for water-vapor-sensitive wavelengths on high mountains, and even in the Antarctic.

The atmosphere can also contribute unwanted foreground emission that raises the detection limit for faint astronomical objects. There are many components (Leinert *et al.* 1998), but most of them play a relatively minor role. Two major ones are scattered moonlight, which can be significant from 0.3 to 1 μm, and the airglow lines (mostly from OH with a contribution

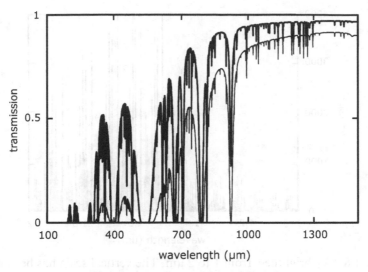

Figure 1.6. Atmospheric transmission in the submillimeter, for precipitable water vapor levels of 0.5 (heavy line) and 2 mm (light line). Prepared using APEX transmission calculator.

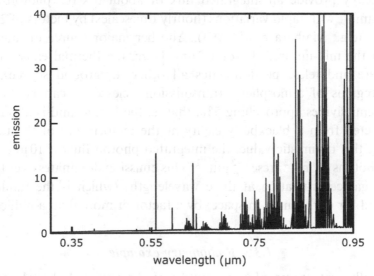

Figure 1.7. Sky brightness in the visible and near-infrared. Based on Rousselot *et al.* (2000).

from O_2) that start in the 0.7–1 μm range (Figure 1.7) and become totally dominant in the near-infrared bands (Figure 1.8; Rousselot *et al.* 2000). In the visible (0.56 μm), a typical dark sky provides about 200 photons $s^{-1} m^{-2}$ arcsec^{-2} μm^{-1} but with the full moon above the horizon this emission is increased by about a factor of 5 (the increase is much greater in the blue

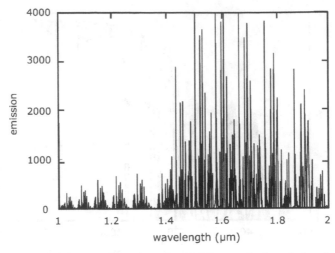

Figure 1.8. Sky brightness from 1 to 2 μm. The vertical scale has been made 100 times larger than in Figure 1.7. Based on Rousselot *et al.* (2000).

and ultraviolet). Where they are strongest, between 1.5 and 1.8 μm, the airglow lines typically provide an integrated flux of about 3×10^4 photons $\mathrm{s}^{-1}\,\mathrm{m}^{-2}$ arcsec$^{-2}\,\mu\mathrm{m}^{-1}$, with rapid variations (hourly timescales) by factors of 2 around this value (e.g., Maihara *et al.* 1993). Another major source of interference occurs in the mid-infrared, beyond 2 μm. Here, the thermal emission of the atmosphere and telescope dominates all other foreground sources. In the clearest regions of atmospheric transmission, telescopes can be built with effective emissivities approaching 5%, that is, the foreground is about 5% of that expected from a blackbody cavity at the temperature of the telescope. Assuming this optimistic value, the integrated photon flux at 10 μm is about 5×10^8 photons $\mathrm{s}^{-1}\,\mathrm{m}^{-2}$ arcsec$^{-2}\,\mu\mathrm{m}^{-1}$. This emission dominates over the zodiacal and galactic radiation at these wavelengths (which is the fundamental foreground for our vicinity in space) by a factor of more than a million.

1.3.3 Radiometry example

We now illustrate some of the points just discussed. A 1000 K spherical blackbody source of radius 1 m is viewed in air by a detector system from a distance of 1000 m. The entrance aperture of the system has a radius of 5 cm; the optical system is 50% efficient and has a field of view half-angle of 0.1°. The detector operates at a wavelength of 1 μm and the light passes through a filter with a spectral bandpass of 0.1 μm. Compute the specific intensities in both frequency and wavelength units. Calculate the corresponding flux densities at the system entrance aperture and the power received by the detector.

The refractive index of air is 1, and the emissivity of a blackbody is also 1. The frequency corresponding to $1\,\mu m$ is $v = c/\lambda = 2.998 \times 10^{14}\,Hz$. Substituting into equation (1.3), we obtain

$$L_v = 2.21 \times 10^{-13}\,W\,m^{-2}\,Hz^{-1}\,ster^{-1}$$

and

$$L_\lambda = 6.2 \times 10^7\,W\,m^{-3}\,ster^{-1}$$

The solid angle subtended by the detector system as viewed from the source is (from equation (1.15) with the entrance aperture area of $7.854 \times 10^{-3}\,m^2$):

$$\Omega = 7.854 \times 10^{-9}\,ster$$

The spectral bandwidth corresponds to $2.998 \times 10^{13}\,Hz$, or to $1 \times 10^{-7}\,m$. The radius of the 0.1° beam at the distance of the source is $1.745\,m$, which is larger than the radius of the source; therefore, the entire visible area of the source will contribute signal. The projected surface area is $3.14\,m^2$ and it will appear to be emitting uniformly since it is Lambertian. The power at the entrance aperture is then the product of the specific intensity, source projected area, spectral bandwidth, and solid angle from the source received by the system; it is $P = 1.63 \times 10^{-7}\,W$.

Because the source falls entirely within the field of view of the detector system, we could have used flux densities in the calculation. The surface area of the source is $12.57\,m^2$. Equations (1.4) then give

$$E_v = 6.945 \times 10^{-19}\,W\,m^{-2}\,Hz^{-1}$$
$$E_\lambda = 208\,W\,m^{-3}$$

Multiplying either by the entrance aperture area and the spectral bandpass yields $P = 1.63 \times 10^{-7}\,W$, as before.

The power received by the detector is reduced by the optical efficiency of 50%, that is, to $8.2 \times 10^{-8}\,W$. From Figure 1.5, the atmosphere is quite transparent at $1\,\mu m$, so we make no further corrections for inefficiencies. The energy per photon (ignoring the slight variation over the bandpass) is, from equation (1.2), $1.99 \times 10^{-19}\,J$, so the detector receives 4.12×10^{11} photons s^{-1}.

1.4 Optical systems

1.4.1 Lenses and mirrors

We use optical elements arranged into a system to control light so we can analyze its information content. We have already implicitly assumed some such arrangement to concentrate light onto a detector. These elements work

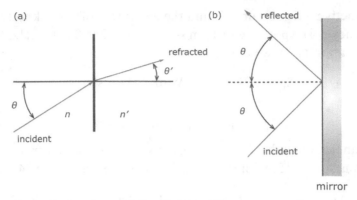

Figure 1.9. (a) Refraction and Snell's Law. The ray passes from the medium with refractive index n into the one with index n'. (b) The law of reflection.

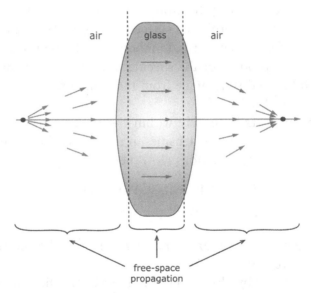

Figure 1.10. Operation of a simple lens.

on two different principles, the law of refraction and the law of reflection (Figure 1.9). Refraction is described by Snell's Law (refer to Figure 1.9(a) for symbol definitions):

$$n \sin \theta = n' \sin \theta' \tag{1.16}$$

The law of reflection simply states that the angle of incidence and angle of reflection are equal, as in Figure 1.9(b).

Figure 1.10 shows a simple lens; it is made of a transparent material with an index of refraction greater than that of the air surrounding it. The surfaces

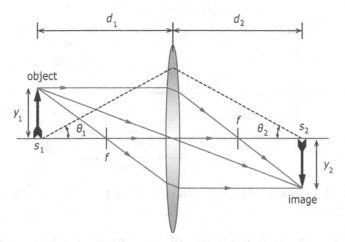

Figure 1.11. Image formation by a positive lens.

are shaped so the rays of light emanating from the point to the left are brought back to a point on the right. This process is imperfect, but we will ignore any such issues for now.

Although real lenses have thickness as shown in Figure 1.10, we can analyze many optical systems ignoring this characteristic. The principles illustrated in Figure 1.11 let us determine the imaging properties of a thin lens. Any ray that enters the lens after passing through its focal point at f is converted by the lens into a ray parallel to its optical axis. Similarly, any ray that passes through the center of the lens is not deflected. This behavior lets us abstract the lens to a simple plane through the middle of the actual lens that acts on the rays. In this approximation (see exercise 1.3),

$$\frac{1}{f} = \frac{1}{d_1} + \frac{1}{d_2} \tag{1.17}$$

and

$$y_1 d_2 = y_2 d_1 \tag{1.18}$$

where f is the focal length and the other terms are defined in Figure 1.11. These equations are described as the thin lens formula and are amazingly useful for a first-order understanding of an optical system containing a number of lenses (and/or mirrors). If our lenses really are thin, then we can compute their focal lengths as

$$\frac{1}{f} = (n-1)\left(\frac{1}{R_1} - \frac{1}{R_2}\right) \tag{1.19}$$

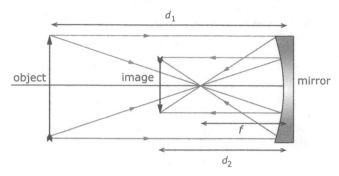

Figure 1.12. A concave mirror can have imaging properties similar to those of a positive lens.

where n is the index of refraction of the lens material (we assume it is immersed in a medium with $n = 1$) and R_1 and R_2 are the radii of curvature of its surfaces, assigned positive signs if the curvature is convex to the left (distances are positive to the right and up; angles are positive rotating counterclockwise from the optical axis).

These concepts lead to another formulation of the principle of preservation of etendue, called the Lagrange Invariant or optical invariant:

$$n_1 \, y_1 \tan \theta_1 = n_2 \, y_2 \tan \theta_2 \tag{1.20}$$

where n_1 and n_2 are the refractive indices. That is, the product of the refractive index, the tangent of the ray angle to a given position on the lens, and the image size is a constant for a perfect optical system. The quantities are illustrated in Figure 1.11. If the ray angle is measured to the edge of the aperture, the preservation of etendue is recovered directly. Although equation (1.18) is usually derived geometrically, it follows immediately from the Lagrange Invariant.

A concave mirror can also form images (Figure 1.12). For a spherical mirror, the focal length is $R/2$ where R is the radius of curvature; however, good imaging generally requires an aspherical surface. Although a little more awkward to demonstrate with simple pictures, the image-forming properties of a mirror can also be described by the thin lens formula, equation (1.17), and its image size by equation (1.18) (the radius of curvature is positive if the surface is convex to the left, distances before reflection are positive to the right and up, and angles are measured counterclockwise from the optical axis).

Multiple mirrors and/or lenses are combined into optical systems. The convention is that these systems are drawn with the light coming in from the left, as in the figures above. The elements (individual mirrors or lenses) of a system can take light from one point to another, forming real images (as in

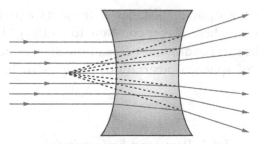

Figure 1.13. A negative lens with concave surfaces forms a virtual image – it looks like the light is coming from a point but it is not.

the cases illustrated above), or they can make the light from a point diverge (e.g., for a lens with concave surfaces, or a convex mirror – see Figure 1.13). A real image is where the light from an unresolved source is brought to its greatest concentration – putting a detector at the position of the real image will record the source. A virtual image can be formed if the optics produces a bundle of light that appears to come from a real image but no such image has been formed – for example, by the negative lens in Figure 1.13, or by a convex mirror that reflects light so it appears that there is a real image behind the mirror. All of this behavior is described by the thin lens formula, if one pays attention to the sign conventions (for lenses, the image distances d_1 and d_2 are both positive for a real image, but a virtual image appears to be to the left of the lens and d_2 is therefore negative).

The thin lens formula and its corollaries provide the simplest description of the imaging properties of an optical system, but they leave out a number of flaws. These issues will be discussed in Chapter 2, but we describe an important design guideline here. One of the goals for an optical system is to form good images over a large field of view. The Abbe sine condition, originally formulated to guide improvement of microscope optics, provides general guidance toward this goal. To meet this condition, any two rays that leave an object at angles α_1 and α_2 relative to the optical axis of the system and that are brought to the image plane at angles α_1' and α_2' must obey the constraint

$$\frac{\sin \alpha_1}{\sin \alpha_2} = \frac{\sin \alpha_1'}{\sin \alpha_2'} \tag{1.21}$$

That is, the sine of the angle with which the ray approaches the image plane is proportional to the sine of the angle with which it leaves the object. In the case of an object at infinite distance, the sine condition becomes

$$\frac{h}{\sin \alpha} = F \tag{1.22}$$

for a ray striking the optic at distance h from its axis and approaching the image at angle α, where F is a constant for all rays independent of h. Systems obeying the sine condition are free of the aberration coma (assuming they also are free of spherical aberration) – topics to be discussed in the next chapter.

1.4.2 Detectors: basic principles

Gathering light on the scale demanded by astronomers is difficult and expensive. Consequently, they demand that it be converted as efficiently as possible into electronic signals they can analyze. However, efficiency is not enough. The detectors have to be very well behaved and characterized; artifacts in the data due to detector behavior can easily mask interesting but subtle scientific results, or create spurious ones. Three basic types of detector meet the goals of astronomers in different applications. The first type is discussed in Chapter 3, while the remaining two are described in Chapter 8.

In *photon detectors*, the light interacts with the detector material to produce free charge carriers photon-by-photon. The resulting miniscule electrical currents are amplified to yield a usable electronic signal. The detector material is typically some form of semiconductor, in which energies of about an electron volt[1] suffice to free charge carriers; this energy threshold can be adjusted from ~0.01 eV to ~3 eV by appropriate choice of the detector material. Photon detectors are widely used from the X-ray through the mid-infrared.

In *thermal detectors*, the light is absorbed in the detector material to produce a minute increase in its temperature. Exquisitely sensitive electronic thermometers react to this change to produce the electronic signal. Thermal detectors are in principle sensitive to photons of any energy, so long as they can be absorbed in a way that allows the resulting heat to be sensed by their thermometers. However, in the ultraviolet through mid-infrared, photon detectors are more convenient to use and are easier to build into large-format arrays. Thermal detectors are important for the X-ray and the far-infrared through mm-wave regimes.

In *coherent detectors*, the electrical field of the photon interacts with a signal generated by a local oscillator (LO) at nearly the same frequency. The interference between the electrical field of the photon and that from the LO produces a signal with a component at the difference frequency, termed the intermediate frequency (IF). The IF signal is put into a frequency range

---

[1] Corresponding to photons of wavelength just longer than 1 μm.

that is compatible with further processing and amplification using conventional electronics.

1.5 Statistics and noise

1.5.1 Error propagation

Signals are subject to errors. Error analysis is a subject in its own right, for which we recommend you acquire a dedicated book (e.g., Wall and Jenkins 2003). We will make use of some very simple relations in our discussion, which we list here.

If two quantities, x and y, have uncertainties that are random and independent, σ_x and σ_y, and the uncertainties are small compared with x and y, then the error in $x + y$ or in $x - y$ is the quadratic sum:

$$\sigma = \sqrt{\sigma_x^2 + \sigma_y^2} \tag{1.23}$$

Under the same assumptions, the error in the product $f = xy$ or ratio $f = x/y$ is

$$\frac{\sigma_{xy}}{f} = \sqrt{\left(\frac{\sigma_x}{x}\right)^2 + \left(\frac{\sigma_y}{y}\right)^2} \tag{1.24}$$

If x is multiplied by a constant, k, which itself has negligible error, then the error in the product is $k\sigma_x$.

1.5.2 Probability distributions

Multiple measurements of a quantity with uncertainties will fall into a distribution around the true value of the quantity. This distribution can take a number of mathematical forms, depending on the circumstances. Three of them are important for us.

The binomial distribution is appropriate when there are exactly two mutually exclusive outcomes of a measurement: success and failure. This distribution describes the probability of observing m successes in n observations (assuming the observations are independent), with the probability of success in a single observation denoted by p:

$$P(m, p, n) = \frac{n!}{m!(n-m)!} p^m (1-p)^{(n-m)} \tag{1.25}$$

The Poisson distribution is derived from the binomial one in the limit of rare events and a large number of trials. It describes the probability of m successes in an observation, when the average number of successes in a series of observations is μ:

$$P(m, \mu) = \frac{e^{-\mu}\mu^m}{m!} \tag{1.26}$$

The normal (often termed Gaussian) distribution is a continuous function in the observed values x:

$$P(x, x_0, \sigma) = \frac{1}{\sigma\sqrt{2\pi}}e^{-(x-x_0)^2/2\sigma^2} \tag{1.27}$$

where x_0 is the center of the distribution (and equal to the mean for a large number of measurements) and σ is the standard deviation (σ^2 is the variance). For large values of μ the Poisson distribution closely resembles a normal distribution with $\sigma = \mu^{1/2}$ and it can be shown generally that the standard deviation of the Poisson distribution is $\mu^{1/2}$. Thus, the standard deviation of N discrete and independent events that follow a Poisson or Gaussian distribution is $N^{1/2}$, where by necessity we take the observed number, N, as the best estimate of the expected average number, μ. Levels of confidence in a measurement are often quoted in units of standard deviations, which, from equation (1.27), can be converted to a probability for random errors of any size relative to the measured value.

Normal distributions are nearly universally used to describe experimental errors. The justification for this approach is the Central Limit Theorem, which starts with n independent values drawn from the same distribution, with expected value μ and variance σ^2. Then the theorem states that the averages of the n values in many repeats of the experiment will tend to be distributed normally with average μ and variance σ^2/n, no matter what the distribution from which the initial values were drawn. Since many experiments roughly satisfy these conditions, it is plausible that their results are distributed normally.

The variance of a series of n measurements with results x_i is

$$\sigma^2 \equiv \frac{1}{n}\sum_{i=1}^{i=n}(x_i - \mu)^2 \tag{1.28}$$

where μ is the expected mean result. We usually do not know μ (a very large number of measurements would be necessary to determine it), but have to estimate it as the mean of the actual measurements. In that case,

$$\sigma^2 = \frac{1}{n-1}\sum_{i=1}^{i=n}(x_i - \bar{x})^2 \tag{1.29}$$

where $\bar{x}$ is the mean of the *measurements*.

Fits of measurements to functions are optimized by minimizing the variance. For example, to fit a set of measurements y_i at x_i instances to a straight line

$$y(x) = a + bx \tag{1.30}$$

a quantity "chi square" is defined:

$$\chi^2 = \sum \left[\frac{1}{\sigma_i^2} (y_i - a - bx_i)^2 \right] \tag{1.31}$$

where σ_i is the standard deviation of the ith measurement and the sum compares the measured values, y_i, with the expected ones, $a - bx_i$ (compare equation (1.29)), weighted by the inverse variance of each measurement. The fit is best when we have minimized χ^2. To find this case, the derivatives of χ^2 with respect to a and b are set equal to zero and the result is solved to find the optimum values of these two parameters:

$$a = \frac{1}{\Delta} \left(\sum \frac{x_i^2}{\sigma_i^2} \sum \frac{y_i}{\sigma_i^2} - \sum \frac{x_i}{\sigma_i^2} \sum \frac{x_i y_i}{\sigma_i^2} \right)$$

$$b = \frac{1}{\Delta} \left(\sum \frac{1}{\sigma_i^2} \sum \frac{x_i y_i}{\sigma_i^2} - \sum \frac{x_i}{\sigma_i^2} \sum \frac{y_i}{\sigma_i^2} \right) \tag{1.32}$$

$$\Delta = \sum \frac{1}{\sigma_i^2} \sum \frac{x_i^2}{\sigma_i^2} - \left(\sum \frac{x_i}{\sigma_i^2} \right)^2$$

1.5.3 Photon statistics

Photons obey Bose–Einstein statistics, so as a result of quantum mechanical effects they do *not* arrive strictly independently of each other. Consequently, the noise is increased from the usual dependence on the square root of the number of photons received (appropriate only for independent events)

$$\langle N^2 \rangle = n \tag{1.33}$$

to

$$\langle N^2 \rangle = n \left[1 + \frac{\varepsilon \tau \eta}{e^{h\nu/kT} - 1} \right] \tag{1.34}$$

where $\langle N^2 \rangle$ is the mean square noise, n is the number of photons detected, ε is the emissivity of the source of photons, τ is the transmittance of the optical system, and η is the detector quantum efficiency (i.e., the fraction of the photons absorbed). Inspection of equation (1.34) shows that the correction is only potentially significant in the Rayleigh–Jeans regime of an emitting blackbody (see exercise 1.4), that is at wavelengths much longer than the peak of the blackbody emission. Because of the efficiency term, $\varepsilon\tau\eta$, in fact the correction is rarely large enough to worry about. Therefore, to the necessary level of accuracy, we can take the noise to just be proportional to the square root of the number of photons detected.

We have now rationalized the usual assumption, that the signal photons follow normal (Gaussian) or Poisson statistics – so in the usual case of reasonably large signals, the ratio of the signal to the noise in the intrinsic signal is just the square root of the number of photons, n:

$$\left(\frac{S}{N}\right)_{\mathrm{ph}} = n^{1/2} \tag{1.35}$$

n can be determined from the power detected using equation (1.2). In a detector, only the absorbed photons produce free charge carriers that eventually contribute to the output. Therefore, n photons produce ηn charge carriers, and in the ideal case the detector can achieve

$$\left(\frac{S}{N}\right)_{\mathrm{det}} = (\eta n)^{1/2} \tag{1.36}$$

In general a detector will fall short of $(S/N)_{\mathrm{det}}$ – there are many mechanisms that add noise. A succinct way to describe the detector performance is the "detective quantum efficiency", DQE. Let n_{in} be the actual input photon signal and n_{out} be an imaginary input signal that would yield the observed S/N if the detector were perfect. That is, $(S/N)_{\mathrm{out}}$ is the observed signal-to-noise ratio, and $(S/N)_{\mathrm{in}}$ is the potential signal-to-noise ratio in the incoming photon stream, which is determined from the intrinsic photon statistics. Then,

$$\mathrm{DQE} = \frac{n_{\mathrm{out}}}{n_{\mathrm{in}}} = \frac{(S/N)_{\mathrm{out}}^2}{(S/N)_{\mathrm{in}}^2} \tag{1.37}$$

DQE is a way to quantify the loss of information content in the photon stream as a result of our detecting it with a less-than-perfect technology. It can include loss of signal through incomplete absorption by the detectors, along with excess noise such as gain fluctuations in the detector and the readout noise of the amplifiers that receive the signals (which in general can be combined with the photon-based noise as in equation (1.23)). It expresses this loss of information in the simplest form, as the fraction of the incoming photons that would be needed to provide the equivalent amount of information if the instrument were perfect.

1.5.4 Other forms of noise

Specific conditions must be met for the above descriptions of noise – for example, random, independent, and identical events. Many sources of uncertainty do not meet these conditions; in these cases many of the simple relations we have described may not hold and the effects of the noise are more difficult to quantify. One example is systematic errors – for example, an offset in the assignment of fluxes to a set of measurements will never be removed by averaging a larger number of the measurements. Another case is where the disturbances are not independent, for example where repetitive electrical interference creates noise in a measurement. Yet another example is when a single cause triggers a number of non-independent events, such as an energetic cosmic ray particle striking a detector and triggering the release of many charge carriers that get collected as a signal; the noise in the number of charge carriers is not given by the square root because they are not created independently.

A more subtle example is confusion, where so many sources are detected that disentangling the resulting clutter of signals becomes problematic. The issue in this case is that the "noise" sources are not identical, but have a spectrum of sizes. The full treatment is complex; we will only give a high-level overview based on a classic paper by Condon (1974). Although confusion can enter in many types of measurement, we will discuss only imaging data. One assumes an image with no other sources of noise (e.g., the result of a very long integration to reduce the random noise sources to a level well below the structure of the sky due to sources). $P(D)$ is defined as the probability for any point in this image to have intensity D. This probability depends on the distribution of fluxes from the sources, which we take to be a power law (in differential source counts):

$$n(S)\, dS = k\, S^{-\gamma}\, dS \qquad (1.38)$$

where S is the flux density, k is a constant, and γ is the power law index. $P(D)$ also depends on the angular resolution with which these sources are observed. For a Gaussian point source image,

$$\Omega_b = \frac{\pi \theta^2}{4 \ln 2} \qquad (1.39)$$

is the geometric solid angle of the beam if its full width at half maximum is θ.

There is a simple specification of a "confusion limit" in source flux density beyond which measurements become unreliable. It is expressed in terms of the number, β, of measurement beam areas, Ω_b, in the image per source detected at this flux density limit, the "beams per source." For reliable measurements, the minimum permissible number of beams per source is a function of the power law slope:

$$\beta = \frac{q^2}{3 - \gamma} \tag{1.40}$$

where q is the factor above the root-mean-square (rms) confusion level at which one can detect individual sources (and hence where one no longer counts them as clutter). For reference, a population of sources uniformly distributed in space and without corrections for cosmological effects yields a "Euclidean" value of $\gamma = 2.5$. For a detection threshold of three standard deviations ($q = 3$), the result from equation (1.40) would be $\beta = 18$. However, the variation in spectral indices for more realistic situations leads to a significant range in γ and hence in β. In addition, different measurement goals can lead to different values of q. Consequently, there is a large range in quoted values of β, from about 10 to 100 or more. When observations reach this range, one should investigate more sophisticated ways to analyze the data that may increase its reliability.

1.6 Exercises

1.1 For $h\nu \ll kT$ (and $\varepsilon = n = 1$), show that

$$L_\nu = \frac{2kT\nu^2}{c^2}$$

This expression is the Rayleigh–Jeans Law.

1.2 From Wien's Displacement Law, the wavelength of the maximum flux density times the temperature is a constant:

$$\lambda_{\max} T = C$$

Derive this expression and show that $C \approx 0.3\,\mathrm{cm\,K}$.

1.3 Derive the thin lens formula. Hint: Take Figure 1.11 and convert it to a set of triangles. Make the lens a single line running through its center, and then draw the three rays going from the tip of the arrow at the object to the tip of the arrow on the image.

1.4 Show that the Bose–Einstein correction to the rms photon noise $\langle N^2 \rangle^{1/2}$ is less than 10% if $(5\varepsilon\tau\eta kT/h\nu) < 1$. Consider a 1000 K source viewed by a detector system with an optical efficiency of 50% and quantum efficiency of 50%. At what wavelength is its peak flux density? At what wavelength does the Bose–Einstein correction become 10%?

1.5 A circular photoconductor of diameter 0.5 mm views through a filter centered at 3.5 µm and with spectral bandwidth of 5% of its center wavelength and 60% average transmission over its nominal band. Suppose it views a circular blackbody source of diameter 1 mm at a distance of 100 m and at a temperature of 1,500 K. Compute the specific intensities in both frequency and wavelength units. Calculate the corresponding flux densities at the system entrance aperture and the power received by the detector.

1.6 A student is measuring a star in a spectral band at 1.55 µm, where the sky brightness is 1500 photons s^{-1} (square arcsec)$^{-1}$. The instrument aperture on the sky is 3 arcsec in diameter and the measurement is obtained by integrating for equal times pointed at the source and at a nearby region of sky and differencing the two readings. The source signal is 100 photons s^{-1}. The net instrument efficiency is 10%. How long does it take to measure the source to a ratio of signal to noise of 10?

1.7 Assume that the noise in a measurement is distributed normally (equation (1.27)). What is the probability that one of many measurements will exceed the average by more than three standard deviations?

1.8 A student is using an infrared detector to measure the flux of photons on one of its pixels. The mean photon rate from the sky background radiation is 30 photons s^{-1} and the photon arrival rate is governed by the Poisson distribution. The signal of interest has a much lower flux of 1 photon s^{-1}. The source can be moved on and off the pixel of interest at any rate. This allows the student to make a differential measurement. That is, she can repeatedly measure the overall signal with and without the signal present and take the difference of the two to derive the photon rate of the signal.

   (a) Write down the signal and noise estimates for this setup for a given integration, t. How long does it take to achieve a signal-to-noise level of 3?

   (b) Now assume that two additional noise sources are present in the detector. Each time the integration is recorded, the result is done so with an uncertainty of 10 photons (this is the "read" noise for a detector). The uncertainty has a Gaussian distribution. The second is cosmic rays; they strike the detector on average once per hundred seconds, but randomly in time. A cosmic ray destroys the information in the read cycle it hits. Calculate the overall signal-to-noise ratio (SNR) now, in terms of the number of cycles, N, and the time per integration, t. You should assume that the detector integrates for the time spent observing the source, and that an equal amount of time is spent observing the scene with no source present. What is the optimum value of t under these considerations, and how long does it take to reach a signal-to-noise level of 3?

1.9 Use Fermat's Principle to derive Snell's Law.

Further reading

Barbastathis, G. (2004). *Optics, MIT Open Course*: http://ocw.mit.edu/courses/mechanical-engineering/2–71-optics-fall-2004/

Condon, J. J. (1974). Confusion and flux density error distributions. *ApJ*, **188**, 279.

Rieke, G. H. (2003). *Detection of Light from the Ultraviolet to the Submillimeter*, 2nd edn. Cambridge: Cambridge University Press.

Wall, J. V. and Jenkins, C. R. (2003). *Practical Statistics for Astronomers*. Cambridge: Cambridge University Press.

2

Telescopes

2.1 Basic principles

Astronomy centers on the study of vanishingly faint signals, often from complex fields of sources. Job number one is therefore to collect as much light as possible, with the highest possible angular resolution. So life is simple. We want to:

1. build the largest telescope we can afford (or can get someone else to buy for us),
2. design it to be efficient,
3. at the same time shield the signal from unwanted contamination,
4. provide diffraction-limited images over as large an area in the image plane as we can cover with detectors, and
5. adjust the final beam to match the signal optimally onto those detectors.

Figure 2.1 shows a basic telescope that we might use to achieve these goals.

Achieving condition 4 is a very strong driver on telescope design. The etendue is conserved (equation 1.12) only for a beam passing through a perfect optical system. For such a system, the image quality is limited by the wavelength of the light and the diffraction at the limiting aperture for the telescope (normally the edge of the primary mirror). Assuming that the telescope primary is circular, the resulting image illumination is

$$I(\theta) \propto \left(\frac{J_1(2m)}{m}\right)^2 \tag{2.1}$$

where $m = (\pi \, (D/2) \, \theta/\lambda)$, $(D/2)$ is the radius of the telescope aperture, λ is the wavelength of operation, and J_1 is a Bessel function of the first kind. The result is the well-known Airy function (Figure 2.2), named after the British astronomer George Biddell Airy. We will illustrate a simple derivation later in this chapter.

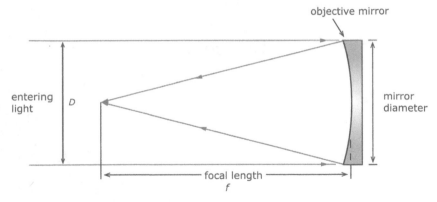

Figure 2.1. Prime-focus telescope.

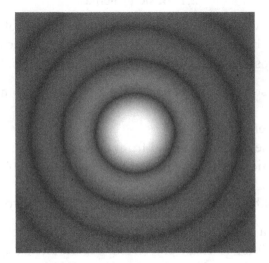

Figure 2.2. The Airy function.

The first zero of the Bessel function occurs at $m = 1.916$, or for small θ, at $\theta \cong 1.22\,\lambda/D$. The classic Rayleigh Criterion states that two point sources of equal brightness can be distinguished only if their separation is $\geq 1.22\,\lambda/D$, that is, if the peak of one image is no closer than the first dark ring in the other (with high signal-to-noise images and computer processing, sources closer than this limit can now be distinguished reliably). The full width at half maximum (FWHM) of the central image is about $1.03\,\lambda/D$ for an unobscured aperture and becomes slightly narrower if there is a blockage in the center of the aperture (e.g., a secondary mirror).

Although we will emphasize a more mathematical approach in this chapter, recall from the previous chapter that the diffraction pattern can be

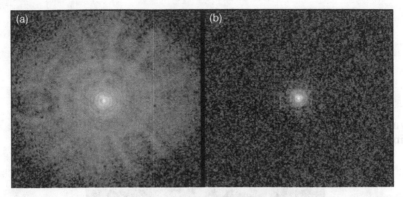

Figure 2.3. The most famous example of spherical aberration, the Hubble Space Telescope images before (left) and after (right) correction. From NASA Images.

understood as a manifestation of Fermat's Principle. Where the ray path lengths are so nearly equal that they interfere constructively, we get the central image. The first dark ring lies where the rays are out of phase and interfere destructively, and the remainder of the bright and dark rings represent constructive and destructive interference at increasing path differences (e.g., the first bright ring occurs at a path difference of one wavelength).

Although the diffraction limit is the ideal, practical optics have a series of shortcomings, described as aberrations. There are three primary geometric aberrations:

1. *Spherical* aberration occurs when an off-axis input ray is directed in front of or behind the image position for an on-axis input ray, with rays at the same off-axis angle crossing the image plane symmetrically distributed around the on-axis image. Spherical aberration tends to yield a blurred halo around an image (Figure 2.3).

   A spherical reflector imaging a source at infinity has strong spherical aberration. It forms a perfect image of a point source located at the center of the sphere, with the image produced on top of the source. However, for a source at infinity, the reflected rays cross the mirror axis at smaller values of f the farther off-axis they impinge on the mirror, where f is the distance along the optical axis measured from the mirror center (see Figure 2.1). The result is shown in Figure 2.4; there is no point along the optical axis with a well-formed image.

2. *Coma* occurs when input rays arriving at an angle from the optical axis miss toward the same side of the on-axis image no matter where they enter

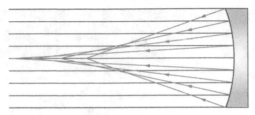

Figure 2.4. The focusing properties of a spherical reflector for a source at infinity.

Figure 2.5. Comatic image displayed to show interference.

the telescope aperture (so long as they all come from the same off-axis angle), and with a progressive increase in image diameter with increasing distance from the center of the field. Coma is axially symmetric in the sense that similar patterns are generated by rays incoming at all azimuthal positions so long as they are at the same off-axis angle. Comatic images have characteristic fan or comet shapes with the tail pointing away from the center of the image plane (Figure 2.5). Optical systems that do not satisfy the Abbe sine condition suffer from coma; a paraboloid is an example (see exercise 2.1).

3. *Astigmatism* (Figure 2.6) is a cylindrical wavefront distortion resulting from an optical system that has different focal planes in one direction perpendicular to the optical axis of the system compared with the orthogonal direction. It results in images that are elliptical on either side of best-focus, with the direction of the long axis of the ellipse changing by 90° going from ahead to behind focus.

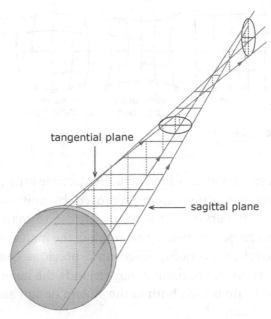

tangential plane

sagittal plane

Figure 2.6. Astigmatism.

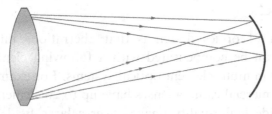

Figure 2.7. Curvature of field.

Two other aberrations are less fundamental but in practical systems can degrade the images:

4. *Curvature of field* occurs when the best images are not formed at a plane but instead on a surface that is convex or concave toward the telescope entrance aperture (see Figure 2.7).

5. *Distortion* arises when the image scale changes over the focal plane. That is, if the system observes a set of point sources placed on a uniform grid, their relative image positions are displaced from the corresponding grid positions at the focal plane. Figure 2.8 illustrates symmetric forms of distortion, but this defect can occur in a variety of other forms.

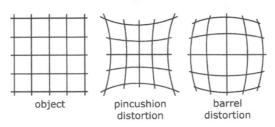

object   pincushion   barrel
          distortion   distortion

Figure 2.8. Distortion.

Neither of these last two aberrations actually degrades the intrinsic images. The effects of field curvature can be removed by suitable design of the following optics, or by curving the detector surface to match the curvature of field. Distortion can be removed by resampling the images and reconstructing them with correct proportions, based on a previous characterization of the distortion effects. Such resampling may degrade the images, but need not do so significantly if care is used both in the system design and in the method to remove the distortion.

Optical systems using lenses are also subject to another form of aberration:

6. *Chromatic aberration* results when different colors of light are not brought to the same focus.

A central aspect of optics design is that aberrations introduced by an optical element can be compensated with a following element to improve the image quality in multi-element optical systems. For example, some high-performance commercial camera lenses have up to 20 elements that all work together to provide high-quality images over a large field. A simpler case applies to an achromatic lens, in which a positive element of one type of glass is combined with a negative element of another with different wavelength dependence of the refractive index to bring two wavelengths to the same focus. An apochromatic lens has additional elements and brings three wavelengths to the same focus.

In addition to these aberrations, the telescope performance can be degraded by manufacturing errors that result in the optical elements not having the exact prescribed shapes, by distortions of the elements in their mountings, and by misalignments. Although these effects are also sometimes termed aberrations, they are of a different class from the six aberrations we have listed. In operation, the images can be further degraded typically to $\sim 1''$ diameter by atmospheric seeing, the disturbance of the wavefronts of the light from the source as they pass through a turbulent atmosphere with refractive index variations due to temperature inhomogeneity (discussed in Chapter 7).

There are a number of ways to describe the imaging performance of a telescope or other optical system. The nature of the diffraction-limited image is a function of the details of the telescope configuration (shape of primary mirror, size of secondary mirror, etc.), so we define "perfect" imaging as that obtained for the given configuration and with all other aspects of the telescope performing perfectly. The ratio of the signal from the brightest point in this perfect image divided into the signal from the same point in the achieved image is the Strehl ratio. We can measure the ratio of the total energy in an image to the energy received through a given round aperture at the focal plane, a parameter called the encircled energy within that aperture. We can also describe the root-mean-square (rms) errors of the delivered wavefronts. There is no simple way to relate these performance descriptors to one another, since they each depend on different aspects of the optical performance. However, roughly speaking, a telescope can be considered to be diffraction-limited if the Strehl ratio is ≥ 0.8 or the rms wavefront errors are $\leq \lambda/14$. This relation is known as the Maréchal criterion. A more general relation between the Strehl ratio, S, and the rms wavefront error, σ, is an extended version of an approximation due to Maréchal:

$$S \approx e^{-(2\pi\sigma/\lambda)^2} \tag{2.2}$$

2.2 Telescope design

We will discuss general telescope design in the context of optical telescopes in the following two sections, and then in Section 2.4 will expand the treatment to telescopes for other wavelength regimes.

In general, telescopes require the use of combinations of optical elements. The basic formulae we will use to describe the functions of these elements can be found in Section 1.4. For more, consult any optics text. However, all of these relations are subsumed for modern telescope design into computer "ray-tracing" programs that follow multiple rays of light through a hypothetical train of optical elements, applying the laws of refraction and reflection at each surface to produce a simulated image. The properties of this image can be optimized iteratively, with assistance from the program. These programs usually can also include diffraction to provide a physical optics output in addition to the geometric ray trace.

Except for very specialized applications, refractive telescopes using lenses for their primary light collectors are no longer employed in astronomy; in addition to chromatic aberration, large lenses are difficult to support without distortion and they absorb some of the light rather than delivering it to the image plane (in the infrared, virtually all the light is absorbed). However, refractive elements are important in other aspects of advanced telescope design, to be discussed later.

Reflecting telescopes are achromatic and are also efficient because they can be provided with high-efficiency mirror surfaces. From the radio to the near-ultraviolet, metal mirror surfaces provide reflection at normal incidence with efficiencies $> 90\%$, and usually even $> 95\%$. At wavelengths shorter than $\sim 0.1\,\mu m$, metals have poor reflection for normal incidence; here, grazing incidence reflection can be used for telescopes, but with significant changes in configuration as discussed in Section 2.4.4.

Although modern computer-aided optical design would in principle allow a huge combination of two-mirror telescope systems, the traditional conic-section-based concepts already work very well. The ability of conic-section mirrors to form images can be described in terms of Fermat's Principle. As the surface of a sphere is everywhere the same distance from its center, so a concave spherical mirror images an object onto itself when that object is placed at the center of curvature. An ellipse defines the surface that has a constant sum of the distances from one focus to another, so an ellipsoid images an object at one focus to the position of its second focus. A parabola can be described as an ellipse with one focus moved to infinite distance. More specifically it is defined as the surface for which reflected rays describe equal distance from points on a plane perpendicular to its axis to its focus. There-fore, it images an object at large distance to its focus. A hyperbola is defined as the locus of points whose difference of distances from two foci is constant. Thus, a hyperboloid forms a perfect image of a virtual image.

The basic telescope types based on the properties of these conic sections are:

1. A *prime-focus* telescope (Figure 2.1) has a paraboloidal primary mirror and forms images directly at the mirror focus; since the images are in the center of the incoming beam of light, this arrangement is inconvenient unless the telescope is so large that the detector receiving the light blocks a negligible portion of this beam.
2. A *Newtonian* telescope (Figure 2.9) uses a flat mirror tilted at 45° to bring the focus to the side of the incoming beam of light where it is generally more conveniently accessed.
3. A *Cassegrain* telescope intercepts the light from its paraboloidal primary ahead of the focus with a convex hyperboloidal mirror. This mirror refocuses the light from the virtual image formed by the primary/secondary to a second focus. Usually the light passes through a hole in the center of the primary and the second focus is conveniently behind the primary and out of the incoming beam.

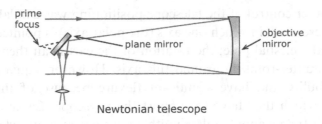

Newtonian telescope

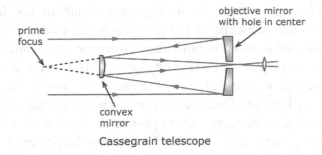

Cassegrain telescope

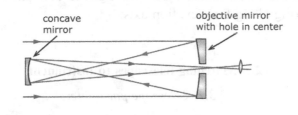

Gregorian telescope

Figure 2.9. The basic optical telescope types.

4. A *Gregorian* telescope brings the light from its paraboloidal primary mirror to a focus, and then uses an ellipsoidal mirror beyond this focus to bring it to a second focus, usually behind the primary mirror as in the Cassegrain design.

Assuming perfect manufacturing of their optics and maintenance of their alignment, all of these telescope types are limited in image quality by the coma due to the failure of their paraboloidal primary mirrors to satisfy the Abbe sine condition (equation (1.22)). A less obvious conic-section combination is the Dall–Kirkham, with an ellipsoidal primary and spherical secondary. Although its optics are relatively easy to make, this design suffers from even larger amounts of coma than for the paraboloid-primary types, and hence is seldom used.

Whatever their optical design, groundbased telescopes are placed on mounts that allow pointing them to any desired point in the accessible sky.

Before computer control of the telescope positioning was available, equatorial mounts were used, in which one axis (the polar axis) is pointed toward the
north or south celestial pole; the rotation of the earth can then be compensated by a counter-rotation around this axis. However, equatorial mounts
tend to be bulky and have significant flexure because of the variety of
directions in which they have to support the telescope. Large modern telescopes use altitude-azimuth (altazimuth or alt-az) mounts where one axis
rotates the telescope around an axis perpendicular to the horizon and
the other points the telescope in elevation. Sidereal tracking then requires
computer-controlled motions in both axes. However, these mounts are much
more compact than equatorial ones and also tend to flex far less.

To avoid flexure in very large instruments, they are often mounted at
specialized foci where the force of gravity is always in the same direction.
Equatorially mounted telescopes accomplished this goal with a Coudé focus,
parallel to their polar axes. A similar benefit can be achieved in an alt-az
mount with a Nasmyth focus, where the light is brought out through the
elevation bearing along the elevation axis.

2.3 Matching telescopes to instruments

2.3.1 Telescope parameters

To match the telescope output to an instrument, we need to describe the
emergent beam of light in detail.[2]

A convenient vocabulary to describe the general matching of the two is
based on the following terms:

The *focal length* is measured by projecting the conical bundle of rays
that arrives at the focus back until its diameter matches the aperture of
the telescope; the focal length is the length of the resulting ray bundle. See
Figure 2.1 for the simplest example. This definition allows for the effects, for
example, of secondary mirrors in modifying the properties of the ray bundle
produced by the primary mirror.

The *f-number* of the telescope is the diameter of the incoming ray bundle,
that is, the telescope aperture, divided into the focal length, f/D in Figure 2.1.
This term is also used as a general description of the angle of convergence of a
beam of light, in which case it is the diameter of the beam at some distance
from its focus, divided into this distance. The f-number can also be expressed
as $0.5 \cot(\theta)$, where θ is the beam half-angle. Beams with large f-numbers

---

[2] During the process of designing an instrument, an optical model of the telescope is often included in the
ray trace of the instrument to be sure that the two work together as expected.

are "slow" while those with small ones are "fast." Although this terminology may seem strange applied to astronomical instrumentation, it originates in photography where large f-numbers are imposed by artificially reducing the aperture of a lens (stopping it down), reducing the amount of light it delivers, and therefore requiring longer exposures – making the photography slow.

The *plate scale* is the translation from physical units at the telescope focal plane, for example mm, to projected angular ones, for example arcsec. As for the derivation of the thin lens formula, we can determine the plate scale by treating the telescope as an equivalent single lens and recalling that any ray passing through the telescope on the optical axis of this lens is not deflected. Thus, if we define the magnification of the telescope, M, to be the ratio of the f-number delivered to the focal plane to the primary mirror f-number, then the equivalent focal length of the telescope is

$$f_{equivalent} = M f_{primary} \tag{2.3}$$

and the angle on the sky corresponding to a distance b at the focal plane is

$$\theta_b \text{ (radians)} = b/f_{equivalent} \tag{2.4}$$

As an illustration of these concepts, equation (1.12) can be used to determine the required f-number of the beam at the focus of a telescope, given the desired mapping of the size of a detector element into its resolution element on the sky. From the Lagrange Invariant (equation 1.20), we can match to any detector area by adjusting the optics to shape the beam to an appropriate solid angle. Simplifying to a round telescope aperture of diameter D accepting a field of angular diameter θ_{in} and a detector of diameter d accepting a beam from the telescope of angular diameter θ_{out},

$$D\theta_{in} = d\theta_{out}$$

Thus, if the detector element is $10\,\mu m = 1 \times 10^{-5}\,m$ on a side and we want it to map to 0.1 arcsec $= 4.85 \times 10^{-7}\,rad$ on the sky on a 10-m aperture telescope, we have

$$10\,m \times 4.85 \times 10^{-7}\,rad = 1 \times 10^{-5}\,m\,\theta_{out}$$

Solving, $\theta_{out} = 0.485$ radians $= 27.8$ degrees. The large angle of incidence onto the detector illustrates the issue of matching the outputs of large telescopes onto modern detector arrays with their relatively small pixels. The optimum projected pixel size depends on the application of an instrument, but see Section 4.4.

The *field of view* (FOV) is the total angle on the sky that can be imaged by the telescope. It might be limited by the telescope optics and baffles, or by the dimensions of the detector (or acceptance angle of an instrument).

A *stop* is a baffle or other construction (e.g., the edge of a mirror) that limits the bundle of light that can pass through it. The *aperture stop* limits the diameter of the incoming ray bundle from the object. For a telescope, it is typically the edge of the primary mirror. A *field stop* limits the range of angles a telescope can accept, that is, it is the limit on the telescope field of view. In addition to these two examples, stops can be used for other purposes such as to eliminate any stray light from the area of the beam blocked by the secondary mirror and other structures in front of the primary mirror.

A *pupil* is an image of the aperture stop or primary mirror. An *entrance pupil* is an image of the aperture stop formed by optics ahead of the stop. The term entrance pupil is also applied to the illumination of the primary mirror itself.

An *exit pupil* is an image of the aperture stop formed by optics behind the stop. For example, the secondary mirror of a Cassegrain (or Gregorian) telescope forms an image of the aperture stop/primary mirror and hence determines the exit pupil of the telescope. The exit pupil is often reimaged within an instrument because it has unique properties for performing a number of optical functions. Because the primary mirror is illuminated uniformly by an astronomical source, the light from a source is spread uniformly over the pupil, making it the ideal place for filters, dispersers in a spectrometer, and masks for a coronagraph, to name a few examples.

2.3.2 An example of telescope design

Consider the Cassegrain telescope in Figure 2.9. Assume its primary mirror is 4 meters in diameter and $f/3$ so its focal length is 12 meters. The secondary mirror is placed 3 meters before the primary mirror focus, where the beam is 1 meter in diameter. It forms images 1 meter behind the vertex of the primary.[3]

Where is the exit pupil?

First we determine the focal length of the Cassegrain secondary mirror from the thin lens formula

$$\frac{1}{f} = \frac{1}{s_1} + \frac{1}{s_2}$$

With $s_1 = -3\,\mathrm{m}$ and $s_2 = 10\,\mathrm{m}$ (it helps to draw the optics as the lens-equivalent to understand the signs), we have $f = -4.2857\,\mathrm{m}$. To locate the exit pupil, we can imagine a tiny light laid on the surface of the primary

---

[3] The vertex is where the optical axis crosses the surface of an optical element; in this case, it is where the primary mirror surface would cross the optical axis if there were not a hole through the mirror.

mirror (near the center) and calculate where its image would lie. We re-apply the thin lens formula, this time with $s_1 = 9$ m and $f = -4.2857$ m, finding that s_2 is -2.903 m. That is, the exit pupil is 2.903 m behind the vertex of the secondary mirror.

Suppose we place a lens with a focal length of 0.25 m, 0.4 m behind the focal plane of the telescope. Using the thin lens formula, this lens will relay the focal plane back by 0.667 m from the lens, or 1.067 m from the original focal plane, while increasing the image scale by a factor of 1.667. Also, the reimaged pupil is located via: $s_1 = 2.903$ m $+ 10$ m $+ 0.4$ m $= 13.303$ m and $f = 0.25$ m, so the pupil is 0.5195 m behind our lens, or 0.148 m in front of the reimaged focal plane.

2.3.3 Image description

The image formed by a telescope departs from the diffraction-limited ideal for many reasons, such as: (1) aberrations in the telescope optics; (2) manufacturing errors; and (3) misalignments. Further degradation is likely to occur in any instrument used with the telescope. Therefore, we need a convenient way to understand the combined imaging properties of the telescope plus instrumentation being used with it.

A simple approach to measuring imaging capability is illustrated by Figure 2.10. One takes an image of black and white alternating bars of ever closer spacing. We can think of the spacing of the bars as the period, P, of the spatial variations, and then $f_{sp} = 1/P$ is the *spatial frequency*. In Figure 2.10, the blurring of the input image results in reduced contrast in the bars as the spatial frequency is increased. We describe the limit where the line pairs can no longer be distinguished as the line pairs per mm limit to the resolution capability of the imaging system. The conventional limit is set at the spatial frequency where the line contrast has been reduced to 4%.

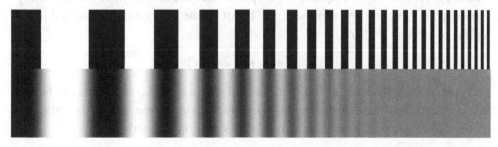

Figure 2.10. Bar chart test of resolution. The upper half shows the object imaged, while the lower half is the image and shows the blurring due to the optical system. Based on material from Norman Koren (n.d.), with permission.

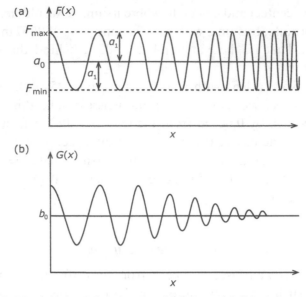

Figure 2.11. Modulation of a signal – (a) input and (b) output.

The *modulation transfer function*, or MTF, makes use of the properties of Fourier transforms to provide a more general description of the imaging properties of an optical system. It falls short of a complete description of these properties because it does not include phase information; the optical transfer function (OTF) is the MTF times the phase transfer function. However, the MTF is more convenient to manipulate and is an adequate description for many applications.

To illustrate the meaning of the MTF, imagine that an array of detectors is used to make an image of a chart similar to Figure 2.10. To make the output easier to interpret, the "bars" are made sinusoidal in density, so they provide a pure spectrum of spatial frequencies (the square wave nature of the conventional resolution chart requires high spatial frequencies to produce the sharp bar edges). We can then represent the input to the array as a sinusoidal input signal of period $P = 1/f_{sp}$,

$$F(x) = a_0 + a_1 \sin(2\pi f_{sp}x) \tag{2.5}$$

Here, x is the distance along one axis of the array, a_0 is the mean height (above zero) of the pattern, and a_1 is its amplitude. These terms are indicated in Figure 2.11(a). The modulation of this signal is defined as

$$M_{in} = \frac{F_{max} - F_{min}}{F_{max} + F_{min}} = \frac{a_1}{a_0} \tag{2.6}$$

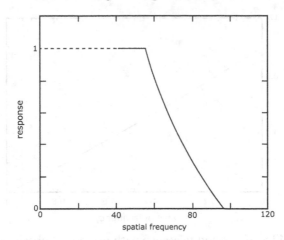

Figure 2.12. The MTF (schematically) corresponding to the function in Figure 2.11(b).

where F_{max} and F_{min} are the maximum and minimum values of $F(x)$.
The resulting output can be represented by

$$G(x) = b_0 + b_1(f_{sp}) \sin(2\pi f_{sp} x) \tag{2.7}$$

where x and f_{sp} are the same as in equation (2.5), and b_0 and $b_1(f_{sp})$ are analogous to a_0 and a_1 (Figure 2.11(b)). The limited ability to image very fine details results in a progressive reduction in the amplitude of the sinusoidal pattern. The dependence of the signal amplitude, b_1, on the spatial frequency, f_{sp}, shows the drop in response as the spatial frequency grows (i.e., it describes the loss of fine detail in the image). The modulation in the image will be

$$M_{out} = \frac{b_1(f_{sp})}{b_0} \leq M_{in} \tag{2.8}$$

The modulation transfer factor is

$$MT = \frac{M_{out}}{M_{in}} \tag{2.9}$$

A separate value of the MT will apply at each spatial frequency. This frequency dependence of the MT is expressed in the modulation transfer function. Figure 2.12 shows the MTF corresponding to the response of Figure 2.11(b).

In principle, the MTF provides a virtually complete specification of the imaging properties of an optical system. Computationally, the MTF can be determined by taking the absolute value of the Fourier transform, $F(u)$, of the image of a perfect point source (this image is called the point spread function

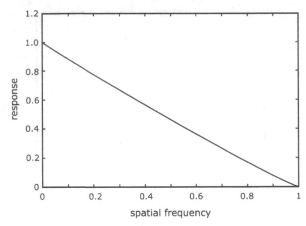

Figure 2.13. MTF of a round telescope with no central obscuration. Spatial frequencies are in units of D/λ.

(PSF)). Fourier transformation is the general mathematical technique used to determine the frequency components of a function $f(x)$ (see, e.g., Press *et al.* 1986; Bracewell 2000). $F(u)$ is defined as

$$F(u) = \int_{-\infty}^{\infty} f(x)\, e^{-j2\pi ux} dx \tag{2.10}$$

with inverse

$$f(x) = \int_{-\infty}^{\infty} F(u)\, e^{j2\pi ux}\, du \tag{2.11}$$

Common examples can be found in Appendix B. The Fourier transform can be generalized in a straightforward way to two dimensions, but for the sake of simplicity we will not do so here. The absolute value of the transform is

$$|F(u)| \;=\; (F(u)\,F^*(u))^{1/2} \tag{2.12}$$

where $F^*(u)$ is the complex conjugate of $F(u)$; it is obtained by reversing the sign of all imaginary terms in $F(u)$. If $f(x)$ represents the PSF, $|F(f_{sp})|/|F(0)|$ is the MTF of the system. The MTF is normalized to unity at spatial frequency 0 by this definition. As emphasized in Figure 2.12, the response at zero frequency cannot be measured directly but must be extrapolated from higher frequencies. As an example, the MTF for a diffraction-limited telescope is shown in Figure 2.13.

Why does the PSF characterize the optical performance so completely? A sharp impulse or point source, represented mathematically by a δ function, contains all frequencies equally (i.e., its Fourier transform is a constant).

Hence the Fourier transform of the image formed from an input sharp impulse fully describes the attenuation of the high spatial frequencies in the image.

The image of an entire linear optical system is the convolution of the images from each element. Convolution refers to the process of cross-correlating the images, multiplying an ideal image of the scene by the PSF for every possible placement of the PSF on the scene to determine how the scene would look as viewed by the optical system. That is, if $O(x)$ (the image) is the convolution of $S(x)$ (the scene) and $P(x)$ (the PSF), we write:

$$O = S \otimes P \tag{2.13}$$

which is shorthand for

$$O(x_0) = \int_0^\infty S(x)P(x - x_0)\, dx \tag{2.14}$$

It is far more convenient to work in Fourier space, where according to the Convolution Theorem,

$$F[O(\varsigma)] = F[S(\varsigma)]F[P(\varsigma)] \tag{2.15}$$

That is, we can avoid the complicated integration in equation (2.14) and simply multiply the transforms to get the transform of the convolution.

By the Convolution Theorem, the MTF (say, of a complex optical system) can be determined by multiplying together the MTFs of the constituent optical elements, and the resulting image is then determined by inverse transforming the resulting MTF. The overall resolution capability of complex optical systems can be more easily determined in this way than by brute force image convolution.

2.3.4 Examples of image behavior

We illustrate the power of these approaches by deriving that a chain of optical elements each of which forms a Gaussian image makes a final image that is also Gaussian, with a width equal to the quadratic combination of the individual widths. We make use of the fact that the Fourier transform of a Gaussian is itself a Gaussian: if

$$f(x) = e^{-\pi x^2} \tag{2.16}$$

then

$$F(u) = e^{-\pi u^2} \tag{2.17}$$

Also, the Fourier transform of $f(ax)$ is

$$\frac{1}{|a|} F\left(\frac{u}{a}\right) \tag{2.18}$$

if $F(u)$ is the transform of $f(x)$. Then, if the image made by the ith optical element is

$$f_i(x) = C_i e^{-\pi(x/a_i)^2} \tag{2.19}$$

its Fourier transform is

$$F_i(u) = C_i a_i e^{-\pi a_i^2 u^2} \tag{2.20}$$

By the Convolution Theorem, the transform of the image resulting from the series of elements is the product of the transforms of each one, or

$$F_{\text{final}}(u) = \prod C_i a_i e^{-\pi a_i^2 u^2} = K e^{-\pi u^2 \Sigma a_i^2} \tag{2.21}$$

To get the image, we do the inverse transform:

$$f_{\text{final}}(x) = C e^{-\pi \left(x/\sqrt{\Sigma a_i^2}\right)^2} \tag{2.22}$$

which is what we set out to demonstrate.

We selected Gaussian image profiles for this example because this shape has the property that its Fourier transform is the same function. Only a relatively small number of functions have analytic Fourier transforms that are easy to manipulate. However, with the use of computers to calculate transforms and reverse transforms for any function, the approach can be generalized to any combination of optical elements and imaging properties.

We now show how the Airy function is derived. To do so, we imagine that we are broadcasting a signal from the telescope (we will encounter this point of view again in Chapter 8 where it is termed the Reciprocity Theorem). The field pattern of the resulting beam on the sky is $E(\phi)$, the Fourier transform of the electric field illumination onto the telescope. By the Van Cittert–Zernicky Theorem (Born and Wolf 1999):

$$E(\phi) = \int E(x) e^{-j2\pi x\phi} \, dx \tag{2.23}$$

and the far-field intensity, that is, the PSF, is the square modulus $(E\,E^*)$ of the field pattern (compare equation (1.9)). We approach the problem in one dimension. We define the box function in one dimension as

$$\text{rect}(x) = 1, x < \tfrac{1}{2}$$
$$= 0, x \geq \tfrac{1}{2} \tag{2.24}$$

with Fourier transform

$$F(u) = \frac{\sin(\pi u)}{\pi u} = \text{sinc}(u) \tag{2.25}$$

The box function represents a uniformly illuminated line of length 1. It can be used to impose limits of integration in equation (2.23) of $\pm D/2\lambda$. From equation (2.25),

$$E(\phi) = \text{sinc}\left(\phi D /_{\lambda}\right) \tag{2.26}$$

The radiated power (the PSF) is proportional to the square of the electric field:

$$P(\phi) \propto \text{sinc}^2\left(\phi D /_{\lambda}\right) \tag{2.27}$$

We now sketch how to apply this approach in two dimensions. We describe a round aperture of radius a functionally in terms of the cylinder function:

$$C(x,y) = 1 \quad for \quad \sqrt{x^2 + y^2} \leq a.$$
$$= 0 \quad for \quad \sqrt{x^2 + y^2} > a \tag{2.28}$$

The Fourier transform is

$$C(u,a) = 2\pi a^2 \frac{J_1(ua)}{ua} \tag{2.29}$$

where $J_1(x)$ is a Bessel function of the first kind. The intensity in the image is

$$I(\theta) \propto \left(\frac{2 J_1(2\pi(D/2)\theta/\lambda)}{2\pi(D/2)\theta/\lambda}\right)^2 \tag{2.30}$$

This result is identical to equation (2.1) describing the diffraction-limited image. The corresponding MTF is

$$T(w) = \frac{2}{\pi}\left[\arccos\left(\frac{\lambda w}{2(D/2)}\right) - \frac{\lambda w}{(D/2)}\left(1 - \frac{\lambda^2 w^2}{4(D/2)^2}\right)^{1/2}\right] \tag{2.31}$$

(see Figure 2.13).

Of course, the images from real telescopes are much more complex because of additional diffraction effects. The secondary supports cause prominent linear features that have symmetry across the center of the image, for example a telescope with three secondary supports spaced at 120° will have six diffraction artifacts spaced at 60°. If the primary mirror is segmented, additional complex artifacts are introduced.

2.4 Telescope optimization

2.4.1 Wide field

The classic telescope designs all rely on the imaging properties of a paraboloid; secondary mirrors relay the image formed by the primary mirror to a second focus. What would happen if no image were demanded of the primary, but the conic sections on the primary and secondary mirrors were selected to provide optimum imaging as a unit? The simplest goal would be to modify the optics to meet the Abbe sine condition. Conceptually, one could envision modifying the secondary mirror of a traditional Cassegrain telescope so that narrow beams launched from the telescope focus to the secondary over a range of angles were reflected to the primary mirror to satisfy equation (1.22), and then modifying the primary mirror to direct these beams in a parallel configuration toward a distant source. This approach yields the Ritchey–Cretién telescope. In this design, a suitable selection of hyperbolic primary and secondary together can compensate for the spherical and comatic aberrations (to third order in an expansion in angle of incidence) and provide large fields, up to an outer boundary where astigmatism becomes objectionable. The change is subtle: the entire modification is in the form of tiny modifications of the mirror shapes. An alternative, the aplanatic Gregorian telescope, has similar wide-field performance.

Still larger fields can be achieved by adding optics specifically to compensate for the aberrations. A classic example is the Schmidt camera, which has a full-aperture refractive corrector in front of its spherical primary mirror. The corrector imposes spherical aberration on the input beam in the same amount but opposite sign as the aberration of the primary, producing well-corrected images over a large field. The Maksutov and Schmidt–Cassegrain designs achieve similar ends with correctors of different types.

However, refractive elements larger than 1–1.5 m in diameter become prohibitively difficult to mount and maintain in figure. For large telescopes, a variety of designs for correctors near the focal plane are possible (e.g., Epps and Fabricant 1997). A simple version invented by F. E. Ross uses two lenses, one positive and the other negative, to cancel each other in optical power but to correct coma. Wynne (1974) proposed a more powerful device with three lenses, of positive-negative-positive power and all with spherical surfaces.

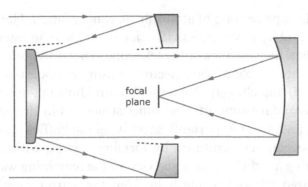

Figure 2.14. Three-mirror wide-field telescope design after Baker and Paul.

A Wynne corrector can correct for chromatic aberration, field curvature, spherical aberration, coma, and astigmatism. Similar approaches are used with all the wide-field cameras on 10-m-class telescopes. These optical trains not only correct the field but also can provide for spectrally active components, such as compensators for the change of refractive index of the air with wavelength (the dispersion), allowing high-quality images to be obtained at the off-zenith pointings and over broad spectral bands (e.g., Fabricant *et al.* 2004) (see Section 4.2).

Even wider fields can be obtained by adding large reflective elements to the optical train of the telescope, an approach pioneered by Baker and Paul; see Figure 2.14. The basic concept is that, by making the second mirror spherical, the light is deviated in a similar manner to the way it would be by the corrector lens of a Schmidt camera. The third mirror, also spherical, brings the light to the telescope focal plane over a large field. The Baker–Paul concept is a very powerful one and underlies the large field of the Large Synoptic Survey Telescope (LSST). With three or more mirror surfaces to play with, optical designers can achieve a high degree of image correction in other ways also. For example, a three-mirror anastigmat is corrected simultaneously for spherical aberration, coma, and astigmatism.

2.4.2 Infrared

The basic optical considerations for groundbased telescopes apply in the infrared as well. However, in the thermal infrared (wavelengths longer than about $2\,\mu$m), the emission of the telescope is the dominant signal on the detectors and needs to be minimized to maximize the achievable signal-to-noise on astronomical sources. It is not possible to cool any of the exposed parts of the telescope because of atmospheric condensation. However, the sky is both colder and of lower emissivity (since one looks in bands where it is transparent) than the structure of the telescope. Therefore, one minimizes the telescope structure visible to the detector by using a modest-sized secondary mirror (e.g., $f/15$), matching it with a small central hole in the primary mirror, building the secondary supports and other structures in the beam with thin cross-sections, and, most importantly, removing the baffles. The instruments operate at cryogenic temperatures; by forming a pupil within their optical trains and putting a tight stop at it, the view of the remaining warm structures can be minimized without losing light from the astronomical sources. To avoid excess emission from the area around the primary mirror entering the system, often the secondary mirror is made slightly undersized, so its edge becomes the aperture stop of the telescope. Only emission from the sky then enters the instrument from beyond the edge of the secondary, and in high-quality infrared atmospheric windows the sky is an order of magnitude less emissive than the primary mirror cell would be.

The emissivity of the telescope mirrors needs to be kept low. If the telescope uses conventional aluminum coatings, they must be kept very clean. Better performance can be obtained with silver or gold coatings. Obviously, it is also important to avoid warm mirrors in the optical train other than the primary and secondary. Therefore, the secondary mirror is called upon to perform a number of functions besides helping form an image. Because of the variability of the atmospheric emission, it is desirable to compare the image containing the source with an image of adjacent sky rapidly, and an articulated chopping secondary mirror can do this in a fraction of a second. Modulating the signal with the secondary mirror has the important advantage that the beam passes through nearly the identical column of air for both chop positions, starting high in the atmosphere and continuing as the beam traverses the telescope to its focus. Therefore, any fluctuations in the infrared emission along this column are common to both mirror positions and tend to cancel in the difference of the images at the two positions. A suitably designed secondary mirror can also improve the imaging, either by compensating for image motion or even by correcting the wavefront errors imposed by the atmosphere (see Chapter 7).

The ultimate solution to the issue of thermal background is to place the infrared telescope in space, where it can be cooled sufficiently that its emission no longer dominates the signals. Throughout the thermal infrared, the foreground signals can be reduced by a factor of a million or more in space compared with the ground, and hence phenomenally greater sensitivity can be achieved. Obviously, space also offers the advantage of being above the atmosphere and hence free of atmospheric absorption, so the entire infrared range is accessible for observation. Astronomy has benefited from a series of very successful cryogenic telescopes in space – the Infrared Astronomy Satellite (IRAS), the Infrared Space Observatory (ISO), the Spitzer Telescope, Akari, the Cosmic Background Explorer (COBE), and the Wide-Field Infrared Survey Explorer (WISE), to name some examples.

2.4.3 Radio

Radio telescopes are typically of conventional parabolic-primary-mirror, prime-focus design, on alt-azimuth mounts. The apertures are very large, so the blockage of the beam by the receivers at the prime focus is insignificant. The primary mirrors have short focal lengths, f-ratios ~ 0.5, to keep the telescope compact and help provide a rigid structure. Sometimes the entire telescope is designed so the flexure as it is pointed in different directions occurs in a way that preserves the figure of the primary – these designs are described as deforming homologously. For example, the 100-m aperture Effelsberg Telescope flexes by up to 6 cm as it is pointed to different elevations, but maintains its paraboloidal figure to an accuracy of $\sim$4 mm. Alternatively, actuators can be used to adjust the panels making up the telescope reflective surface according to pre-determined corrections, as is done with the Greenbank Byrd Telescope. Another approach is taken by the Arecibo Telescope, which has a spherical primary to allow steering the beam in different directions while the primary is fixed. The spherical aberration is compensated near the prime focus of the telescope in this case.

Telescopes for the mm- and sub mm-wave regimes are generally smaller than cm-wave ones and usually are built in a Cassegrain configuration, often with secondary mirrors that can be chopped or nutated over small angles to help compensate for background emission. They can be considered to be intermediate in design between longer-wavelength radio telescopes and infrared telescopes.

Radio receivers are uniformly of coherent detector design; only in the mm- and sub mm realms are bolometers used for continuum measurements and low-resolution spectroscopy. The operation of these devices will be described in more detail in Chapter 8. For this discussion, we need to understand that all such receivers are limited by the Antenna Theorem, which states that they are sensitive only to the diffraction-limited images and to only a single polarization. This behavior modifies how the imaging properties of the telescope are described. The primary measure of the quality of the telescope optics is *beam efficiency*, the ratio of the power from a point source in the central peak of the image to the power in the entire image. The rings of the Airy pattern and other diffraction structures appear to radio astronomers as potential regions of unwanted sensitivity to sources away from the one at which the telescope is pointed. The sensitive regions of the telescope beam outside the main Airy peak are called sidelobes.

2.4.4 Extreme ultraviolet and X-ray

A challenge in developing telescopes for the extreme ultraviolet (UV) and X-rays is that reflection at normal incidence does not work with the types of surfaces used for this function in the radio through the visible. However, certain materials have indices of refraction in the 0.01–10 keV range (0.1–0.0001 μm) that are slightly less than 1. Consequently, at grazing angles these materials reflect by total external reflection. In the extreme UV, angles of up to 10° can be used as in the Extreme Ultraviolet Explorer (EUVE), but at the high-energy end of this energy range the angle of incidence must be < 1°. We will concentrate on the X-ray range below.

X-ray telescopes are built around reflection by conic sections, just as radio and optical ones are. However, the necessity to use grazing incidence results in dramatically different configurations from those for shorter wavelengths. In addition, the images formed by grazing incidence off a paraboloid fail dramatically to meet the Abbe sine condition and have severe coma. However, two reflections can yield better images: one off a paraboloid and the other off a hyperboloid or ellipsoid. For example, a Wolter Type-1 geometry uses a reflection off a paraboloid followed by one off a hyperboloid (Figure 2.15). A Wolter Type-2 telescope uses a convex hyperboloid for the second reflection to give a longer focal length and larger image scale, while a Wolter Type-3 telescope uses a convex paraboloid and concave ellipsoid.

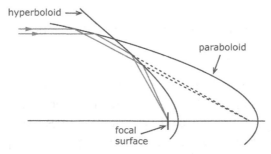

Figure 2.15. A Wolter Type-1 grazing incidence X-ray telescope – vertical scale exaggerated for clarity.

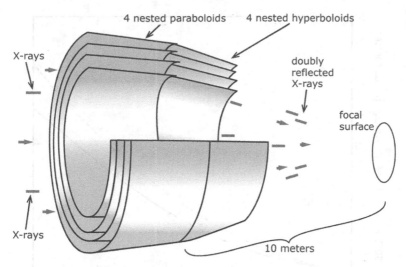

Figure 2.16. The Chandra telescope. Based on image from Chandra CXC.

The on-axis imaging quality of such telescopes is strongly dependent on the quality of the reflecting surfaces. The constraints in optical design already imposed by the grazing incidence reflection make it difficult to correct the optics well for large fields and the imaging quality degrades significantly for fields larger than a few arcminutes in radius. The areal efficiency of these optical trains is low, since their mirrors collect photons only over a narrow annulus. It is therefore common to nest a number of optical trains to increase the collection of photons without increasing the total size of the assembled telescope.

As an example, we consider Chandra (Figure 2.16). Its Wolter Type-1 telescope has a diameter of 1.2 m, within which there are four nested optical trains that together provide a total maximum collecting area of 1,100 cm^2 (i.e., ~10% of the total area within the entrance aperture). The focal length is 10 m, the angles of incidence onto the mirror surfaces range from 27 to 51 arcmin, and the reflecting material is iridium. The telescope efficiency is reasonably good from 0.1 to 7 keV. The Chandra design emphasizes angular resolution. The on-axis images are 0.5 arcsec in diameter but degrade by more than an order of magnitude at an off-axis radius of 10 arcmin (Figure 2.17).

The Chandra design can be compared with that of XMM-Newton, which emphasizes collecting area. Its individual telescopes were electroformed, allowing very thin shells that could be nested efficiently, at a cost of surface accuracy compared with the Chandra Zerodur$^®$ (a glass-ceramic) shells. Because of the lower-quality mirror surfaces, the on-axis images are an order of magnitude larger in diameter than for Chandra. In compensation,

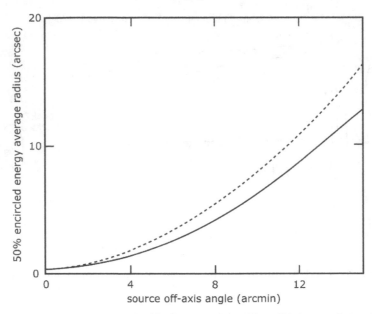

Figure 2.17. The Chandra images degrade rapidly with increasing off-axis angle. The solid line is at 1.49 keV and the dashed one at 6.40 keV.

XMM-Newton has three modules of 58 nested optical trains, each of diameter 70 cm and with a collecting area of 6000 cm² (~50% of the total enclosed in the three entrance apertures). The range of grazing angles, 18–40 arcmin, is smaller than for Chandra, resulting in better high-energy (~10 keV) efficiency.

If the requirement for optimal imaging on-axis is dropped, a family of telescope designs is possible that has more uniform images over a large field. An image size of less than 5″ (FWHM) over a one degree diameter field is possible for a telescope of similar size to the Chandra and XMM-Newton ones (Burrows *et al.* 1992). These designs no longer use conic-section mirrors, but instead the mirror shapes are based on polynomials that are varied to optimize the imaging for a specific application.

For reflecting surfaces, Chandra and XMM-Newton used single materials (e.g., platinum, iridium, gold) in grazing incidence. This approach works only at increasingly grazing incidence with increasing energy, reducing the effective area that can be achieved. Higher-energy photons can be reflected over larger angles using the Bragg effect. Figure 2.18 shows how a crystal lattice can reflect by constructive interference, when

$$m\lambda = 2d\sin\theta \tag{2.32}$$

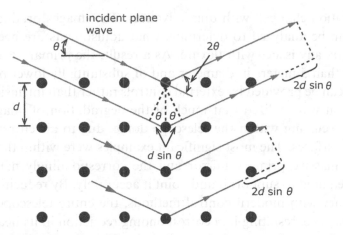

Figure 2.18. Bragg effect. The dots represent the atoms in a regular crystal lattice. Constructive interference occurs at specific grazing incidence reflection angles.

With identical layers, the reflection would be over a restricted energy range. By grading multiple layers, broad ranges are reflected. The top layers are made with large spacing (d) and reflect the low energies, while lower layers have smaller d for higher energies. A stack of up to 200 layers yields an efficient reflector.

The Nuclear Spectroscopic Telescope Array (NuStar, to be launched in 2012) uses this approach to make reflecting optics. It has a Wolter-type design with multilayer surfaces deposited on thin glass substrates, slumped to the correct shape. There are 133 reflecting shells, with a maximum diameter of $\sim$40 cm, grazing angles of up to 16 arcmin, and a focal length of 10 meters. This design provides a collecting area of $\sim$800 cm^2 at 10 keV and $\sim$60 cm^2 at 78 keV. More advanced versions of this technology are being developed for very large future missions, such as the International X-ray Observatory (IXO).

2.5 Modern optical-infrared telescopes

2.5.1 10-meter-class telescopes

For many years, the Palomar 5-m telescope was considered the ultimate large groundbased telescope; flexure in the primary mirror was thought to be an insurmountable obstacle to construction of larger ones. The benefits from larger telescopes were also argued to be modest. If the image size remains the same (e.g., is set by a constant level of seeing), then the gain in sensitivity with a background limited detector goes only as the diameter of the telescope primary mirror.

This situation changed with dual advances. The images produced by the telescope can be analyzed to determine what adjustments are needed to its primary to fix any issues with flexure. As a result, the primary can be made both larger than 5 meters in diameter and of substantially lower mass, since the rigidity can be provided by external controls rather than intrinsic stiffness. In addition, it was realized that much of the degradation of images due to seeing was occurring within the telescope dome, due to air currents arising from warm surfaces. The most significant examples were within the telescope itself – the massive primary mirror and the correspondingly massive steel structures required to support it and point it accurately. By reducing the mass of the primary with modern control methods, the entire telescope could be made less massive, resulting in a corresponding reduction in its heat capacity and a faster approach to thermal equilibrium with the ambient temperature. This adjustment is further hastened by aggressive ventilation of the telescope enclosure, including providing it with large vents that almost place the telescope in the open air while observing. The gains with the current generation of large telescopes therefore derive from both the increase in collecting area and the reduction in image size. A central feature of plans for even larger telescopes is sophisticated adaptive optics systems to shrink their image sizes further (see Chapter 7).

Three basic approaches have been developed for large groundbased telescopes. The Keck Telescopes, Gran Telescopio Canarias (GTC), Hobby-Eberly Telescope (HET), and South African Large Telescope (SALT) exemplify the use of a segmented primary mirror. The 10-m primary mirror of Keck is an array of 36 hexagonal segments, each 0.9 m on a side, and made of Zerodur®. The positions of these segments relative to each other are sensed by capacitive sensors at the segment edges. A specialized alignment camera is used to set the segments in tip and tilt and then the entire mirror is locked under control of the edge sensors. The segments must be adjusted very accurately in piston for the telescope to operate in a diffraction-limited mode. The alignment camera allows for adjustment of each pair of segments in this coordinate by interfering the light in a small aperture that straddles their edges.

The Very Large Telescope (VLT), Subaru, and Gemini telescopes use a thin monolithic plate for the optical element of the primary mirror. We take the VLT as an example – it is actually four identical examples. They each have 8.2-meter primary mirrors of Zerodur® that are only 0.175 meters thick. A VLT primary is supported against flexure by 150 actuators that are controlled by analysis of images of stars at an interval of a couple of times per minute.

The Multiple Mirror Telescope (MMT), Magellan Telescopes, and Large Binocular Telescope (LBT) (primaries respectively 6.5, 6.5, and 8.4 m in diameter) are based on a monolithic primary mirror design that is deeply relieved in the back to reduce the mass and thermal inertia. Use of a polishing lap with a computer-controlled shape allows the manufacture of very fast mirrors; for example, the two for the LBT are $f/1.14$. Actuators at the backs of these mirrors make small adjustments in their shapes. The LBT is developing pairs of instruments for the individual telescopes, such as prime focus cameras (with a red camera at the focus of one mirror and a blue camera for the other), or twin spectrographs on both sides. The outputs of the two sides of the telescope can also be combined for operation as an interferometer. The design with the primary mirrors on a single mount eliminates large path-length differences between them and gives the inter-ferometric applications a uniquely large field of view compared with other approaches.

2.5.2 Wavefront sensing

To make the adjustments that maintain their image quality, all of these telescopes depend on frequent and accurate measurement of the image changes resulting from flexure and thermal drift. A common way to make these measurements is the Shack–Hartmann sensor, which divides the wave-front at a pupil using an array of small lenslets (Figure 2.19). A "perfect" optical system maintains wavefronts that are plane or spherical. In Figure 2.19, the situation for a perfect plane wavefront is shown as a dashed line going into the lenslet array. Each lenslet images its piece of the wavefront onto the charge coupled device (CCD; the imaging process is shown as the paths of the outer rays, not the wavefronts) directly behind the lens, on its optical axis. For the plane input wavefront, these images will form a grid that is uniformly spaced.

Aberrations, flexure, and thermal drift impose deviations on the wave-fronts. An example is shown as a solid line in Figure 2.19. Each lenslet will see a locally tilted portion of the incoming wavefront. As a result, the images from the individual lenslets will be offset relative to the optical axis of the lens and displaced when they reach the CCD. A simple measurement of the positions of these images can then be used to calculate the shape of the incoming wavefront, and hence to determine the flaws in the optics from which it was delivered. These can be corrected by a combination of adjust-ments on the primary mirror and motions of the secondary one.

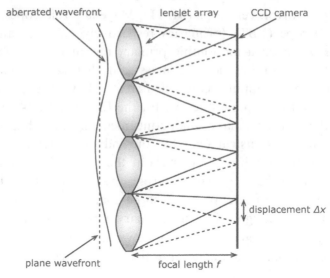

aberrated wavefront　　　　　lenslet array　　　CCD camera

displacement Δx

plane wavefront　　　　　　　focal length f

Figure 2.19. Principle of operation of a Shack–Hartmann sensor.

2.5.3 Telescopes of the future

The methods developed for control of the figure of large primary mirrors on the ground have been adopted for the James Webb Space Telescope, in this case so the 6.5-m primary mirror can be folded to fit within the shroud of the launch rocket. The demands for very light weight have led to a segmented primary mirror of beryllium. After launch, the primary is to be unfolded and then a series of ever more demanding tests and adjustments will bring it into proper figure. Periodic measurements with the near-infrared camera will be used to monitor the primary mirror figure and adjust it as necessary for optimum performance. The overall design is a three-mirror anastigmat, with a fourth mirror for fine steering of the images.

There are a number of proposals for groundbased telescopes of roughly 30-meter aperture. Given the slow gain in sensitivity with increasing aperture for constant image diameter, all of these proposals are based on the potential for further improvements in image quality to accompany the increase in size. These gains will be achieved with multi-conjugate adaptive optics (MCAO), as discussed in Chapter 7.

One proposal, the Thirty Meter Telescope (TMT), would build on the Keck Telescope approach. Its primary mirror would have 492 segments. The European Southern Observatory (ESO) is planning the European Extremely Large Telescope (E-ELT), a 42-m aperture segmented mirror design. The Giant Magellan Telescope is based on a close-packed arrangement of seven 8.4-m mirrors, shaped to provide one continuous primary mirror surface. Its collecting area would be equivalent to a 21-m single round primary.

All of these projects face a number of technical hurdles to work well enough to justify their cost. Their dependence on the development of multi-conjugate adaptive optics is one example. In addition, their large downward looking secondary mirrors are a challenge to mount, because they have to be "hung" against the pull of gravity, a much more difficult arrangement than is needed for the upward-looking primary mirrors. For the segmented designs, the electronic control loop to maintain alignment will be very complex. All of them will be severely challenged by wind, which can exert huge forces on their immense primary mirrors and structures. Nonetheless, we can hope that the financial and technical problems will be surmounted and that they will become a reality.

2.6 Exercises

2.1 Show that a parabola does not satisfy the Abbe sine condition.

2.2 Use the Fourier transforms discussed in this chapter to derive the Fourier transform of triangle function.

2.3 A spare WFC3 infrared array has turned up on eBay for a great price and you are designing a telescope to use it for large field imaging at 1.55 microns (short H band, 1.4–1.7 microns). The array has 1024×1024 pixels, each 18 microns square. Suppose you are going to buy a 1.5-m telescope for your survey. [Hint: read the first page of Chapter 7 and use the fact that r_0 is proportional to $\lambda^{6/5}$.]

   (a) What projected pixel size on the sky is appropriate? (Consider whether the telescope will be limited by image motion or by image blurring by seeing.)

   (b) What field of view does this provide for the entire array?

   (c) What type of optical design should the telescope have?

   (d) What effective focal length should it have to provide the correct image scale?

2.4 In the system in exercise 2.3, can you just put the detector array at the focus of the telescope, or do you need to worry about the thermal background? Take the airglow emission to be 1,500 photons s^{-1} (square arcsec)$^{-1}$ and assume the telescope is at 290 K with an effective emissivity of 0.20.

2.5 Design an achromatic lens corrected for 486 and 589 nm, with a crown glass positive element and a flint glass negative one (let them be air spaced so you do not have to match curvatures of the facing surfaces). Make your lens 20 cm in diameter and with a 4-m focal length, and use the thin lens formula to compute the focal lengths required for the two elements. The refractive indices can be taken as

$$n(\lambda) = A + \frac{B}{\lambda^2}$$

where for the crown glass, $A = 1.5220$ and $B = 0.00459\,(\mu m)^2$ whereas for the flint, $A = 1.7280$ and $B = 0.01342\,(\mu m)^2$.

2.6 The Steward Observatory 2.3-m telescope is of Ritchey–Cretién design with a scale at the focal plane of $10''/\text{mm}$. What is the f-number of the telescope? Suppose it takes 100 seconds to measure a seeing-limited star image of average brightness x within $1''$ diameter. How long will it take to measure a source $1'$ in diameter with the same average surface brightness, to the same signal-to-noise ratio per $1''$ diameter? Now suppose the telescope were four times smaller in aperture, with no other design changes. How long would it take to measure the star to the same signal-to-noise ratio? How long to measure the extended source?

2.7 Prove the following theorems about Fourier transforms.

   (a) Convolution Theorem: If $F(s)$ is the Fourier transform of $f(x)$ and $G(s)$ is the Fourier transform of $g(x)$, then $f(x) \otimes g(x)$ has the Fourier transform $F(s)G(s)$.

   (b) Similarity Theorem: If $F(s)$ is the Fourier transform of $f(x)$, then $|a|^{-1}F(sa)$ is the Fourier transform of $f(ax)$.

   (c) Autocorrelation Theorem: If $F(s)$ is the Fourier transform of $f(x)$, then the autocorrelation of f,

$$\int_{-\infty}^{\infty} f^*(u)f(u+x)\,du$$

   has the Fourier transform $|F(s)|^2$.

   (d) Shift Theorem: If $F(s)$ is the Fourier transform of $f(x)$, then the transform of $f(x-a)$ is $e^{-2\pi jas}\,F(s)$.

Further reading

Bely, P. Y. (ed.) (2003). *The Design and Construction of Large Optical Telescopes.* Berlin, New York: Springer.

Bracewell, R. N. (2000). *The Fourier Transform and its Applications*, 3rd edn. Boston, MA: McGraw-Hill.

Press, W. H. , Teukolsky, S. A., Vetterling, W. T., and Flannery, B. P. (2007). *Numerical Recipes*, 3rd edn. Cambridge: Cambridge University Press.

Sacek, Vladimir (n.d.), Notes on amateur telescope optics: http://www.telescope-optics.net/

Schroeder, D. J. (2000). *Astronomical Optics*, 2nd edn. San Diego, CA: Academic Press.

Wilson, R. N. (2007). *Reflecting Telescope Optics*, 2nd edn. Berlin, New York: Springer.

3

Detectors for the ultraviolet through the infrared

3.1 Basic properties of photodetectors

For nearly a century, photography was central to huge advances in astronomy. Photographic plates were the first detectors that could accumulate long integrations and could store the results for in-depth analysis away from the telescope. They had three major shortcomings, however: (1) they have poor DQE; (2) their response can be nonlinear and complex; and (3) it is impossible to obtain repeated exposures with the identical detector array, an essential step toward quantitative understanding of subtle signals. The further advances with electronic detectors arise largely because they have overcome these shortcomings.

Modern photon detectors operate by placing a bias voltage across a semiconductor crystal, illuminating it with light, and measuring the resulting photo-current. There are a variety of implementations, but an underlying principle is to improve the performance by separating the region of the device responsible for the photon absorption from the one that provides the high electrical resistance needed to minimize noise. Nearly all of these detector types can be fabricated in large-format two-dimensional arrays with multiplexing electrical readout circuits that deliver the signals from the individual detectors, or pixels, in a time sequence. Such devices dominate in the ultraviolet, visible, and near- and mid-infrared. Our discussion describes: (1) the solid-state physics around the absorption process (Section 3.2); (2) basic detector properties (Section 3.3); (3) infrared detectors (Section 3.4); (4) infrared arrays and readouts (Section 3.5); and charge coupled devices (CCDs – Section 3.6). This chapter also describes image intensifiers as used in the ultraviolet, and photomultipliers (Section 3.7). Heritage detectors that operate on other principles are discussed elsewhere (e.g., Rieke 2003, Kitchin 2008).

3.2 Photon absorption

For the simplest photodetectors, the detection process begins when a photon entering a crystal of semiconductor is absorbed by freeing an electron from its bonds, allowing it to move freely through the detector volume. The free electron can carry an electrical current. This type of photoconductivity is termed intrinsic, because the photon energy required is an intrinsic property of the detector material.

A similar process occurs for isolated atoms, where the absorption can only occur between the discrete energy levels allowed for the electrons and the result of the absorption is to lift an electron to a higher energy level. In the case of a solid material, atoms are packed close together and, by the exclusion principle, their electrons are not allowed to share a single energy level. Instead, the energy levels are split into multiple closely spaced levels that we can treat as continuous energy bands (see Figure 3.1).

In this picture, the behavior of different types of material is largely due to the energy differences between the low-lying, or valence, bands representing the bound energy states and the high-lying, or conduction, bands where the electrons are unbound and their motions are responsible for conducting currents. Insulators require a large energy to elevate an electron to the conduction band. Room temperature is too low to elevate a significant number, and hence these materials do not conduct electricity well unless they are at high temperature. In metals, the conduction and valence bands are adjacent and free electrons exist in large numbers with little energy input, making the materials conductive under nearly any conditions. Semiconductors are in between. Their bandgaps match the energies of near-ultraviolet,

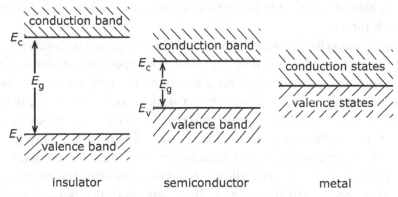

Figure 3.1. Bandgap diagrams for insulators, semiconductors, and metals. In these diagrams, energy increases along the *y*-axis, while the *x*-axis schematically shows one spatial dimension in the material.

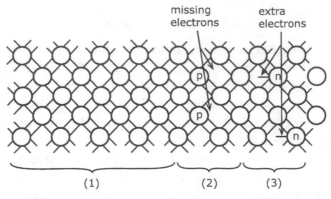

Figure 3.2. Crystal structure of (1) intrinsic, (2) p-type, and (3) n-type semiconductors.

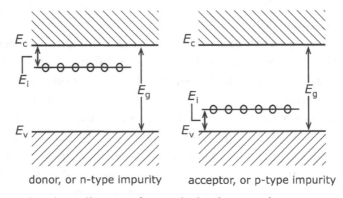

Figure 3.3. Bandgap diagrams for extrinsic photoconductors.

visible, and near-infrared photons. Hence, intrinsic absorption in semiconductors is almost always the first stage in the detection process across this wavelength range.

For wavelengths longer than the near-infrared, materials with suitably small bandgaps do not make satisfactory detectors, but doped semiconductors can be used. Dopants are impurities that do not supply the right number of bonding electrons to complete the semiconductor crystal; p-type dopants are missing a bond within the crystal lattice, while n-type ones have an extra electron: see Figure 3.2. The terminology *semiconductor:dopant* is used, for example Si:As or Ge:Ga. Extra energy levels are added to the band diagram to indicate these dopants (Figure 3.3). As they indicate, it takes relatively little energy, called the excitation energy, to free one of the n-type partially bonded electrons. Similarly, a small amount of energy can cause the bonds to shift in the p-type material to cause the empty bond to move through the crystal (in which case

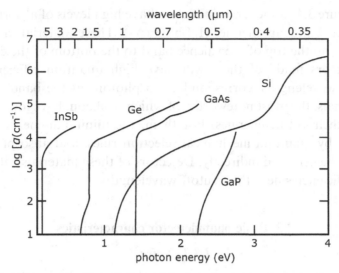

Figure 3.4. Absorption coefficients for some intrinsic photoconductors.

it is called a hole and treated as a positively charged mobile particle). The resulting photoconductivity is termed extrinsic, because the dopants that make it possible are not a fundamental constituent of the crystal material.

The efficiency of the detection process depends on the extent of the photon absorption in the detector. The absorption coefficient is $a(\lambda)$ (in cm^{-1}) and has a characteristic cutoff at the bandgap energy (intrinsic material; see Figure 3.4) or the excitation energy (extrinsic). The absorption of a flux S of photons passing through a path dl is

$$\frac{dS}{dl} = -a(\lambda)\, S \tag{3.1}$$

with solution at depth l

$$S = S_0\, e^{-a(\lambda)l} \tag{3.2}$$

where S_0 is the flux that penetrates the surface to the bulk material (e.g., after reflection losses at the surface). The absorptive quantum efficiency, η_{ab}, is the portion of this flux absorbed in the detector:

$$\eta_{ab} = \frac{S_0 - S_0\, e^{-a(\lambda)d_1}}{S_0} = 1 - e^{-a(\lambda)\, d_1} \tag{3.3}$$

where d_1 is the thickness of the detector. The net absorption must allow for losses such as reflection and incomplete collection of the signals from freed electrons, so the realized quantum efficiency, η, is generally less than the value for absorption alone.

From Figure 3.4, some semiconductors have high levels of absorption right down to their bandgap energies (InSb, GaAs). These materials allow direct transitions from the top of the valence band to the bottom of the conduction band. Detectors made of them will have high quantum efficiency up to their cutoff wavelengths, corresponding to photons at the bandgap energy. Other materials, the most noteworthy of which is silicon, have low absorption just above their cutoff energies. For them, the minimum-energy transition is forbidden by quantum mechanical selection rules, and absorption at this energy must be achieved indirectly. Detectors of these materials will have low quantum efficiencies near their cutoff wavelengths.

3.3 Basic photodetector characteristics

3.3.1 Noise

In addition to the basic noise associated with the input photons from both the source and background, photodetector noise includes a contribution from the Brownian motion of free charge carriers, called Johnson or Nyquist noise:

$$\langle I_J^2 \rangle = \frac{4kT\,df}{R} \tag{3.4}$$

where $\langle I_J^2 \rangle$ is the average square noise current, df indicates the electronic frequency bandwidth, k is Boltzmann's constant, and T and R are the temperature and resistance. Minimizing this contribution requires operating the detector with very high resistance and/or at very low temperature. With an integrating amplifier, the same process manifests itself as kTC noise, to be discussed below. Excess noise may also result from dark current, the flow of charge carriers when the detector is shielded from light. Low operating temperatures reduce dark current to some minimum level but it still may contribute noise. The readout amplifier may contribute additional noise in a practical circuit, for example, "read noise." All of these mechanisms generally satisfy the conditions for combining their effects via equation (1.23). Additional noise may be contributed by the discreteness of the digitization of the signal; in a well-designed system, this component should be small and unbiased, so it is also often treated as a Gaussian distributed random process.

3.3.2 Linearity/dynamic range

As a detector accumulates signal, its output might look like Figure 3.5. We have assumed the input flux is constant and that the detector integrates the signal with time. In that case, initially the collected signal grows linearly with

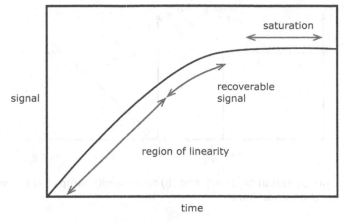

Figure 3.5. Response of an integrating detector to a constant signal.

integration time, but eventually it reaches a hard limit (termed saturation). Where the output is a linear function of the input signal, it is easy to interpret what the detector is telling us. When the detector is saturated, all information about the input flux is lost (other than that it is too large!). With care, information can be derived in the nonlinear regime before saturation. The dynamic range is the total range of signal over which useful information is yielded.

3.3.3 Time response

The time response of the detection process is set by phenomena such as the time required for the photo-generated charge carriers to recombine or be collected, so the detector returns to the state it was in before it was exposed to light. In addition, detectors are electrical devices with resistance, R, and capacitance, C, leading to an exponential response time

$$\tau_{RC} = RC \tag{3.5}$$

and a voltage response to a sharp input pulse at time $t = 0$ of

$$V_{\text{out}} = 0, \qquad t < 0$$
$$V_{\text{out}} = \frac{V_0}{\tau_{RC}} e^{-t/\tau_{RC}} \qquad t \geq 0 \tag{3.6}$$

3.3.4 Spectral response

Over what range of wavelengths does the detector respond to light? For an ideal photon detector, there is a characteristic answer as illustrated in Figure 3.6.

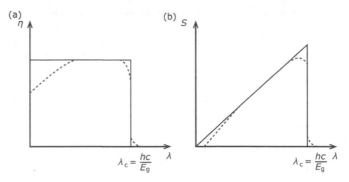

Figure 3.6. (a) Quantum efficiency and (b) responsivity (amps out per watt in) for an idealized photodetector.

The detector does not absorb at energies below its bandgap (or excitation) energy, corresponding to a cutoff wavelength of

$$\lambda_c = \frac{hc}{E_g} = \frac{1.24\mu m}{E_g(eV)} \tag{3.7}$$

where E_g is the bandgap (or excitation) energy. In an ideal photon detector, each absorbed photon creates one free charge carrier, so the detector responsivity, S (in amps of current out per watt of photons in), rises linearly with wavelength to λ_C. A number of effects in real detectors act to round off this ideal response curve (dashed lines).

3.3.5 Detector arrays and resolution

Semiconductor detectors can be manufactured into large-format detector arrays, and these devices are the basis of almost all astronomical instruments in the ultraviolet through the mid-infrared. The capability of a detector array for spatial resolution over its area is expressed to first order as its number of pixels, or detector elements. However, for a more complete description a number of other issues need to be included. At some level, signal spills from one pixel to its neighbors – this behavior is termed crosstalk and is usually described as the fractional response of the neighboring pixels compared with that from the pixel actually receiving the signal. In addition, the pixels may not cover the entire array area; the ratio of pixel area to the total is the fill factor.

The resolution achieved by detector arrays can be better characterized using the methods described in Section 2.3.3. We will elaborate on how the scale of the image projected onto a detector array affects the achievable resolution in Section 4.4.

3.4 Photon detector types for the infrared

In this section, we describe the operation of two types of photon detector: Si:As IBC devices; and photodiodes. We discuss readouts and arrays in Section 3.5 and charge coupled devices (CCDs) in Section 3.6. This organization has been adopted because, conceptually, the infrared arrays are simpler than CCDs.

Extrinsic absorption photon detectors (for wavelengths longer than ~5 μm) must operate under two diametrically opposed requirements. They need to have very high resistance to suppress noise currents, such as Johnson noise. At the same time, they need to have a high impurity level for good photon absorption, which drives the resistance down. Photoconductive detectors can be made by applying contacts to opposite sides of a suitable piece of extrinsic semiconductor, placing a bias between them, and monitoring the current across the device. However, in this case the doping of the semiconductor must be kept low to suppress dark current, and as a result reasonably high quantum efficiencies require that the detectors be thick (up to a few millimeters), making construction of fine-pitch detector arrays impossible. In addition, at low signal levels, these devices act like capacitor–resistor circuits with huge values of the resistance, resulting in response characteristics with very long time constants that bedevil calibration.

The solution is to separate the detector regions responsible for the electrical characteristics (low doping, high resistance) and photon absorption (heavy doping, reduced resistance).

3.4.1 Impurity Band Conduction (IBC) detectors

Silicon Blocked Impurity Band (BIB) or Impurity Band Conduction (IBC) – they are the same thing – are the detectors of choice for wavelengths between about 5 and 35 μm. We consider the most common type, Si:As IBC detectors, which respond to about 28 μm. These devices separate the absorbing and high resistance regions as shown in Figure 3.7. A thin blocking layer of high-purity and high-electrical-impedance silicon is grown together with an infrared-active layer, so heavily doped with arsenic that nearly total absorption requires a path of only about 30 μm. The active layer is sufficiently heavily doped that the electrons are forced to occupy a band of closely spaced energy levels, the impurity band (as Fermions they cannot share the same quantum mechanical state); consequently, the impedance of this layer is low. Still heavier doping is used for the degenerate – electrically conducting – buried contact layer. The electrical connection to this layer is made through the

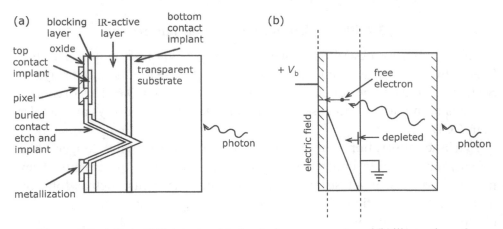

Figure 3.7. A Si:As IBC detector: (a) physical arrangement; and (b) illustration of operating principles.

V etched region to one side of the detector array. The second contact is put on the blocking layer. A thicker layer of silicon supports the device. When a voltage is placed across the contacts, the free charge carriers are driven out of the infrared-active layer, and it is said to be 'depleted'. As a result, it is in a suitable state for measurement of photo-generated signals; it has no spurious free electrons to add to the signal, and no holes where photo-generated charges might get trapped. Because it has few free charge carriers, it has a sufficiently high resistance that a significant electric field develops across it as well as in the high-impedance blocking layer. When infrared photons are absorbed in this depletion region, they free electrons that are driven by this field to the contact where they are collected to produce the photo-current that is a measure of the detected photons.

But how do these free electrons get through the high-purity blocking layer? Why don't the thermally generated carriers in the impurity band flood the contact? The answer is in the solid-state 'trick' in Figure 3.8. The blocking layer has few impurities, so the carriers in the impurity band are blocked there (hence 'BIB'). However, a photon-generated free carrier has been promoted into the silicon conduction band, which is continuous from the IR-active layer through the intrinsic one, so the electron can traverse the intrinsic layer readily.

The critical issue with this detector type is to adjust the parameters so it works! Arsenic is the dominant, 'majority' impurity (since we put it there at high concentration); it is n-type. There will be a much lower level of minority impurities, of p-type, in the IR-active layer. These p-type impurities will tend to acquire electrons from the arsenic, leaving them as negative charge centers.

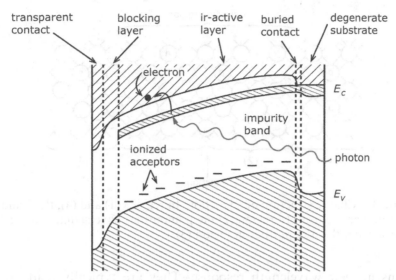

Figure 3.8. Band diagram for a Si:As IBC detector. The p-type impurities that have captured electrons are called inoized accepters.

The resulting negative space charge tends to neutralize the electric field in the As-doped, infrared-absorbing layer. Photo-electrons are only collected where this field has been established, that is, in the depleted portion of the layer. The thickness, w, of the depletion region is:

$$w = \left[\frac{2\kappa_0 \varepsilon_0}{q N_A}|V_b| + t_B^2\right]^{1/2} - t_B \tag{3.8}$$

where N_A is the density of ionized p-type impurity atoms, t_B is the thickness of the blocking layer and V_b is the bias voltage, q is the electronic charge, κ_0 is the dielectric constant (11.8 for silicon), and ε_0 is the permittivity of free space. The larger N_A, the smaller is w and the lower the quantum efficiency of the detector. One might try to combat this problem by increasing the bias voltage, but the result is incipient avalanching that increases the noise. Therefore, it is critical to minimize the concentration of the 'minority' p-type impurities.

State of the art semiconductor processing allows control of N_A to be less than 10^{12} cm^{-3}. An acceptable arsenic concentration is 3×10^{17} cm^{-3}. For arsenic in silicon, the absorption cross section is 2.2×10^{-15} cm^2, so the absorption length is about 15 μm. Assuming $N_A = 10^{12}$ cm^{-3} and $V_b = 1$ V, $w = 32$ μm from equation (3.8), so a high quantum efficiency detector can be built. Si:As IBC detectors have good quantum efficiency (~60%) at the longer wavelengths (10–26 μm), but the absorption falls toward shorter wavelengths. Because the photon absorption is not complete, they can have strong fringing – periodic

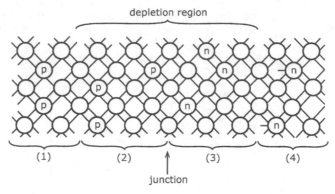

Figure 3.9. Charge structure of a junction. In regions (1) and (4), the p and n impurities are neutral. In region (2) the p impurities have acquired an extra electron from the n impurities in region (3).

variations in their wavelength response. They are typically read out with simple source-follower integrating amplifiers (to be discussed below) and have a modest and readily correctable nonlinearity in such a circuit.

Very-high-performance IBC detector arrays up to 1k × 1k pixels in size have been manufactured for infrared space missions where the thermal backgrounds are very low and the full potential detector performance can be realized. However, in use on the ground, the fully optimized arrays are overwhelmed by the high thermal backgrounds. Very rapid readout of the detector/amplifier unit is required to avoid saturation, and as a result there are compromises in the sizes of the arrays, the read noise, and possibly other parameters.

3.4.2 Solid-state photomultiplier

A modification of the detector architecture described above can enhance the avalanching gain, providing a fast pulse whenever a photon is absorbed. Where rapidly varying infrared signals are to be observed, this solid-state photomultiplier (SSPM) can have unique advantages.

3.4.3 Photodiodes

To make a diode, one dopes adjacent regions in a semiconductor with opposite type impurities. At the resulting "junction," charges migrate over a narrow region to fill all the open bonds. This situation defines a depletion region – no free charge carriers. The resulting charge sheets on either side of the junction create an electric field. See Figure 3.9.

This situation can be described as follows. Electrons in a semiconductor are in a Fermi–Dirac distribution relative to the energy levels:

$$f(E) = \frac{1}{1 + e^{(E-E_F)/kT}} \tag{3.9}$$

where $f(E)$ is the probability that an electron will be in a state of energy E and E_F is the Fermi level, defined by $f(E_F) = 0.5$. The significance of the Fermi level is clear in a metal, with a continuous range of available energy levels. At low enough temperature that thermal excitation can be ignored, the electrons fill all the energy levels up to the Fermi level. As the temperature increases, electrons are lifted from levels below E_F to ones above, with a distribution given by equation (3.9). In a semiconductor, the Fermi level lies between the valence and conduction bands. When dopants are added to the semiconductor, they shift the Fermi level: n-type dopants move it toward the conduction band and p-type toward the valence band. This statement reflects the fact that, for example, in n-doped material it takes less energy to free an electron – raise it into the conduction band – than with intrinsic material.

More or less by definition of the Fermi level, current will flow in a semiconductor to bring it to the same energy throughout the crystal. The valence and conduction bands must then shift with different doping levels in the crystal, resulting in built-in electric fields. The field across a diode junction is an example, leading to a "contact potential," V_0. This potential drives free charge carriers out of the region of the junction between the two doping levels, creating the depletion region.

When a photon is absorbed in a way that the free charge carrier it produces can reach the depletion region of the diode, the field maintained by the contact potential drives the carrier across the junction and the resulting current can be sensed to produce a signal. Photons are generally not absorbed in significant numbers in the junction region because it is very thin. To be detected, the charge carriers they free must diffuse through the material until they are captured in the contact potential (Figure 3.10). Diffusion describes the tendency of free charge carriers to spread through the material due to thermal motions. It is characterized by the diffusion coefficient,

$$D = \frac{\mu k T}{q} \tag{3.10}$$

where μ is the "mobility" of the free charge carriers (and is a characteristic of the detector material) and T is the temperature. The distance over which the free electrons can travel is characterized by the diffusion length, L:

$$L = \sqrt{D\,\tau} \tag{3.11}$$

Table 3.1. *Common photodiode materials*

Material	Cutoff wavelength (μm)	Transition type
Si	1.1	indirect
Ge	1.8	indirect
InAs	3.4	direct
InSb	6.8	direct
HgCdTe	~1.2– ~15	direct
GaInAs	1.65	direct
AlGaAsSb	0.75–1.7	direct

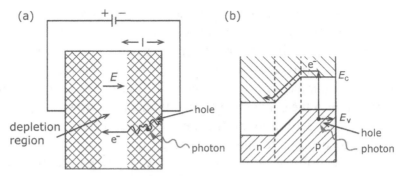

Figure 3.10. Detection of a photon in a diode.

The recombination time, τ, is how long the charge carriers remain free before being recaptured. For the diode to have good quantum efficiency, the thickness of the layer overlying the junction must be less than a diffusion length. This requirement rules out operation with *extrinsic* absorption, since the absorption lengths are too long. Photodiodes are made of intrinsic materials such as those listed in Table 3.1. For materials in this table with a range of λ_c, the bandgap can be controlled by changes in composition, such as by increasing the relative amount of Hg compared with Cd in HgCdTe to reduce the bandgap.

   Photodiodes must operate at low temperature to achieve low noise and dark current. The design of a photodiode must therefore take into account the temperature dependence of the diffusion length. In the regime of doping and temperature of interest D is proportional to T (equation 3.10). The recombination time goes roughly as $T^{1/2}$. Hence, L is proportional to $T^{3/4}$. To build detector arrays, the light is brought in through the substrate carrying the diodes, in an arrangement called back-illumination. To allow the charge carriers to diffuse to the junction, it is necessary that this back substrate layer be made very thin, in some cases no more than 10–20 μm! Two general approaches to making large arrays of very thin back illuminated diodes are: (1) to grow the diodes on the back of a transparent carrier; or (2) to thin them after they have been attached to a strong substrate, in this case the silicon wafer carrying the readouts.

The current conducted by a diode can be expressed to first order by the diode equation:

$$I = I_0(e^{qV_b/kT} - 1) \tag{3.12}$$

where I_0 is the "saturation current" and V_b is the bias voltage on the junction. As indicated by this equation, high impedance is achieved when the diode is modestly "back-biased" – that is, when a voltage is put across it with the polarity that causes it to add to the internal field across the junction and to increase the size of the depletion region. If the diode is forward biased, the depletion region shrinks and substantial current is conducted. The diode equation does not include the condition for large back bias, where the field across the depletion region is large enough that free electrons are accelerated strongly and can free additional ones, producing avalanche gain; a single absorbed photon can produce a small shower of free charge carriers driven across the junction. With still higher back bias, the diode "breaks down" and becomes highly conducting and not useful as a detector. Detectors are usually operated with a modest back bias, although we will discuss devices that utilize avalanche gain in Section 3.4.4.

With its two charge sheets separated by a thin layer of dielectric at the junction, a diode approximates a classic parallel plate capacitor with capacitance

$$C_J = \kappa_0 \varepsilon_0 \frac{A}{w} \tag{3.13}$$

where w is the width of the depletion region, κ_0 is the dielectric constant of the material, ε_0 is the permittivity of free space, and A is the area of the detector. The width decreases with increased doping and increases with increased back bias. Small capacitance is desirable to minimize the noise in reading out the diode.

Near-infrared photodiode arrays show excellent quantum efficiency (~90%). Because of the very high absorption efficiency, there is much less spectral fringing than with Si IBC devices. The diodes are usually read out by simple source-follower amplifiers, as described below. A characteristic of the source-follower circuit is that the detector is de-biased as signal is accumulated. As this process occurs, the width of the depletion region in the photodiode decreases and an integrating amplifier/detector system becomes less responsive because of the increased capacitance. However, this form of nonlinearity is highly repeatable and can be compensated readily in data processing. These attributes are available in very-high-performance detector arrays up to 2k × 2k format (with still larger formats in development) operating from 0.6 to 5 µm with detectors made in either InSb or HgCdTe. These devices are the detectors of choice for the near-infrared. At wavelengths longer than 8–9 µm, however, control of the material properties to minimize dark current becomes very difficult and photodiodes are no longer competitive with Si:As IBC detectors.

3.4.4 Diode variants

So far, we have described photodiodes based on a simple junction between oppositely doped regions of semiconductor. Some variants are described below:

PIN diodes

Some of the limitations discussed above can be removed with a thicker depletion region – it gives lower capacitance, and if the photon absorption occurs mostly in the depletion region, the limitations due to charge diffusion into the junction are removed. The thicker absorption region can improve the quantum efficiency just short of the bandgap for indirect transition absorbers. All of these gains can be achieved by growing the diode with an intrinsic region between the p- and n-type doped ones, hence "P-I-N" or "PIN" diode.

Avalanche diodes

If the back bias across a photodiode is increased sufficiently, the field in the intrinsic material becomes so large that the charge carriers gain enough energy to break more bonds, freeing more charge carriers and leading to a large gain in the device – that is, many electrons are produced when a single photon is absorbed. To work well, the gain must be similar for all detected photons. Therefore, the structure of the diode is modified to allow the bias to establish a depletion region in the intrinsic material so photo-electrons are driven toward the junction in a manner similar to the structure of the IBC detector and the gain occurs just before the junction. The avalanche process adds noise to the signal, but it can be useful if one needs a fast detector. This approach also can yield diodes that pulse count on single photons. Where a very fast detector is needed, pulse counting has great advantages over measuring a detector current because the read noise on the signal is eliminated. Of course, the inverse problem is that where the photon rate is high, or one wants to read out many detectors in an array, pulse counting is far more complex to implement than measuring photo-currents.

Schottky diodes

A junction between a metal and a semiconductor produces an asymmetric potential barrier that acts as a diode. These devices can be used as infrared detectors, although they have low quantum efficiencies.

3.5 Infrared arrays

3.5.1 Readout circuits

Conceptually, the "easiest" way to produce an array of Si IBC detectors or infrared photodiodes is to make the detectors and amplifiers separately and then join them together. In practice, this isn't easy at all – it requires making more than four million solder connections for a 2048 × 2048 array. None-theless, this challenge has been met: the detectors are manufactured in an optimized material with each one placed in a grid; indium bumps are grown on contacts to each of these detectors; amplifiers are made in a matching grid on a silicon wafer; indium bumps are grown on them; and the two grids are squeezed together to cold weld the indium and connect the detectors to the amplifiers. The process is described as "bump bonding," or "flip chip." The device, illustrated in Figure 3.11, is a "direct hybrid array."

Each pixel is given its own complete amplifier, built from a small number of metal-oxide-field-effect-transistors (MOSFETs) (Figure 3.12). In the n-channel device shown, two diodes are formed by n-type regions implanted into a p-type substrate, with contacts through the insulating SiO_2 layer. These implants are termed source and drain, the latter biased positive. A third electrode, the gate, is placed between them. Because the source and drain diodes are back-to-back, the device is normally non-conductive; in this condition, it is "turned off." If a positive voltage is put on the gate, it attracts negative charge carriers, forming an n-type channel through which current can flow. The amount of current depends strongly on the gate voltage.

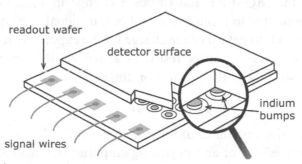

Figure 3.11. Direct hybrid infrared array. Figure adapted from *The Last of the Great Observatories* by George Rieke. © 2006 The Arizona Board of Regents. Reprinted by permission of the University of Arizona Press.

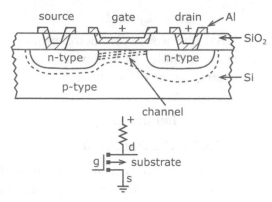

Figure 3.12. A MOSFET. Various channel depths are shown, corresponding to different gate voltages.

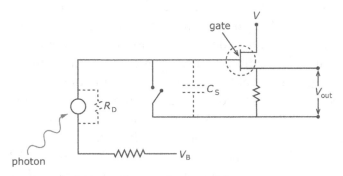

Figure 3.13. Source follower integrating amplifier.

One type of readout amplifier is shown in Figure 3.13. With the switch open as shown, the current through the detector causes charge to collect on the integrating capacitor, C_S, which is the combination of the detector capacitance and the input (gate) capacitance of the MOSFET. As charge collects, it changes the gate voltage $V_g = q/C_S$, which in turn modulates the signal through the channel of the MOSFET and causes a change in V_{out}. Once sufficient charge has accumulated to produce a useful output signal, the amplifier is reset by closing the switch to get rid of the charge on the capacitor. It can be shown through a series expansion of the resistive-capacitive (RC) response that the output is linear so long as the integration time,

$$t_{integration} \ll 2R_D C_S \tag{3.14}$$

with R_D the effective resistance of the detector.

The output waveform of an integrating amplifier therefore shows a linear ramp as charge accumulates, until the reset switch is closed (Figure 3.14). External electronics are used to select voltage values on this ramp and convert

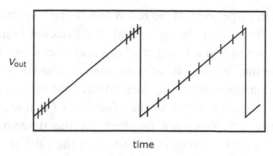

Figure 3.14. Two ways of reading out an integrating amplifier.

them to a digital number (DN, or DU for digital unit) for storage in a computer and later processing. The voltage can be read just before and after reset, with the signal given as $V(t_{before}) - V(t_{after})$. This strategy has the advantage that it is not affected by low-frequency noise in the amplifier, since the signal is extracted over a short time interval. However, it is subject to kTC (or reset) noise, which in units of electrons is

$$\langle Q_N^2 \rangle = \frac{kTC_S}{q^2} \qquad (3.15)$$

This type of noise is fundamental, since it involves the thermodynamic exchange of potential energy (stored on the capacitance) and kinetic energy (Brownian motion of the charge carriers). For a MOSFET with an input capacitance of $0.1\,\text{pF}$ at $T = 150\,\text{K}$, the noise is nearly 100 electrons. Therefore, it is more common to sample the signal at the beginning (t_1) and end (t_2) of the integration ramp. This strategy can avoid kTC noise because the time constant for changes in the charge on C_S is $\tau_S = R_D C_S$, and if $t_2 - t_1 \ll \tau_S$, then the noise electrons are "frozen" on the integrating capacitor during the integration. If $t = t_2 - t_1$ is the time from the read at the start of the integration to that at the end, the resulting contribution of the kTC noise is

$$\text{read noise} = \sqrt{\frac{kTC_S}{q^2}(1 - e^{-t/\tau_S})} \qquad (3.16)$$

Although the starting voltage for an integration ramp will vary from integration to integration due to kTC noise, the effect is automatically subtracted out in the difference between the end and beginning readings.

3.5.2 An example

As an example to illustrate the components of detector noise, suppose a simple source follower integrating readout has a net gate capacitance of

10^{-12} F and is being operated at 40 K. What is the expected level of kTC noise, in electrons rms? Now imagine that the detector being read out has a resistance of 10^{17} ohms and is being read out and reset every 100 seconds, but in a strategy where the signal is the measurement at the end of the integration ramp minus the measurement at the beginning, that is, where the charge is "frozen" between readouts. What is the effective kTC noise in this circumstance? If there are 2.5 photons/s incident on the detector, the quantum efficiency is 0.80, its dark current is 0.1 e/s, and the output amplifier has an electronic noise of 20 electrons rms for a simple reading, what is the total noise?

By simple substitution in equation (3.15), $\langle Q_N^2 \rangle = 21{,}578$, so the kTC noise is the square root of this value or 147 electrons rms. The RC time constant of the readout/detector is 10^{-12} F times 10^{17} ohms or 10^5 seconds. From equation (3.16), the effective reset noise with the readout strategy described is

$$147(1 - e^{-t/RC}) = 0.15 \text{ electrons rms}$$

The photon signal will produce 2 photo-electrons per second, or 200 over a 100-second integration; the dark current yields 10 additional electrons. The resulting noise is equal to the square root of the collected charge. To get the total noise, we add the components (kTC, photon, and amplifier) quadratically,

$$\text{total noise} = \sqrt{0.15^2 + (210) + 20^2} = 24.7 \text{ electrons rms}$$

The noise is dominated by that associated with the amplifier, with a significant addition from the photon noise and virtually nothing from the kTC noise. The signal-to-noise ratio at the output is $200/24.7 = 8.1$. The signal-to-noise ratio at the input photons is $250/\sqrt{250} = 15.81$. From equation (1.37), the detective quantum efficiency is 0.26. The dominance of the amplifier over the photon noise has degraded the potential detector performance from a quantum efficiency of 0.8 to a much lower DQE.

Can we do better? Probably yes. The amplifier noise can be reduced by making multiple reads of the output and averaging them. Figure 3.14 illustrates two ways to implement multiple sampling to reduce the high-frequency noise. In the first ramp, a number of samples are taken at the beginning and end, but otherwise the pattern is identical to our previous discussion. This pattern is sometimes called "Fowler sampling." In the second ramp, sampling is continued at a constant rate while the ramp accumulates – hence "sample up the ramp." The slope can then be fitted by least squares (equations (1.32)). Fowler sampling has the advantage of delivering the lowest noise, at least in principle. Sampling up the ramp allows recovery of most of the signal if the integration is disturbed, for example by a cosmic ray hitting the detector;

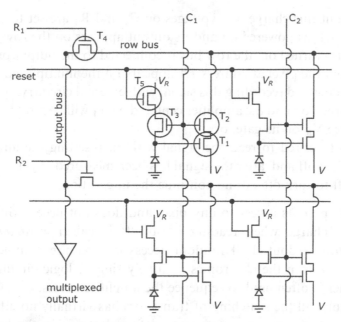

Figure 3.15. Simple multiplexed array readout.

it also allows extracting a valid measurement from the first few samples on a source so bright it saturates in the full integration. Initially, a strategy like Fowler sampling can reduce the noise almost as fast as the square root of the number of samples, but soon other noise sources begin to enter. A plausible gain in our example would be to get from 20 electrons to 6. We would then have a total noise from the detector system of 15.69 electrons, dominated by the photon noise, and a DQE of 0.65.

3.5.3 Multiplexing

To allow many-pixel arrays, the amplifiers' outputs are "multiplexed," meaning that the signals from them are switched successively to the array output. An array circuit diagram with all the necessary switching to read the signals out is shown in Figure 3.15. MOSFET T_1 operates as a source follower as in Figure 3.13. The signal is integrated on its gate and passed through T_2 (rather than the source resistor in Figure 3.13) to the row bus. However, the other MOSFETs allow the signal to be read out at specific times rather than continuously. The sequence to do so is as follows:

1. When integrating, the voltage on C_1 is set to turn off T_2 and T_3, so T_1 is also turned off. The voltage on R_1 is also set to turn off T_4. The photo-current places charge on the gate capacitance of T_1 even though the MOSFET is turned off.

2. To read out this charge, the voltages on C_1 and R_1 are set to turn on T_2 and T_3, so T_1 is powered on and its output appears on the row bus.
3. If T_4 is also turned on, the row bus is connected to the output bus and the signal from the pixel at address 1,1 appears on the multiplexed output.
4. After reading and recording this signal with external circuitry, if desired T_5 can be turned on and the amplifier centered on T_1 will be reset by imposing the voltage V_R on its gate, or
5. if it is just desired to read the signal without resetting the amplifier, T_5 is left turned off and, after the signal has been measured, T_2, T_3, and T_4 are turned off to turn off T_1 and continue the integration.

This readout permits access to any pixel and does not necessarily reset the accumulated charge when reading it out. It is called a *random access, non-destructive* readout. Full random access can require complex control electronics. To keep the electronics relatively simple, logic circuitry on the readout wafer is often used to sequence the amplifier access.

Remarkably, all the switching of transistors has virtually no effect on the final signal, which emerges accurate to a few electrons in the best arrays. Even more remarkable, the DC stability of these circuits is so good that they can integrate for many minutes without drifting so far as to compromise the accuracy of the signal measurement. That is, the strategy of freezing the charge between amplifier reads can be implemented even for time intervals of thousands of seconds between reads.

3.6 Charge coupled devices (CCDs)

3.6.1 Basics

Charge coupled devices were an elegant solution to constructing detector arrays before the miniaturization of integrated circuitry allowed dedicating an amplifier to each detector. They are still the state of the art for imaging in the visible because they have a number of advantages compared with the approach that must be used for infrared arrays, including simpler fabrication.

Consider a wafer of silicon with a thick oxide insulator layer and an electrode deposited on the oxide. In Figure 3.16, the silicon is doped p-type to reduce the concentration of free electrons. A positive voltage has been put on the electrode, or "gate." The voltage forms a depletion region in the silicon and attracts any free electrons into the potential well against the oxide. If light is allowed to penetrate the silicon and free electrons, then the rate at which charges are collected is a measure of the level of illumination.

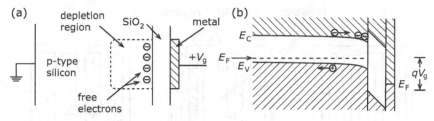

Figure 3.16. CCD charge collection under an electrode.

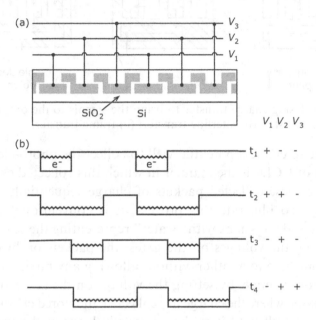

Figure 3.17. (a) Three-phase charge transfer and (b) its control sequence. At time step t_1, the charge is collected under a single electrode by the positive voltage. When the voltage on the neighboring electrode is set to the same voltage, the collection well expands. When the first electrode is set to a negative voltage, the charge transfer is completed. These steps are repeated to pass the charge along a column of the array.

The illumination can be supplied from the right in the picture, through and around the electrodes, in which case the CCD is described as front illuminated. If the light comes from the left and avoids the electrodes, the device is back illuminated. The detector is a fancy form of intrinsic photoconductor with integral charge collection.

The following discussion is based on a three-phase device as drawn in Figure 3.17. Each pixel then has three gates that are controlled by the array clocking circuitry in parallel with the similar gates for the other pixels. While the CCD is exposed to light, one (or two) gate(s) for each pixel is

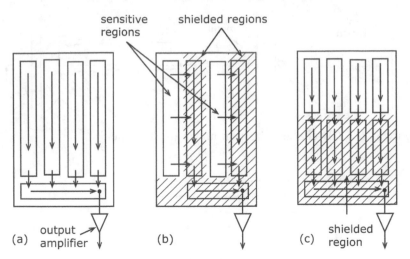

Figure 3.18. Using charge transfer to bring the signals to the output amplifier: (a) line transfer; (b) interline transfer; (c) frame transfer.

biased positive to create a potential well to collect the photo-electrons. The
unique aspect of CCDs is the manner in which this collected charge is read
out by passing these collected packets of charge sequentially through the
detector array. To illustrate this process, it is conventional to draw the
potential wells as depressions with "water" representing the sea of electrons.
By manipulating the voltages on the gates, the packets of "water" can be
passed from one gate to another without allowing any mixing of the separate packets. For example, by setting the voltage on the adjacent gate equal
to that for the one where the charge has collected, the stored charge will diffuse
into the adjacent well until it is shared equally between the two. Then, by
making the charge on the first gate more negative, its well is made smaller
forcing the electrons to complete their migration to the second electrode. By
repeating this sequence, the electrons are moved across the array. Not only do
the collected electrons not mix, but the presence of a depletion region between
them (under the third gate) means that the rest of the array is electrically isolated
from each charge packet.

The three charge-transfer architectures in Figure 3.18 represent three ways
of delivering the charge packets to an output amplifier. Each deals in a
different way with the continued creation of free charge carriers as the array
is exposed to light:

(a) In line transfer devices, the problem is ignored; precise exposures require
a shutter that can be closed while the CCD is read out.
(b) In interline transfer devices, the charge packets are moved at the end
of an exposure to a neighboring set of gates that are shielded from light.

These gates can be read out as desired. These devices generally have low net efficiency because of the real estate occupied by the shielded gates.

(c) In frame transfer devices, the whole image is transferred to a shielded region of the array where it is read out. The efficiency of the light-sensitive region is not compromised.

In all cases, the line of gates that feeds the output amplifier is called the output register.

During the transfers, doped regions along the transfer direction – called *channel stops* – prevent the charge from spreading orthogonally. A performance liability of CCDs is the tendency of strong signals to spill into adjacent wells, producing "blooming" – images that are extended along the direction of charge transfer. A form of beefed-up channel stop called an anti-blooming gate can intercept the extra charge and conduct it away before it spills into the adjacent well, but at a price in terms of fill factor, well depth, and effective quantum efficiency.

For good performance in a CCD, the charge transfer efficiency (CTE $= 1 - \varepsilon$, where ε is the fraction of charge lost in a transfer) must be very high. Poor CTE leads to crosstalk between pixels and to excess noise. To consider the noise issue, if N_0 charges are transferred then on average (εN_0) are left behind on each transfer, and the uncertainty in the number left behind is $\sqrt{(\varepsilon N_0)}$. In addition, $\varepsilon N_0 \pm \sqrt{(\varepsilon N_0)}$ charge carriers from the preceding packet may join the packet. Thus, if there are n transfers to get to the output amplifier, the charge transfer noise is:

$$N_{\mathrm{CTE}} = (2\varepsilon n\, N_0)^{1/2} \qquad (3.17)$$

Consider a 2048×2048 three-phase CCD with CTE $= 0.9999$ ($\varepsilon = 0.0001$), and an average signal level of $N_0 = 1000$. The number of transfers from the most remote pixel to the output amplifier is (roughly) 12 288, and $N_{\mathrm{CTE}} = 50$ electrons (larger than the $\sqrt{n}$ noise from the signal!). Thus, a large-format high-performance CCD must have a CTE approaching 0.999 999!

A number of mechanisms drive the charge packet transfers: (1) electrostatic repulsion among the electrons in the well; (2) fringing fields from neighboring electrodes; and (3) diffusion. Poor charge transfer can result if the device is read out too quickly to allow these mechanisms to drive *all* the electrons from one electrode to the next. The net result is that reading out a large CCD sufficiently slowly to preserve the CTE can take many seconds. The readout may be further slowed by the speed of the circuitry that digitizes the output voltages.

Another source of poor CTE arises in surface channel CCDs such as in Figure 3.16, where the charge is collected and transferred at the

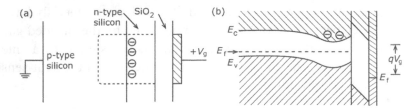

Figure 3.19. Buried channel CCD. Panel (a) shows the doping pattern, while panel (b) is a band diagram.

silicon–silicon-oxide gate interface. The open crystal bonds (traps) at this interface can capture the charge carriers and delay their transfer so that they rejoin charge packets that come by later. To circumvent this problem, low-noise CCDs are made with buried channels, in which a weak junction is used to move the potential well away from the oxide interface and the charge packets can be moved from gate to gate entirely within the silicon crystal. Figure 3.19 shows the concept, with panel (b) demonstrating how the rule that the Fermi level must remain constant through the material explains the bending of the conduction band that produces a "buried" well to collect the photo-electrons. If the wells are overfilled, the charge carriers contact the oxide and the device becomes surface channel. As a result the well capacity is reduced compared with surface channel devices. Nonetheless, with well depths of the order of 10^5 electrons and read noises of only a few electrons, these devices have large dynamic range.

Exposure to ionizing radiation can permanently damage the crystal lattice in a CCD, producing open bonds that can trap electrons and release them slowly. This process also reduces the CTE; it causes CCDs operating in space to degrade slowly. If the CCD is operated below 70–100 K, the buried channel "freezes out," that is, the charge carriers no longer have enough energy to detach themselves from bonds and migrate to establish the buried channel. As a result, the device becomes surface channel, with resulting problems with charge transfer and noise. This is one reason why CCD readouts are not used with infrared detectors, given their low operating temperatures.

The charge transfer structure allows an elegant solution to the kTC noise issue. The CCD electrodes can be used to pass the charge packets over a floating CCD gate, which couples them capacitively to the output MOSFET gate (see Figure 3.20). That is, when a cloud of electrons is transferred to lie under the electrode at V_b, its field repels electrons from the floating CCD gate resulting in additional charge appearing on the MOSFET gate, where

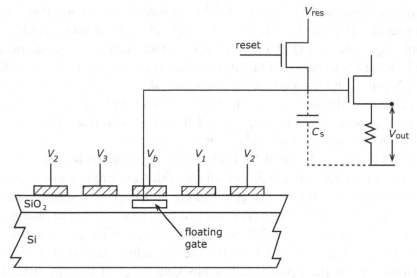

Figure 3.20. CCD readout amplifier.

it can be sensed and recorded as with any other source-follower amplifier. The sequence is:

1. The charge on the MOSFET gate is removed by closing the reset switch and then opening it.
2. A reading is taken of the amplifier output.
3. Then the CCD gate voltages are manipulated to pass a charge packet over the floating CCD gate, which transfers a charge into the amplifier *while maintaining high impedance to the rest of the world for the FET gate capacitance.*
4. A reading is taken to measure this additional charge.

Thus, the conditions for freezing the thermally driven charge on the capacitor are satisfied.

Lower noise can be achieved (but with even longer readout times) by reading the signal a number of times, for example by using CCD structures to pass the charge into a series of amplifiers, or back and forth between two amplifiers. Through a combination of slow readout, buried channels, high-quality material, and multiple read strategies, CCDs can achieve read noises of only a couple of electrons.

3.6.2 Other aspects of CCD performance

1. UV performance: the absorption coefficient of silicon is so high in the blue and UV that the photons are absorbed right at the back surface of the

device, away from the field created by the electrodes. There, they can fall into surface traps at the silicon-oxide layer (all silicon grows an oxide layer upon exposure to air). To instead drive them across the device into the wells, a variety of steps are taken, such as: (a) physically thinning the CCD to ~20 μm – thinning also reduces crosstalk, since the photo-electrons have less chance to diffuse into the "wrong" well; and (b) back surface charging in which special coatings are applied to the back surface that can repel photo-electrons from the oxide layer.

The UV performance of CCDs is also compromised by the very rapid changes of the refractive index of silicon with wavelength short of 300 nm, making anti-reflection coatings difficult. However, if the quantum efficiency at longer wavelengths is not a priority, it is possible to obtain good quantum efficiencies (~70%) over the 200–300 nm range. In addition, a perverse disadvantage of CCDs for the ultraviolet is that their excellent performance in the optical makes blocking spectral leakage from these longer wavelengths difficult. Filters capable of blocking this response have poor transmission in their ultraviolet passbands, leading to poor DQEs for the detection systems.

2. Although the CCD is basically a linear transfer device, a clever design developed at Lincoln Laboratory allows two-dimensional charge shifting in a four-phase device.

3. The CCD readout sequence, called clocking, can be modified to combine charge packets, a process called pixel binning. If the output register is not clocked, the charge in a number of pixels along a column can be transferred into a single output register well. The register contents can then be output to a final output summing well allowing a number of register wells to be combined before reading the signal.

4. Time-delay-integration (TDI): the CCD charge transfer process lends itself naturally to clocking charge in one direction at a set rate. This capability can be useful in applications where images drift across the detector array at a constant (relatively slow) rate – the charge generated by a source can be moved across the CCD to match the motion of the source. As a result, the CCD can integrate efficiently on a moving scene without physically moving anything to track the motion.

5. Deep depletion or fully depleted: silicon has low absorption efficiency in the near-infrared (~0.9 μm; see Figure 3.4). The brute-force way to get good absorption in this spectral range with a CCD would be just to make the absorbing region thicker, but there are bad side-effects like reduced charge collection and increased crosstalk. These problems can be partially mitigated by using high-purity silicon for the absorbing layer, allowing the

field from the electrodes to penetrate to a depth of 40–50 μm in a deep depletion CCD. Much greater depletion depths can be established by adding an electrode to the back surface and establishing a voltage across the absorbing region that drives the photo-electrons towards the gates and their potential wells, creating "fully depleted" CCDs. They can have absorbing layers a few hundreds of microns thick and achieve quantum efficiencies of ~90% at 0.9 μm, with usable response beyond 1 μm (Oluseyi *et al.* 2004). The price is higher dark current, more susceptibility to cosmic rays, and generally greater difficulty in fabrication.

6. L3 technology: E2V, Inc. supplies CCDs that can apply a high bias voltage on extra registers that feed the normal CCD output register. This voltage is close to the point of avalanching. The result is a small gain per transfer, 1–2%; after many transfers, the gain is significant. This style of operation can be useful when clocking the CCD quickly, since the larger charge packets allow faster operation of the output register without compromising the effective noise.

7. CCDs can be operated in a subarray mode by clocking the output register very quickly, not worrying about CTE, until the pixels of interest are about to be read out. Because the columns clock much more slowly than the output register, good CTE can be preserved for these pixels, and the output register can be slowed to provide low noise for the region of interest.

3.6.3 *Some alternative optical detectors*

Direct hybrid PIN silicon diodes

High-performance arrays can be manufactured in the same fashion as hybrid infrared arrays, but using PIN silicon diodes for the detectors. These devices have excellent red sensitivity, but they have liabilities because of the complex processing required for any bump bonded array.

CMOS imagers

Another spinoff from infrared arrays, CMOS (complementary-symmetry metal-oxide semiconductor) imagers are fabricated with a silicon diode as part of the circuit for each amplifier in a readout similar to those hybridized onto infrared detectors. CMOS imagers can be produced in standard integrated circuit foundries and hence are relatively cheap. They can replace CCDs in many applications. Since they are more-or-less conventional integrated circuits, it is easy to add circuitry to them that carries out various signal-processing functions. In addition, they do not require charge transfer with its attendant issues and they are more tolerant to ionizing radiation than CCDs are.

However, for use at low light levels, CMOS imagers have their own list of problems. Since the amplifiers compete for space on the wafer with the diodes, fill factors (see Section 3.3.5) range from ~70% downwards, depending on the pixel size and the complexity of the amplifier. In devices not designed to be cooled, the fill factor can be improved with an array of tiny lenses, one over each sensor. Back-illumination is being explored as a solution, but it takes the devices out of the integrated circuit mainstream and thus loses one of the advantages of these detectors. They may also have higher noise, local peaks in dark current, and poorer pixel-to-pixel uniformity.

3.7 Image intensifiers

Image intensifiers have largely been supplanted by CCDs in the near-ultraviolet (~0.3 μm) out to 0.9 μm. However, they are very competitive farther into the ultraviolet when large field imaging detectors are desired and where silicon-based detectors are difficult to make with high quantum efficiencies. Their detection process is based on the vacuum photodiode as shown in Figure 3.21. When the photocathode absorbs a photon with energy greater than its bandgap, an electron is lifted into the conduction band. This electron can diffuse to the surface. If it arrives with sufficient energy, it can escape (photoelectric effect) into the vacuum space of the tube. This process is less efficient than just falling into the junction field as in a semiconductor photodiode; that is, the escape probability is significantly less than 1, contributing to the lower quantum efficiency (10–30% is typical). Once the electron gets into the vacuum, it is driven across by the field and interacts (perhaps

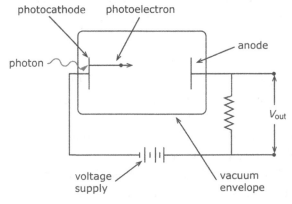

Figure 3.21. A vacuum photodiode.

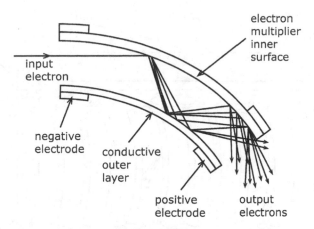

Figure 3.22. Operation of a microchannel.

focused by electron optics – charged surfaces that create an appropriate electric field) with an output device at the anode – a phosphor or array of electrodes, for example. The vacuum plays the role of the depleted region at the junction of the semiconductor photodiode, and the cathode and anode play roles analogous to the overlying semiconductor layers.

A popular version that can be quite compact and allows efficient readout takes the output to a microchannel plate (MCP). Microchannels are thin tubes of lead-oxide glass with inner diameters of 2–25 μm and length-to-diameter ratios of 40–60 (see Figure 3.22). They are supplied in large-scale arrays or plates, with typically 40% of the area occupied by glass and the remaining 60% active microchannels (i.e., they reduce the net quantum efficiency to about 60% of that of the photocathode). The inside surfaces of the microchannels are coated with a layer of PbO, which acts as an electron multiplier, and a large voltage (≥ 1000 V) is maintained along them. When an electron is accelerated by this voltage and impacts the PbO, a number of electrons are released. These electrons are in turn accelerated into a wall of the microchannel, producing more electrons, and so forth. For straight microchannels, the achievable gain is limited to ~1000 because positive ions produced from electron collisions with residual gas can migrate "upstream," collide with the walls, and initiate an unwanted new electron cascade. Therefore, the tubes are curved (called C-plate MCPs), or multiple MCPs are stacked with alternating tilts (Z-plate or chevron MCPs) and in these configurations MCPs are capable of gains of 10^7 to 10^8.

A variety of approaches have been developed to receive the outputs of MCPs. In a generation II image intensifier, they are conveyed to a phosphor that emits a greatly amplified version of the input image. This approach is

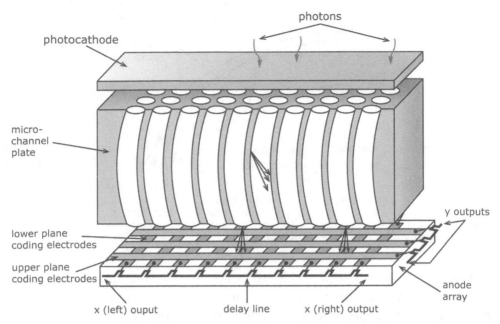

Figure 3.23. The GALEX image intensifier with microchannel and delay line readout.

used in the XMM-Newton Optical Monitor, where a CCD receives the phosphor output. The approach has a number of disadvantages, including the problem of light from the output leaking back to the input to produce a source-induced background signal, which is very difficult to calibrate and remove. Therefore, purely electronic outputs are preferred. The challenge is to provide multiplexing of a pulsed signal; the type of electronic multiplexers used with infrared arrays and CCDs are much too slow for this application. Nonetheless, there are a number of solutions.

As an example, Figure 3.23 shows schematically the GALEX (Galaxy Evolution Explorer) near-ultraviolet (177 nm–283 nm) device with its crossed delay line readout (the full array has 4k × 4k pixels). When the photocathode releases an electron, it enters one of the microchannels. The gap between the photocathode and the entrance to the microchannels is only 300 μm, so the positional information is retained (this approach is called proximity focusing). Orthogonal electronic charge collectors are placed at the output of the MCP that convey the signal to external delay lines interconnecting all of the collectors for each coordinate. By measuring the time interval between the emergence of the two signals at the opposite ends of a delay line, it is possible to locate where the signal originated along its coordinate and thus where the original photon hit the photocathode.

Other types of intensifier extract the signal in different ways: (1) overlying orthogonal wire grids in a multi-anode microchannel array (MAMA – used, for example, in the Space Telescope Imaging Spectrograph (STIS)); and (2) dual delay lines, where the charge collectors are shaped so that the relative signal strengths provide the fine position information in one dimension, with delay lines as in the crossed delay line readout for the other (as used in the Far Ultraviolet Spectroscopic Explorer (FUSE) mission and far-UV portion of the Cosmic Origins Spectrograph (COS) instrument on HST).

The far-ultraviolet (134 nm–179 nm) detector in GALEX is similar in concept, except that the photocathode is deposited directly on the inside of the microchannel tubes. A similar approach can be used in the X-ray (e.g., the High Resolution Camera (HRC) on Chandra). Both of these examples use CsI for the photocathode, but other materials can also be useful, such as KBr and CsTe. However, none of these materials yields quantum efficiencies above the 10–30% range. Photocathodes with much higher quantum efficiency (up to ~80%) are under development using GaN and its alloys, especially AlGaN. These materials also have reduced response in the visible and hence ease substantially the problem of blocking the visible part of the spectrum; for example, the bandgap for GaN is 3.4 eV.

3.8 Photomultipliers

Photomultipliers were the first truly electronic detector type to be used widely in astronomy. With only modest (if any) cooling, they provide low dark current and are sufficiently sensitive to allow counting individual photons, that is, their read noise can be equivalent to one input electron. However, their quantum efficiencies are relatively low compared with CCDs, and they cannot be built easily into large-format arrays.

A photomultiplier can be viewed as a single-pixel image intensifier, as shown in Figure 3.24. The entire apparatus is enclosed in a vacuum-tight envelope and evacuated. At one end of this envelope a transparent faceplate admits light to a transparent photocathode similar to that for an image intensifier. The released electrons are guided by additional electrodes and accelerated into the first dynode of the multiplier chain, made of a material that produces a number of output secondary electrons for every high-energy electron that strikes it. A 100–200 V difference from one dynode to the next is maintained from a single high-voltage power supply feeding a chain of resistors that divide the voltage dynode-to-dynode. The voltages accelerate the electrons, and at each dynode more are produced to yield a large net gain by repetitive avalanche multiplication. In principle, as with all avalanche

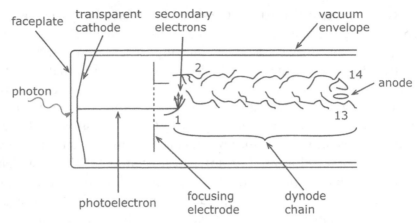

Figure 3.24. A photomultiplier.

multipliers, the statistics of the first few stages where only a few electrons are produced can result in substantial gain fluctuations. This issue can be mitigated with a type of dynode surface with "negative electron affinity (NEA)" that can produce 10–20 output electrons per input (but cannot be used for the later stages because it is damaged by high currents). The final pulse of perhaps 10^6 electrons is collected at the anode. Output pulse widths of a few nanoseconds can be achieved. The signal can be recorded as the average anode current (i.e., smoothing the pulses), or by pulse counting. The latter method is usually preferred because it provides greater immunity to extraneous noise and to variations in pulse heights due to statistical differences in the amount of multiplication.

3.9 Exercises

3.1 Suppose a simple source follower integrating readout has a net gate capacitance of 10^{-12} F and is being operated at 40 K. What is the expected level of reset (kTC) noise, in electrons rms? Now imagine that the detector being read out has a resistance of 10^{17} ohms and is being read out and reset every 100 seconds, but in a strategy where the signal is the reading just before the $(n + 1)$th readout minus the reading just after the nth readout, that is, where the charge is "frozen" between readouts. What is the effective kTC noise in this circumstance?

3.2 Compare the signal-to-noise ratio achieved with two methods of multiple sampling of the output of an integrating amplifier, dividing 24 samples equally between the beginning and end of the integration ramp and distributing them spaced equally in time up the ramp. Assume each sample has the same noise of σ.

3.3 Suppose you are making a measurement using a photomultiplier with very large dynode gain (so differences due to dynode gain can be ignored). The signal you are measuring corresponds to an average of 100 events per second. In addition, there is an excess noise process that produces an average of 1 pulse per second that is 100 times as large as a photon-generated pulse. All the events arrive with a Gaussian distribution in time (i.e., the uncertainty in the number of pulses received in a time interval can be taken to be the square root of the expected number). Compare the signal-to-noise ratio that will be achieved in 100 seconds by current measurement (where in effect the power in all the pulses is added together) and by pulse counting, where each event is counted individually.

3.4 Consider a backside-illuminated CCD of thickness 15 µm and operating at a wavelength of 0.9 µm. How many microns of thickness variation change the fringes from bright to dark (i.e., change the interference from constructive to destructive)? How large a change of wavelength does the same thing? Estimate the amplitude of the fringing if all the light incident on the electrodes is reflected and the absorption coefficient of the silicon at this wavelength is $800 \, \text{cm}^{-1}$.

Hint: the interference intensity between two beams of intensity I1 to I2 is proportional to $2(\text{I1 I2})^{1/2} \cos\theta$, where θ is the phase difference between them.

3.5 A photographic plate has grains that become developable upon exposure to four or more photons. The plate is illuminated at a level that corresponds to an average of one photon per grain, with the exposure following a Poisson distribution. After development, the signal is read out with a light system / detector that illuminates a plate area containing 100 grains on average. The grains are distributed non-uniformly such that the number in the area being read out has a one-standard-deviation variation of 25 (i.e., 25% of its average value). Compute the detective quantum efficiency of the plate in this application.

Further reading

Csorbe, I. P. (1985). *Image Tubes*.Indianapolis, IN:Howard Simms.

Janesick, J. R. (2001). *Scientific Charge-Coupled Devices*. Bellingham, WA: SPIE Press.

Joseph, C. L. (1995). UV Image Sensors and Associated Technologies. *Exp. Ast.*, **6**, 97.

Rieke, G. H. (2003). *Detection of Light from the Ultraviolet to the Submillimeter*, 2nd edn. Cambridge: Cambridge University Press.

Rieke, G. H. (2007). Infrared detector arrays for astronomy. *ARAA*, **45**, 77.

Timothy, G. J. (2010). Microchannel plates for photon detection and imaging in space. In *Observing Photons in Space*, ed. M. C. E. Huber, A. Pauluhn, J. L. Culhane, J. G. Timothy, K. Wilhelm, and A. Zehnder. ISSI Scientific Reports Series, ESA/ISSI.

4

Optical and infrared imaging; astrometry

4.1 Imagers

Imagers capture the two-dimensional pattern of light at the telescope focal plane. They consist of a detector array along with the necessary optics, electronics, and cryogenic apparatus to put the light onto the array at an appropriate angular scale and wavelength range, to collect the resulting signal, and to hold the detector at an optimum temperature while the signal is being collected. Imaging is basic to a variety of investigations, but is also the foundation for the use of other instrument types that need to have their target sources located accurately. In this chapter we discuss the basic design requirements for imagers in the optical and the infrared and guidelines for obtaining good data and reducing it well. We finish with a section on astrometry, a particularly demanding and specialized use of images.

4.2 Optical imager design

A simple optical imager consists of a CCD in a liquid nitrogen dewar[4] or cryostat with a window through which the telescope focuses light onto the CCD (see Figure 4.1). Broad spectral bands are isolated with filters, mounted in a wheel or slide to allow different ones to be placed conveniently over the window. Although this imager is conceptually simple, good performance requires attention to detail. For example, if the filters are too close to the CCD, any imperfections or dust on them will produce large-amplitude artifacts in the image.

---

[4] A dewar is analogous to a thermos bottle. It maintains a vacuum inside its case to provide a high degree of thermal isolation. A cryogenic liquid (e.g., liquid nitrogen) is stored in a vessel isolated from external heat by this vacuum surround. A similar design might use a cold head from a mechanical refrigerator in place of the cryogenic liquid.

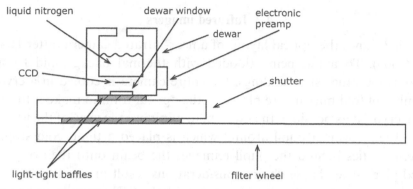

Figure 4.1. Optical camera schematic.

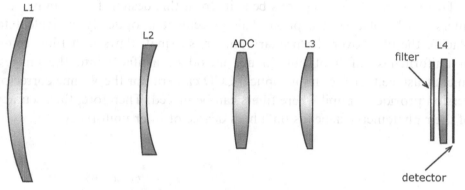

Figure 4.2. Wynne-type field correcting optics.

Other components can improve the quality of the data. A shutter can provide accurate and uniform exposures. For instance, a curtain can initiate the exposure by opening across the CCD, and a second curtain can close off light in the same direction to end the exposure, providing the same exposure time across the entire CCD. More ambitious instruments use optical systems to provide good images over a large field that can be filled by a mosaic of detectors. There are a broad variety of possible corrector designs. Figure 4.2 shows a generic example, based on the concept proposed by Wynne (Section 2.4.1) that is used widely. The first three lenses (L1, L2, and L3) work together to correct the telescope aberrations over the field; lens 4 flattens the field to match the detector. An atmospheric dispersion corrector (ADC) counters the chromatic effects due to atmospheric refraction. Without the ADC, images taken at small elevation angles will look like tiny spectra, with the blue end pointing to the zenith because the atmospheric refraction bends blue light more than red light. In this design, the ADC is moved as the telescope elevation changes to provide a good correction.

4.3 Infrared imagers

Figure 4.3 shows the optical layout of a near-infrared camera (after Hodapp *et al.* 2003). To avoid being flooded with thermal background from the telescope and other surroundings, the entire camera is cooled in a cryostat. A number of fold mirrors are placed in the optical train to make it fit into as small a cryostat as possible. In addition, to minimize the view of the telescope, the field lens forms a pupil around which is placed a tight, cold stop. The "camera" optics behind the pupil reimages the beam onto the array at the desired pixel scale. These design considerations result in an instrument configuration changed substantially from that for the CCD camera, although the two instruments take very similar-appearing data.

There are some serendipitous benefits from this design. For example, the filters can be placed at a pupil. This placement is optically equivalent to placing the filter over the primary mirror, so small flaws in a filter result in uniform loss of light but do not introduce artifacts into the images. In contrast, neither the simple optical CCD camera nor the Wynne corrector version provides a pupil where filters can be placed. Therefore, the accuracy of their photometry depends on a high degree of filter uniformity.

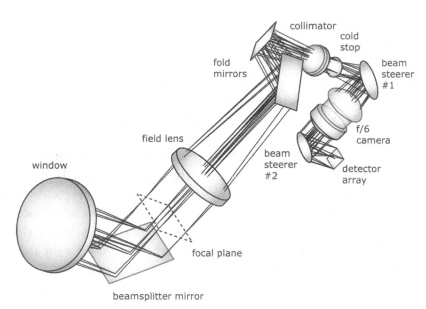

Figure 4.3. Layout of a near-infrared camera. After entering through the dewar window, the light is focused by the field lens plus collimator on to the pupil where the cold stop is placed. The remaining optics relay the focal plane to the detector array. © PASP

However, all those optics impose other issues:

1. Cos$^N\theta$ effects refer to a general drop in signal with increasing off-axis angle (θ). In general, optical systems are most efficient on-axis. For example, anti-reflection coatings are generally designed for normal incidence and their effectiveness is reduced for off-axis rays. The efficiency of a detector array is likely to fall off with increasing off-axis angles of incidence. Vignetting may reduce the response at the edges of the field.
2. Ghost images result when light reflected from refractive optics gets back to the array and appears as point-like or extended images, depending on the geometry. Even simple optical imagers can produce ghosts due to reflections from the dewar window and/or filter.

4.4 Nyquist sampling

A basic requirement in imager design is to set the equivalent angular size of the array pixels to extract the maximum possible information. Pixels that project to angles that are too small will underutilize the detector array and reduce the size of the field of view unnecessarily. Pixels that project to angles that are too large will lose information on small angular scales. From Section 2.3.1, we know how to control the projected pixel sizes; now we will determine what size we should aim for. Fortunately, there is a rule that says "how small" is small enough that essentially no details are lost in an image.

We can determine the required pixel size by determining the MTF of a detector array and seeing under what conditions the spatial frequency spectrum from the telescope is preserved. We only have to compute the MTF of a pixel, since the full suite of pixels will have the same impact on the result. The function

$$\operatorname{rect}(x) = 1 \quad \text{for} \quad |x| \leq \tfrac{1}{2}$$
$$= 0 \quad \text{otherwise}$$

is the response function for a pixel of width 1. We need to modify it for a pixel of width w:

$$\operatorname{rect}\left(\frac{x}{w}\right) = 1 \quad \text{for} \quad \left|\frac{x}{w}\right| \leq \tfrac{1}{2}$$

Using equations (2.18) and (2.25), the corresponding Fourier transform is

$$\boldsymbol{F}(w) = \frac{1}{|1/w|} \frac{\sin(\pi f_\text{S}/(1/w))}{\pi(1/(1/w))} = |w| \frac{\sin(\pi w f_\text{S})}{\pi w f_\text{S}}$$

where f_s is the spatial frequency.

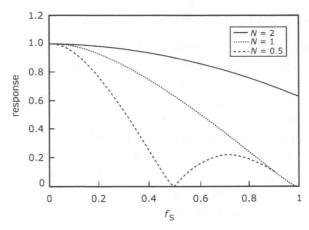

Figure 4.4. MTFs for sampling at $N = 0.5$, 1, and 2 pixels per spatial period. The spatial frequencies are in units of D/λ.

The MTF is $\sqrt{(F\,F^*)}$, normalized to 1 at $f_s = 0$. Since

$$\frac{\sin(x)}{x} \to 1 \quad \text{as} \quad x \to 0$$

it follows that

$$\text{MTF} = \left|\frac{\sin(\pi w f_S)}{\pi w f_S}\right| \tag{4.1}$$

Some samples are shown in Figure 4.4. We can identify $f_s = 1$ with the natural cutoff frequency of a telescope, D/λ cycles/radian. Since $f_s = 1/P_s$ where P_S is the spatial period, the MTF at D/λ for sampling with two pixels across the spatial period is

$$\text{MTF} = \left|\frac{\sin(\pi/2)}{\pi/2}\right| = 0.64$$

That is, the attenuation of high spatial frequencies is modest. With one pixel per period the loss at high spatial frequencies is already severe. That is, we have shown that two pixels across the image diameter (FWHM $\sim\lambda/D$) is a good goal for the pixel size that does not lose any spatial information.

This result is often described as Nyquist sampling, after the Nyquist or Sampling Theorem. This theorem states that a bandwidth-limited signal with maximum frequency f and period $P = 1/f$ can be completely reconstructed from time samples at a time interval of $P/2$ (strictly at slightly smaller than $P/2$).

The situation with finite pixels is analogous but not completely identical with the assumptions in proving this mathematical result (e.g., the pixels have significant size compared with their spacing whereas the Sampling Theorem applies to samples instantaneous in time). Furthermore, a number of people derived the result independently of Nyquist, but the "Nyquist" terminology applied to imaging is firmly entrenched.

Images where the pixels are smaller than half the image FWHM are described as being oversampled. There are situations where oversampling is beneficial. One reason is that real arrays fall short of the ideal (see Section 4.5 below) and some of their flaws can be overcome by finer sampling. Undersampled images have pixels larger than half the FWHM. Besides the irretrievable loss of information at high spatial frequencies, there is a related issue with undersampling, called aliasing. With undersampled pixels, the source distribution that fits a detector array output image is not uniquely determined. Consider the case with ½ pixel per FWHM of the image (i.e., two image FWHMs per pixel) in Figure 4.4. The null at frequency $D/2\lambda$ occurs because a signal containing only this spatial frequency has a full period across the pixel width, so it cancels its own signal. An example of a signal contributing to the secondary peak at higher frequencies would be one 1.5 times higher than the null frequency; it would have 1.5 periods across the pixel and so it would only partially cancel itself. However, at frequency D/λ there is another null because there are now two spatial periods across the pixel. The high spatial frequency signals no longer have unique signatures and can get confused with source structures at lower spatial frequency. There may be no unique distribution of sources that can be associated with an image. It may be easier to visualize this problem in image space; if the pixels seriously undersample the input image, then a number of arrangements of diffraction-limited images distributed over pixels could produce identical images. Fortunately, for mild degrees of undersampling, if we take multiple exposures offset in pointing by a fraction of a pixel size we can mitigate aliasing. However, Figure 4.4 shows that this strategy does not recover all the information; once the pixels are as large as λ/D, the highest spatial frequencies are lost completely from the images.

A problematic case occurs with seeing-limited images. Our derivation of the required sampling assumed that there is a unique spatial frequency cutoff associated with the telescope aperture. Seeing-limited images behave in a more complex manner. For them, one might oversample to try to extract information at spatial scales finer than the simple "seeing limit" (Chapter 7).

4.5 Imager data reduction

4.5.1 Benefits and issues

Imagers have huge advantages over single detector instruments for nearly all astronomical observations. They permit very accurate position determination and enable astrometry, as discussed in Section 4.7. For photometry they:

- allow centering on the source and setting other parameters of extraction of photometry after the fact
- enable use of small apertures in crowded fields and rejection of backgrounds for improved sensitivity
- allow differential photometry relative to other sources in the field for accurate measurements under non-optimum conditions
- enable removal of foreground stars that might interfere with the signal from an extended source
- enable use of custom extraction apertures with shapes optimized to the shape of the source
- allow construction of multiple-color images for comparison of the behavior of a source in all the colors precisely.

However, to gain these advantages, there are number of steps that are required:

- Various issues with detector arrays need to be anticipated in data-taking.
- The data reduction strategy needs to allow for elimination of problems such as transient signals.
- Calibration must take into account the differing properties of the detectors in the array.

Some of the optical phenomena that affect array data are:

- *Inter-pixel gaps and intra-pixel response*: Since array pixels are discrete, the sensitivity may have minima between pixels (Figure 4.5). These "gaps" can have significant effects if the pixels do not sample the PSF well. It is also possible that the response varies over the face of a pixel. Figure 4.6 shows the dependence of the signal on the centering of a source on a NICMOS Camera 3 pixel – for this example, 1 pixel $\sim 1.5\ \lambda/D$. None of these effects will be detected through normal flat fielding image processing.
- *Crosstalk*: Small amounts of charge can transfer from one pixel in an array to the next, blurring the image. This issue is much more prominent if the pixels undersample the intrinsic image.
- *Fringing/channel spectra*: Arrays are based on thin parallel plate components. When the absorption in the detectors is low, interference within the

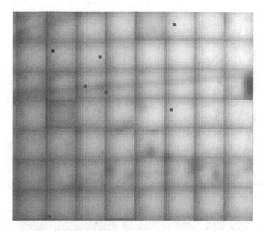

Figure 4.5. Interpixel response gaps; to first order, no charge is lost but the gaps indicate where charge can migrate to the "wrong" pixel. Image courtesy of Tim Hardy and John Hutchings.

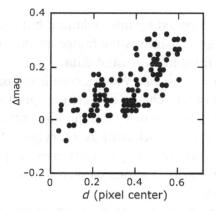

Figure 4.6. Signal vs. centering for a NICMOS array. Adapted from Hook and Fruchter (2000).

material causes fringing (Figure 4.7). The nature of the fringing is a sensitive function of the spectral content of the illumination. Therefore, fringing can best be corrected by generating response (or flat field) images with the identical illumination conditions as those for the science data.

There are also a number of electrical issues (see Figer *et al.* (2003) for illustrative examples – the last four in our list apply primarily to infrared and CMOS arrays):

- *Hot and dead pixels*: Some pixels have anomalously high dark current (termed "hot") and some do not respond at all ("dead"). In either case, they return no useful data. The analogous CCD fault is a dead column. The missing

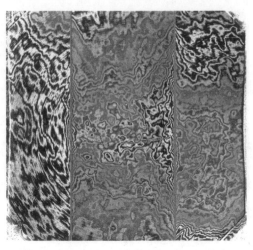

Figure 4.7. Fringing in a CCD at 0.95 μm. Image by Bryan Miller, credit Gemini/AURA, by permission.

information can be supplied by interpolation, but it is much better to use an observing strategy that moves the image on the array so the effects of these pixels can be filled in with valid data.

- *Cosmic ray hits, other transients*: When a cosmic ray passes through a pixel, it generates a cloud of free charge carriers that produce a transient signal.
- *Nonlinearity and soft saturation*: Modest nonlinearity can be corrected with a suitable fit to the detector behavior as a function of signal strength. Soft saturation occurs when the output continues to grow with increasing signal, but the nonlinearity is strong enough to make it difficult to recover accurate measurements.
- *Latent images*: Many electronic arrays retain an image at the 0.1–1% level on the next readout after a bright source has been observed (assuming the two readout cycles are identical). If a short exposure on a bright source is followed by a long exposure (~10 minutes) on a faint source, then the total latent charge collected can be much larger than 1% since many readout intervals similar to that on the bright source are combined. The images are from charge trapped at surface or interface layers in the detector; they usually decay over about 10 minutes. In addition, much longer decay times can result if the array has been saturated by an extremely strong signal.
- *MUX glow*: The array readout transistors are sources of light through electroluminescence. This glow can be picked up by the array, contributing to a lack of flatness in the images and also contributing noise. Standard reduction procedures correct for the MUX (multiplexer) glow signals but the statistical noise from these signals cannot be removed.

- *Electrical crosstalk/ghosting*: Various effects can produce secondary images. A common example is inadequate drive power on the output amplifiers of an array with multiple outputs. In this situation, when a bright source requires a lot of drive power to push its signal out of the array the supply to the other output amplifiers can be affected.
- *Pedestal effects*: Readouts sometimes have electrical offsets that appear as structure in images. Typically, these effects can be removed by standard methods with no degradation in the final images.

4.5.2 Taking good data

Addressing these issues in the final reduced images requires care in taking the data. The first rule of array imaging is repetition.

- Multiple images allow systematic identification of outlier signals due to cosmic rays and other transients.
- They also allow replacing areas compromised by cosmic rays, latent images, hot or dead pixels, ghosting, etc. with *real data*.
- By dithering the pointing on the sky between exposures (moving the telescope slightly so that the images fall on different parts of the array), the sky signal itself can be used to flatten the image (as discussed below). If the sky dominates the signal, then fringing effects are removed to first order, along with many other potential contributors to non-flatness. These benefits require at least three dither positions and more are better.
- Properly sampled images are another form of repetition – more than one pixel contributes to the signal. Accurate photometry benefits from spreading the light over multiple pixels (which can also be done with dithering).

The second rule of imaging photometry is "don't change anything." Artifacts like MUX glow, pedestals, and many others will disappear from the reduced data virtually completely if all the input data – science and calibration frames – are taken in identical ways (e.g., constant exposure times and readout cadences, plus constant temperatures and backgrounds).

4.5.3 Calibration

We now address what is sometimes called the inverse problem – working with the observations to remove artifacts and generate a high-fidelity version of the photon input that yielded the measurements. A more mundane term is to calibrate the data, which we assume is in the form of a two-dimensional field of digital units (DU, or DN for digital numbers) representing the outputs of all of the array pixels. In discussing this process, we assume that the data consist of repeated exposures of the field, moving the source on the array between exposures (dithering). This is the most common way to take imaging data (because it works well). We also assume that there is an offset frame (sometimes called bias frame) with a very short exposure and no signals, plus a dark current frame with a long exposure (ideally the same as for the frames on the science field) and no signals. We will describe how to generate a response frame from the data.

The first step in reducing the data is to pre-process the images to get rid of artifacts. In the case of persistent latent images, it may be necessary to generate an ad hoc response image from the images with the impressed latent and use it to make a correction. One might also have to identify electrical and optical ghost images and other such effects and fix them by hand or with custom routines. Modern infrared arrays often include non-active reference pixels that are electrically identical to those that detect photons. They can be used to correct the images for slow drifts in the electronics and other such effects, although how useful they are depends on the array and the use it is being put to. Their use may be hard-wired into the reduction pipeline, or one may need to experiment or seek advice on whether they improve the quality of the data. For CCDs, the same benefit is obtained by overscanning the array, that is, continuing the readout clocking beyond the area of the real pixels. The electrical artifacts in the overscan data are a measure of those impressed on the data from the readout process. As appropriate, the reference pixel or overscan information can be used to correct all the images.

Cosmic ray hits can be identified in three basic ways. First, if one is sampling up the ramp on an integrating detector output, a cosmic ray will cause a discontinuous jump that can be identified and excised from the data in this first stage of processing. The second approach is to obtain a large number of exposures and reject the cosmic rays by comparing the independent images. For example, one can take the median for each position on the sky, rather than the mean. That way, an outlier or two (due to a cosmic ray hit or other stochastic event) will automatically be excluded from the data. A related procedure is called sigma clipping. One calculates the mean and standard deviation for the values for repeated measurements of a position, then rejects those falling more than some number of standard deviations from the mean and recalculates. Both approaches are examples of robust statistics; that is, they are approaches in which the results are not strongly perturbed by a few anomalous readings, but which also impose no bias on the final value. A third approach is possible if an image is significantly oversampled; it may be possible to identify cosmic ray hits as producing an "image" that is impossibly small in size.

To produce good images the data must be processed to take into account and correct for the differing properties of the detectors in the array, that is, pixel-to-pixel variations in: (1) amplifier offset; (2) dark current; and (3) response. The first two calibration frames needed for this process (offset and dark current) should already be available. A response frame (sometimes called a flat field) can be generated as the median average of the dithered frames on sky – as a result of the dithering, sources will disappear because they do not appear at the same place on any two frames.

Image data reduction then consists of:

1. Subtract the offset frame from the data, dark, and response frames to obtain data′, dark′, and response′
2. Scale dark′ to exposure time of data and response and subtract from them to get data″ and response″
3. Divide data″ by response″

That is:

$$\text{data}' = \text{data} - \text{offset}$$
$$\text{dark}' = \text{dark} - \text{offset}$$
$$\text{response}' = \text{response} - \text{offset}$$
$$\text{data}'' = \text{data}' - \alpha\,\text{dark}'$$
$$\text{response}'' = \text{response}' - \beta\,\text{dark}'$$
$$\text{final frame} = \text{data}''/\text{response}''$$

Here α and β are the scaling factors to adjust the dark frame to be equivalent to the dark that would be obtained with the same exposure as the frame being reduced. If the data frame has a uniform exposure, then the product of these reduction steps will be a uniform image at a level corresponding to the ratio of the exposure on the data frame to the exposure on the response frame (exposure = level of illumination multiplied by the exposure time). Sources will appear on top of this uniform background.

There will be one such reduced frame for each position of the telescope (at least for each dither position and if multiple exposures were taken at a given position, the telescope may have drifted in pointing so each exposure may be usefully treated as a separate reduced frame). These frames need to be shifted to a common pointing and combined. In general, the images are unlikely to have been taken with exactly an integral number of pixel widths difference in pointing. The simplest solution is just to assign the pixel values to the nearest point on the master grid of the image. An improvement is to use bilinear interpolation among the four nearest positions to assign the values appropriately for the surrounding pixel positions. Various other interpolation functions such as sinc, spline, or polynomial can produce better performance under specific conditions. Finally, after shifting all the frames, another median average will eliminate bad pixels and cosmic rays, while gaining signal-to-noise ratio on the source image. This final step automatically fills in the positions of bad pixels with valid data from other pointings of the telescope, so the impact of the bad pixels is only a local reduction of the total integration time in the final image.

We have gone through one possible calibration sequence. In general, the image reduction software will include standard or recommended procedures to generate the necessary calibration frames from the data, and to shift and add all the science frames into one high-quality image. In addition, for simplicity, we have specified the minimum number of calibration frame types. Additional types that may be useful include: (1) dome flats obtained with the telescope pointing at an illuminated screen; they provide an alternative response frame set with high signal-to-noise ratio but with the disadvantages that the spectral character and the illumination do not match the sky well; (2) twilight flats obtained in the evening or morning when the sky is bright enough to provide high signal-to-noise ratio in short exposures; they overcome the issues with dome flats to some extent; (3) shutter shading exposures, flats obtained with a range of integration times, especially short ones, so the effect of a shutter on the exposure over the field can be determined; and (4) orientation exposures obtained with the telescope drive off, so the precise orientation of the field on the sky can be determined from the direction of the star trails.

For well-sampled images, the approaches described above work well. For undersampled images, or where one is being exceptionally careful to preserve the image characteristics, a different class of combination works better, exemplified by the "drizzle" algorithm (Fruchter and Hook 2002). A fine grid is imagined to be projected onto the sky, representing an ideal array of pixels for the image. The data for the observed pixels are projected onto this grid as in Figure 4.8. This projection not only reflects the true pixel sizes, but also includes any distortions from the optics, rotations of the instrument with regard to the ideal grid, and any other issues that result in the projection of the real data not coinciding perfectly with the ideal array. The signals are divided according to the overlapping areas of the actual pixels and the ideal ones represented by the ideal grid. For illustration in Figure 4.8, the observed pixel area was set to exactly four times the area of a "pixel" in the ideal grid, but it is distorted in shape and rotated. The algorithm would assign 25% of the signal received by this imager pixel to the ideal one 2 down and 3 from the left (since it overlaps entirely), and would distribute the remaining 75% of the signal to the other ideal pixels in proportion to the

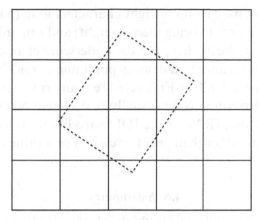

Figure 4.8. Basic operation of drizzle algorithm. The solid lines represent the ideal grid of pixels projected onto the sky. The dashed line outlines the projection onto the sky of an imager pixel.

overlapping areas with the projected imager pixel. Thus, if one of the ideal pixels overlaps with 15% of the projected real pixel, 15% of the signal from that real pixel is assigned to the ideal one. Once this process has been completed over the entire image, the signals over the ideal grid of pixels represent the source field as it might have been imaged starting with the ideal array.

4.5.4 How to carry the measurements around

Once the images are reduced, they need to be recorded. In the late 1970s astronomers developed the Flexible Image Transport System, FITS, as an archive and interchange format for astronomical data files. Since 2000 FITS has also become the standard format for on-line data that can be directly read and written by data analysis software. FITS is much more than just an image format (such as JPG or GIF) and is primarily designed to store scientific data sets consisting of multidimensional arrays and two-dimensional tables containing rows and columns of data.

A FITS file consists of one or more Header + Data Units (HDUs), where the first HDU is called the "Primary Array." The primary array contains an N-dimensional array of pixels. This array can be a 1-D spectrum, a 2-D image or a 3-D data cube. Any number of additional HDUs, called "extensions," may follow the primary array. They are followed by an optional "Data Unit."

Every "Header Unit" is formatted in ASCII and consists of any number of 80-character records that have the general form:

$$\text{KEYNAME} = \text{value/comment string}$$

The keyword names may be up to eight characters long (padding characters are added to shorter ones to bring them to eight) and can only contain capital letters, the digits 0–9, the hyphen, and the underscore character. The value of the keyword may be an integer, a floating point number, a character string, or a Boolean value (the letter T or F). There are many rules governing the exact format of keyword records so it is usually best to rely on standard interface software like CFITSIO, IRAF or the IDL astro library to construct or parse the keyword records rather than directly reading or writing the raw FITS file.

4.6 Astrometry

Well-reduced images invite us to think of the positions of astronomical objects, a direction that leads us to the topic of astrometry – the procedures used to set up systems of positions and to measure the locations of individual objects accurately within these systems.

4.6.1 Coordinate systems

An astrometric coordinate system can be envisioned as a grid projected up into the sky upon which the positions of celestial objects are measured. Such a grid has a fundamental great circle and a secondary one (a great circle is the intersection of a plane running through the center of a sphere with the surface of that sphere). There are four such grids in common use:

1. *Horizon*: The fundamental circle runs around the horizon and the secondary one runs from it over the zenith. This system is the basis of the altitude-azimuth, alt-az, coordinates used to point most large telescopes. As shown in Figure 4.9, the great circle passing through the zenith and north and south celestial poles defines the zero point of azimuth where it intersects the horizon circle to the north. Any object on the celestial sphere lies on a great circle perpendicular to the horizon circle, and the azimuth (A) for this object is the angular distance measured eastwards from the zero point to the first intersection of its great circle with the horizon one. The altitude (a) of the object is measured along this great circle from the horizon circle, $+$ if it is above the horizon and $-$ if it is below.

2. *Equatorial*: Although the horizon system is convenient for telescopes, it has the disadvantage that the coordinates of any object depend on the place and time of the observation. Another system is needed in which the position of a celestial object remains at fixed coordinates. Equatorial coordinates fulfill this role (Figure 4.10); the fundamental circle is the

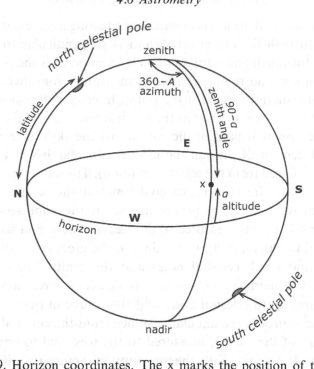

Figure 4.9. Horizon coordinates. The x marks the position of the source. Azimuth runs from north through east, a projection that is not clear in this diagram, so we show 360 – A.

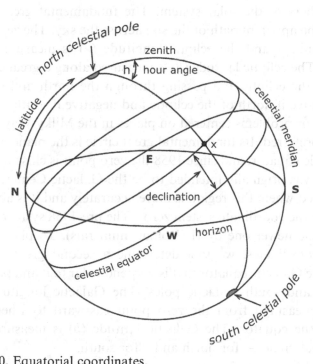

Figure 4.10. Equatorial coordinates.

celestial equator and declination is measured along a secondary great circle that runs through the object (at x) and is perpendicular to the celestial equator, + for north and − for south. The zero point is the position of the Vernal Equinox, and right ascension is measured from there eastward to where the declination circle for the object intersects the celestial equator. The Vernal Equinox is defined as the point where the celestial equator and ecliptic (the apparent path of the sun across the sky) intersect in March (i.e., the placement of the sun the moment in March when it is directly overhead as viewed from the equator at noon). Thus, the zero is roughly at midnight (within the vagaries of civil time) at the Autumnal Equinox. The units of right ascension (α) are hours, minutes, and seconds of time while declination (δ) is measured in degrees, minutes, and seconds of arc. The celestial meridian, or local meridian, is the great circle along which lie the north and south celestial poles and the zenith (the point directly overhead). The meridian of a source is the great circle along which lie the north and south celestial poles and the source in question. The hour angle of the source is the angular distance from the celestial meridian to the meridian of the source, measured to the west and in hours, minutes, and seconds. It is equivalently the time until the source transits the celestial meridian (negative hour angle) or the time since it transited (positive).

3. *Ecliptic*: This coordinate system is the natural one to use when dealing with members of the solar system. The fundamental great circle is the ecliptic – the apparent path of the sun across the sky. The zero point is the Vernal Equinox and the ecliptic longitude (λ) is measured from there eastward. The ecliptic latitude (β) is measured along a great circle perpendicular to the ecliptic and passing through the north and south ecliptic poles: positive if north of the ecliptic and negative if south.

4. *Galactic*: For problems centered on places in the Milky Way, the Galactic system is preferred. Its fundamental great circle is the plane of the Galaxy, the Galactic equator, and since 1958 the zero point is close to the Galactic Center (it was originally intended to be the Galactic Center, but we have since learned where this region is more accurately and it is about 5 arcmin away from the coordinate system zero). The pre-1958 system is designated by I and the newer one by II (Roman numerals). To place an object in Galactic coordinates, we first determine a second great circle passing through the Galactic equator that is perpendicular to it and passes through the north and south Galactic poles. The Galactic longitude (l) of the source is measured from the zero point eastward to where this circle intersects the equator. The Galactic latitude (b) is measured along the second great circle, + for north and − for south.

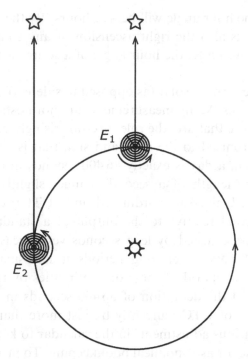

Figure 4.11. Time and local meridian.

4.6.2 Time

To make use of these coordinate systems, we need to synchronize watches – that is, to place the celestial objects in the sky as a function of some time system. A way to do so is to determine when an object is on the local meridian.

Our clocks are set to run (approximately) on *solar* time (sun time). But for astronomical observations, we need to use *sidereal* time (star time). Consider the earth at position E_1 in Figure 4.11. The star shown is on the meridian at midnight by the clock. However, three months later when the earth reaches position E_2, the same star is on the meridian at 6 p.m. by the clock. We define one rotation of Earth as one sidereal day, measured as the time between two successive meridian passages of the same star. Because of the earth's orbital motion, this is a little shorter than a solar day. (In one year, the earth rotates 365 times relative to the sun, but 366 times relative to the stars, so the sidereal day is about 4 minutes shorter than the solar day.)

The local sidereal time (LST) is the sidereal time at a particular location. It is zero hours when the Vernal Equinox is on the observer's local meridian, and by definition of the hour angle, the LST is thus the hour angle of the Vernal Equinox – that is, if the Vernal Equinox is on the local meridian now, in two hours it will be two hours (30 degrees on the celestial equator) west of the meridian and the hour angle will be + 2 hours. By the definition of right ascension, the LST is also the right ascension of any source that is on the local meridian. Equivalently, the hour angle of a source is the LST minus its right ascension.

We also use some form of solar (as opposed to sidereal) time as part of our measurement strategies. Many measurements are not hostage to precise time keeping, but, for those that are, the rules are surprisingly complex. Universal time (UT) is synchronized to the stars and sun, that is, the rotation of the earth. By convention, a day is exactly 86 400 seconds in duration, with the implication that the length of a second changes slightly with changes in the rate of rotation. Universal coordinated time (UTC) solves this problem by defining the second relative to the output of a standard atomic clock. However, it must be adjusted by leap seconds very occasionally to keep it coordinated with UT, so over long periods it has small discontinuities. Terrestrial time (TT) gets rid of these discontinuities by running strictly by the atomic clock and the definition of 86 400 seconds in a day; hence it is slowly getting ahead of UTC (currently by just more than a minute). Over long periods, the various adjustments in the calendar to keep it synchronized are also troublesome for astronomical bookkeeping. To provide a continuous system, the Julian day has been defined, starting at zero at noon (UT) on January 1, 4713 BC. The modified Julian date (MJD) is equal to the Julian date minus 2 400 000.5; the extra half-day puts the transfer from one day to the next at midnight rather than noon. A Julian date converter can be found at http://aa.usno.navy.mil/data/docs/JulianDate.php

4.6.3 Coordinate transformations

Although each of the coordinate systems has its use, they do pose the problem of transforming from one to another. Here are some formulae for that purpose. First, to transform from azimuth, A, and altitude, a, to hour angle, h, and declination, δ, for an observer at latitude on the earth of ϕ:

$$
\begin{aligned}
\cos \delta \sin h &= \cos a \sin A \\
\sin \delta &= \sin \phi \sin a - \cos \phi \cos a \cos A \\
\cos \delta \cos h &= \cos \phi \sin a + \sin \phi \cos a \cos A
\end{aligned} \tag{4.2}
$$

The inverse goes from hour angle and declination to azimuth and altitude:

$$
\begin{aligned}
\cos a \sin A &= -\cos \delta \sin h \\
\sin a &= \sin \phi \sin \delta + \cos \phi \cos \delta \cos h \\
\cos a \cos A &= \cos \phi \sin \delta - \sin \phi \cos \delta \cos h
\end{aligned} \tag{4.3}
$$

From equatorial to ecliptic coordinates, where the obliquity (inclination of the equator of the earth against the ecliptic) is $\varepsilon = 23°26'21.448''$, the transformation is

$$\cos \beta \cos \lambda = \cos \delta \cos \alpha$$
$$\cos \beta \sin \lambda = \cos \delta \sin \alpha \cos \varepsilon + \sin \delta \sin \varepsilon \qquad (4.4)$$
$$\sin \beta = -\cos \delta \sin \alpha \sin \varepsilon + \sin \delta \cos \varepsilon$$

where λ and β are the ecliptic longitude and latitude, respectively. The inverse transformation from ecliptic to equatorial is:

$$\cos \delta \cos \alpha = \cos \beta \cos \lambda$$
$$\cos \delta \sin \alpha = \cos \beta \sin \lambda \cos \varepsilon + \sin \delta \sin \varepsilon \qquad (4.5)$$
$$\sin \delta = \cos \beta \sin \lambda \sin \varepsilon + \sin \beta \cos \varepsilon$$

The previous two sets of transformations are relatively simple mathematically because all the systems are centered on the earth. The conversion to Galactic coordinates does not have this attribute and is more complex. There are a number of web-based coordinate transformation calculators that can be used, for example, http://nedwww.ipac.caltech.edu/forms/calculator.html or http://heasarc.gsfc.nasa/gov/cgi-bin/Tools/convcoord/convcoord.pl or one can find details in Lang (2006) or Kattunen *et al.* (2007).

4.6.4 Defining coordinate systems

The most accurate celestial positions are obtained through very long baseline interferometry (VLBI) in the radio, accurate to a milliarcsecond or better. Therefore, in 1997 the International Astronomical Union adopted the International Celestial Reference System (ICRS), based on VLBI coordinates for 212 extragalactic radio sources. Because these objects are very distant, they should have no significant proper motions and the definition should remain in place indefinitely. We will discuss VLBI position determination in Chapter 9. The ICRS is transferred in the optical to 118 218 stars, all with accurate measurements of positions and proper motions based on the Hipparcos satellite data.

4.6.5 World Coordinate System

As astronomy becomes more and more panchromatic, it has become necessary to have an efficient method to place an image of a field accurately on the sky and in the appropriate equatorial coordinates, so that it can be matched with identifications at other wavelengths. To implement this capability, suitable information is now placed in the FITS header of many types of astronomical data. The standard format for the position-defining keywords in the FITS header is defined as the World Coordinate System (WCS). A common example is to link each pixel in an astronomical image to a specific direction on the sky (such as right ascension and declination). The basic idea is that each axis of the image has a coordinate type, a reference point given by a pixel value, a coordinate value, and an increment. A rotation parameter may also exist for each axis. The FITS WCS standard defined 25 different projections that are specified by the CTYPE keyword. For a complete description of the FITS/WCS projections and definitions see Greisen and Calabretta (2002), Calabretta and Greisen (2002), and Greisen *et al.* (2006). There are a number of software packages that help the astronomer to access the astrometric information using the WCS of the image or to write the WCS of an image to the header. A few of the most commonly used packages are WCStools, WCSLIB, IRAF, and packages in the astronomy IDL library.

If an adequate WCS does not exist for an image the basic steps are:

1. Read in the FITS image and its header.
2. Find all the stars in the image.
3. Find all stars in a reference catalog in a region of the sky where the image header says the telescope is pointing.
4. Match the reference stars to the image stars.
5. Using one of the above WCS software packages perform a fit between the matched stars' pixel positions and the reference positions. Write the resulting WCS information to the header.

Unless very accurate pointing information can already be associated with the data, obtaining the necessary information often requires conducting an automated search to match objects detected in the image with a catalog of objects on the sky. If this search is to be fast, it cannot proceed by brute force. One strategy is to sort the objects in the image in order of decreasing brightness and then to match them with a list similarly sorted of catalog objects in the same region of sky. Once a match has been achieved, it is usually necessary to correct the image data for distortions and other effects that might make the coordinates less accurate away from the specific region of the match.

4.6.6 Changes in celestial coordinates

Unfortunately, we are not done. The linking of the equatorial coordinate system to the celestial equator and poles means that the grid of the system shifts due to a number of motions of the earth. In addition, to use the system accurately there are additional effects to be taken into account. Fortunately, all of the items listed below are well understood and with care can be compensated sufficiently well that they do not interfere with obtaining accurate positions for any objects we wish to observe.

Precession

Because the earth is not perfectly spherical, the gravitational fields of the moon and sun exert a torque on it. The result is that it precesses like a spinning top, its axis describing cones with a half-angle of about 23.5° centered on the north and south ecliptic poles. A precessional cycle takes about 26 000 years. The gravity of the giant planets also drives a change in the obliquity (tilt) of the poles of the earth over a range of about 21.5° to 24.5° and with a period of ~41 000 years. Of course, these effects also change the direction of the celestial equator, and parameters that depend on it drift. For example, the zero of the equatorial system is set by the intersection of the celestial equator and the ecliptic, which is currently drifting at about 50 arcsec per year.

Therefore, coordinate systems defined by the Vernal Equinox must be specified for a certain date. The specified year is called the Equinox (not epoch, as is commonly stated). We currently usually use coordinates for Equinox J2000.0, but one will find coordinates for equinoxes of 1900, 1950, and so forth. Calculators such as http://nedwww.ipac.caltech.edu/forms/calculator.html or http://heasarc.gsfc.nasa/gov/cgi-bin/Tools/convcoord/convcoord.pl are convenient for converting from one equinox to another.

Nutation

On top of precession, the tidal forces from the sun and the moon cause a number of smaller, short-period motions imposed on the much larger circle traced by the earth's axis due to precession. The largest is 18.6 years long with an amplitude of ~9″, but a number of additional periodic terms make the motion relatively complex. These terms are called nutation (after Latin for "nodding"). Like precession, they can be determined and compensated accurately.

Parallax

Of course, one of the primary goals of astrometry is to measure parallax and determine stellar distances. For nearby stars, this effect (up to ~1″) must be accounted for in any accurate position determination. The Hipparcos satellite has been used to measure accurate parallaxes for virtually all nearby stars that are reasonably bright (Section 4.7).

Proper motion

Nearby stars also move measurably across the sky – in about 500 cases at a rate of 1 arcsec per year or more. These proper motions require updating the stellar coordinates to the current date to have an accurate position for the star. The Hipparcos satellite data have been used with earlier astrometry to provide measurements of proper motions (Section 4.7). Where there is a long-time baseline or under circumstances permitting very accurate positional measurements, they can be measured by other means also, including to much fainter levels than are reached by Hipparcos.

Refraction

The index of refraction of air under standard conditions is about 1.0003 and diminishes with reduced pressure. Therefore, light from outside the atmosphere that enters obliquely is bent slightly. Objects that are really 35 arcmin below the horizon will appear (in visible light) to be right on the horizon (if we could see that clearly). However, the effect decreases rapidly with increasing elevation and is less than $2'$ above $30°$. The refractive index is smaller in the near-infrared (and at longer wavelengths), reducing this effect.

Aberration

Because of the finite speed of light, the apparent position of an object is displaced in proportion to the transverse velocity of the earth moving through space relative to the vector of the beam of light from the object. Since the direction of the motion of the earth relative to the vector of the light from the object varies over the year, much of this effect is periodic over a year. Because the motion of the earth is so much smaller than the speed of light, the size of the effect can be estimated as the ratio of the speed of the earth in its orbit, 2.98×10^4 m/s, to the speed of light, that is, yielding a maximum of 20.5 arcsec (for light arriving perpendicular to the orbital motion – the effect is smaller for other arrival directions). Aberration should not be confused with parallax, which is proportional to the diameter of the orbit of the earth, 2.99×10^{11} m, divided by the distance to the object. Aberration occurs with the same amplitude for all objects in the same direction, since it depends only on the observer's instantaneous transverse velocity relative to the direction of the incoming light.

4.7 Astrometric instrumentation and surveys

Astrometry is an important branch of science in its own right (besides providing coordinates for us all). It is the foundation of our distance scales, a basic way to identify members of populations of stars, has provided fundamental evidence for the existence of a supermassive black hole in the Galactic Center, and is a promising approach to search for planets around other stars, just to name a few examples.

Stellar positions are measured from images of the sky. The accuracy of the position measurement of an unresolved source (e.g., a star) can be estimated roughly as the full width at half maximum of its image divided by the signal-to-noise ratio.[5] Standard accuracy limits for photographic astrometry based on combining the results of multiple observations of the same field are about 10 milliarcsec (mas).

To reach this level of accuracy requires that the telescope be very stable. In addition, from our discussion of issues like inter-pixel gaps and intra-pixel response variations (and their analogs for photographic plates), it is very desirable that the image scale be large enough that the image of a star is spread over many pixels. A century ago, the best solution was long-focus refracting telescopes, but as the engineering of reflectors improved they became more than competitive. Although specialized astrometric telescopes are often used for large programs (e.g., the 155-cm telescope with a flat secondary of the US Naval Observatory), with care good results can also be obtained with ones of more conventional design.

In the nineteenth century, visual astrometry was a central topic in astronomy. A number of specialized telescopes and instruments were developed to allow accurate measurements, such as transit telescopes (sometimes called meridian circles) to mark the passage of a star over the meridian, heliometers (telescopes with split lenses to measure angular distances from their neighbors), and various micrometer-adjustable sighting devices integrated with eyepieces.

An ambitious program was initiated in the late nineteenth century, the Carte du Ciel, to obtain all-sky astrometry using the newly available high-sensitivity photographic plates and 18 identical telescopes, each with a 30 cm aperture. In fact, the project proved too ambitious and observations dragged out for more than 50 years, by which time the product was becoming obsolete (254 printed volumes in various formats). This effort was replaced by Hipparcos and Tycho (described below), the latter of which has a similar limiting magnitude (about 11th) and number of stars to the Carte du Ciel. The work invested in the photographic effort has assumed new importance, however, because it provides a long baseline for determining accurate proper motions.

---

[5] This guideline clearly breaks down at low signal-to-noise ratio (if you do not detect a source you cannot locate it at all!). It also fails at very high signal-to-noise ratio; the underlying reason is that the further one pushes the position below the FWHM, the more demanding are the requirements on understanding the structure of the image. Fortunately, it is not required that the image be perfect (aberration-free), nor even that images being compared be identical, just very well understood.

Another common source for accurate positions is the Palomar Optical Sky Survey (POSS), which in particular has been used to provide guide stars for the Hubble Space Telescope (HST). Guide Star Catalog Version 2.2 provides all-sky measurements to accuracies of 200–250 mas and down to about 19th magnitude.

More recent astrometric data have been obtained with electronic imagers. One noteworthy example is the Sloan Digital Sky Survey (SDSS) in the optical. The SDSS uses a 2.5-m Ritchey Cretién telescope with a well-corrected 3° diameter field. Toward the edges of this field there are twenty-two 400×2048 pixel CCDs optimized for astrometry. These detectors avoid saturation on stars as bright as SDSS magnitude $r \sim 8$ (through faster readout and neutral density filters), and can detect stars as faint as $r \sim 17$ well. Therefore, they include a huge number of stars from the Tycho-2 catalog (described below) and other astrometric catalogs to establish the overall reference frame, and then extend this frame to their detection limit. The SDSS photometry CCDs saturate at $r \sim 14$, so the astrometric reference can be transferred to them using stars between 14th and 17th magnitude, and the photometry CCDs extend the astrometry to about $r \sim 22$.

We describe the SDSS reduction steps, since they are typical of position determination with digital imaging data. To start, the CCD data are run through a standard reduction computer pipeline, which carries out the steps we described earlier in this chapter to obtain high-quality images. Positions are then measured off these fully reduced images. First, the images are smoothed to minimize noise artifacts; to avoid degrading the resolution, the smoothing length is adjusted according to the image sizes on the data frame. The pipeline divides each physical pixel into 3×3 subpixels, and quartic interpolation is used to estimate the position of the image peak within these subpixels. This result is compared with a first moment position calculation (the sum of the signals times the distance from some fiducial point divided by the sum of the signals). A number of additional steps correct these estimates for possible biases. The resulting astrometric accuracy is 50–100 mas, the latter at the sensitivity limit of $r \sim 22$ (Pier *et al.* 2003).

Another survey that provides accurate positions is 2MASS (Two Micron All Sky Survey). The 2MASS survey uses 2 arcsec pixels but with multiple sightings of each source. It has proven possible to use these sightings to obtain accurate astrometry (errors of ~ 80 mas relative to Tycho-2) from the composite images, down to K magnitude of about 14. This accuracy is achieved by modeling the positions of the Tycho-2 stars as detected by the 2MASS cameras to identify and correct a number of error sources, such as wandering of the telescope pointing and drifts in the image distortion. These methods work because 2MASS had a large instantaneous field of view (8.5 arcmin on a side) and was scanned rapidly (~1 arcmin per second) so very large areas were covered quickly compared with the time over which the potential error sources changed. Thus, a large number of astrometric calibration stars could be fitted together to obtain an accurate astrometric solution and determine the necessary corrections to the true positions.

Since about 1995, measurements with the Hipparcos satellite have been the ultimate astrometric reference source for the optical range. The instrument concept was very different from the historical imaging approach. Hipparcos had a Schmidt telescope that included a beam-combining mirror that superimposed two fields of view about 1° in size and 58° apart on the sky. The detector was an image-dissector-scanner, basically a photomultiplier with the ability to place its sensitive field anywhere over the large sensitive area that filled the usable focal plane of the telescope. A fine grid was placed over the field viewed by this detector, with alternating opaque and transparent bands. The satellite was put into a slow roll causing any star image to conduct a controlled drift across the grid. The region around a known star was isolated with the image-dissector-scanner, with a field diameter of 38 arcsec. The resulting modulation of the star signal as the telescope scanned it over the grid produced an oscillating signal. A similar signal was produced by a star in the second field of view, differing by 58° in placement on the sky. The phase difference between the two signals was analyzed for an accurate determination of the apparent angle between the two stars. These relative positions were ambiguous at the level of the period of the grid, 1.2 arcsec, but previous measurements of the stellar positions just to a modest fraction of an arcsecond permitted this ambiguity to be removed. Positions of more than 100 000 stars were measured (complete to magnitude $V = 7.3$) to an accuracy of ~1–3 mas in this way, with proper motions (based on comparison with earlier astrometry such as the Carte du Ciel) typically accurate to 1–2 mas/yr. Comparing these numbers, it is clear that the current-epoch accuracy of the Hipparcos coordinates has degraded significantly due to the uncertainties in proper motion – the epoch of observation is 1991.25, so based on a 2 mas uncertainty, by 2011 typical errors were about 40 mas.

The satellite carried a second instrument that gathered astrometric data to an accuracy about 25 times lower, but on more than 2 million stars. It also obtained homogeneous B and V photometry. These measurements are contained in the Tycho-2 catalog (the current best reduction), which is 95% complete to $V = 11.5$ with positional accuracies of 10–100 mas (better for brighter stars) and proper motions typically accurate to 1–3 mas/yr.

The next step for space astrometry is the European Space Agency GAIA mission (Lindegren 2005). GAIA goes back to image analysis, but in a grand way. As with Hipparcos, a key element is for the spacecraft to roll slowly, and for images from two widely separated areas on the sky to be brought to a single focal plane. At the focal plane, GAIA has a large number of CCDs, aligned so that the stellar images can be tracked across the detector through time-delay-integration (TDI – Section 3.6.2). The readout rate is relatively slow, just matching the arrival of the charge packets at the CCD output register; as a result, the read noise can be kept low. Because GAIA uses so many detectors and in a high-performance mode, it can achieve a huge gain over Hipparcos. It is expected to reach 20th magnitude with positional errors of a few hundred micro-arcsec (µas) and to make measurements to about 4 µas at 12th magnitude, the sensitivity limit of the Tycho catalog (Perryman *et al.* 2001).

4.8 Exercises

4.1 The index of refraction of air can be estimated from the Cauchy formula:

$$n(\lambda) = 1 + (272.6 + \tfrac{1.22}{\lambda^2}) \times 10^{-6}$$

Take the net atmospheric refraction at 40 degrees elevation and 0.6 microns to be 1 arcmin (it is a weak function of pressure and temperature). Use this curve to estimate the results of dispersion between 0.4 and 0.8 microns – how much will the near-infrared and far blue images be separated for a source 40 degrees above the horizon?

4.2 You want to observe a star at −10 degrees declination through no more than two air masses from a telescope at latitude 20 degrees north. On the most favourable nights, how long can you observe the star?

4.3 Astrometry is sometimes conducted through narrow spectral filters because of concerns about systematic effects due to the colors of stars. Take the situation described in exercise 4.1 (40° elevation) and consider two stars, one of temperature 10 000 K and the other of 3500 K. Approximate the stellar spectra by blackbodies. Compute the maximum filter width that can be used centered at 0.6 µm such that the positional shift between these two stars is less than 5 mas.

4.4 The first moment can be used to determine the centroid of a distribution of measurements (e.g., the measurements of an image by the pixels of an array):

$$\text{centroid}(x) = \frac{\Sigma\, x_i I_i}{\Sigma\, I_i}$$

where the I_i are the signals at positions x_i. Use this approach to evaluate the impact of inter-pixel gaps and sampling on centroiding accuracy. Assume a Gaussian image, and that the array pixels have uniform response over 80% of their widths, but the response is only 0.9 times the central value for the outer 10% at either side. Place the image half-way between the center and edge of one of the pixels and calculate the centroiding error as a function of the pixel sampling. For 1 arcsec FWHM seeing, what level of sampling is required to keep the centroiding error below 10 mas? If such a narrow band is not acceptable (e.g., to reach faint stars), how could you modify the observing strategy to allow a broader one?

Further reading

Kovalevsky, J. (2002). *Modern Astrometry*, 2nd edn. Berlin, New York: Springer.

Kovalevsky, J. and Seidelmann, P. K. (2004). *Fundamentals of Astrometry*. Cambridge: Cambridge University Press.

Lindegren, L. (2005). The astrometric instrument of GAIA: Principles. In *Proceedings of the GAIA Symposium, "The Three-Dimensional Universe with GAIA,"* ed. C. Turon, K. S. O'Flaherty and M. A. C. Perryman. European Space Agency SP-576.

Smart, W. M. and Green, R. M. (1977). *Textbook on Spherical Astronomy*, 6th edn. Cambridge: Cambridge University Press.

Starck, J.-L. and Murtagh, F. (2006). *Astronomical Image and Data Analysis*. Berlin, New York: Springer.

5

Photometry and polarimetry

5.1 Introduction

Most of what we know about astronomical sources comes from measuring their spectral energy distributions (SEDs) or from taking spectra. We can distinguish the two approaches in terms of the spectral resolution, defined as $R = \lambda/\Delta\lambda$, where λ is the wavelength of observation and $\Delta\lambda$ is the range of wavelengths around λ that are combined into a single flux measurement. Photometry refers to the procedures for measuring or comparing SEDs and is typically obtained at $R \sim 2$–10. It is discussed in this chapter, while spectroscopy (with $R \geq 10$) is described in the following one.

In the optical and near-infrared, nearly all the initial photometry was obtained on stars, whose SEDs are a related family of modified blackbodies with relative characteristics determined primarily by a small set of free parameters (e.g., temperature, reddening, composition, surface gravity). Useful comparisons among stars can be obtained relatively easily by defining a photometric system, which is a set of response bands for the [(telescope)-(instrument optics)-(defining optical filter)-(detector)] combination. Comparisons of measurements of stars with such a system, commonly called colors, can reveal their relative temperatures, reddening, and other parameters. Such comparisons are facilitated by defining a set of reference stars whose colors have been determined accurately and that can be used as transfer standards from one unknown star to another. This process is called classical stellar photometry. It does not require that the measurements be converted into physical units; all the results are relative to measurements of a network of stars. Instead, its validity depends on the stability of the photometric system and the accuracy with which it can be reproduced by other astronomers carrying out comparable measurements.

In other circumstances, it is necessary to convert measurements with a photometric system into physical units. This situation applies in general for non-stellar targets and, since such targets are dominant away from the optical and near-infrared, it is the rule for most wavelength regions. The initial steps are similar to those for classical stellar photometry: a photometric system must be defined, a network of standard sources is established, and measurements are made relative to these standard sources. However, the outputs of the standards must be known in physical units, so the intercomparisons can be expressed in such units.

Polarimetry is an additional avenue to learn about astronomical sources. The instrumental approaches stem from those used in photometry but for good performance require a high degree of specialization. They are described at the end of this chapter.

5.2 Stellar photometry

5.2.1 Source extraction and measurement

To provide a specific framework for our discussion, we will assume for now that we are working with optical or infrared array data. Given a well-reduced image, as described in the preceding chapter, photometric information can be extracted using a number of alternative software packages,[6] often optimized for a particular type of observation. McLean (2008) gives an overview of some of these packages, but any plan for actually using one of them needs to start with its own documentation. Here we concentrate on the underlying principles. There are two basic ways to conduct photometry: aperture photometry and point spread function (PSF) fitting.

Aperture photometry

A digital aperture is placed over the source image and the counts within the aperture are summed.[7] To determine the contribution to these counts from

---

[6] Examples of analysis systems appropriate for optical and infrared data include the National Optical Astronomy Observatory (NOAO) Image Reduction and Analysis Facility (IRAF) with applications for data from specific facilities layered on top of it, the ESO-MIDAS system, and Starlink (maintained at the Joint Astronomy Centre (UK Science and Technology Facilities Council)). In the radio, popular packages are the Astronomical Image Processing System (AIPS) developed at the National Radio Astronomy Observatory (NRAO) for reducing radio interferometric data, its update and expansion as the Common Astronomy Software Applications (CASA) package, and the Continuum and Line Analysis Single-Dish Software (CLASS) developed at IRAM (Institut de Radioastronomie Millimétrique). In the X-ray, there are the Chandra Interactive Analysis of Observations (CIAO), Xanadu, and specific packages for the XMM-Newton instruments. In addition, there are a host of applications written in the Interactive Data Language (IDL). Basically, there are many choices, allowing selection of a specific package optimized both for the envisioned application and for the access to support of the package.

[7] As with many other topics in this book, we make this procedure sound much easier than it is: to place a round aperture in a field of square pixels requires some kind of interpolation or counting rule to determine how to include signal from pixels that are only partially within the aperture.

the sky (allowing them to be subtracted), either the aperture is placed on nearby blank sky and the counts summed, or the sky level is determined over an annulus around the source position. This procedure is the same as is used with a single detector aperture photometer, but it is done in the computer after the data have been obtained (see Stetson 1987, 1990 for more information). There are a number of considerations in applying it effectively:

- The aperture is placed on the basis of a centroiding routine; for example, if I_i is the intensity in an image at x_i, then the center of the image can be determined as the central moment:

$$x_c = \frac{\Sigma \, x_i I_i}{\Sigma \, I_i} \tag{5.1}$$

(often it is helpful to subtract a uniform background from the I_i's before carrying out this calculation). Equation (5.1) applies to a single coordinate; a similar calculation provides the position in the orthogonal one. Programs like DAOFIND or DOPHOT can be used to locate all the sources in a field to allow efficient photometry of multiple sources in an array image.

- An aperture diameter must be defined. It is not necessary to get nearly all the signal, so long as one is measuring point sources and all of them are done in the same way. Small apertures are advantageous in crowded fields. For faint sources, there is a relatively small aperture size that optimizes the signal-to-noise ratio even with no nearby sources. One should experiment with different aperture sizes to determine which is best for any application. In the end, the measurements will be corrected for the signal lost from the aperture by use of measurements in multiple apertures on a bright, isolated source – this factor is the "aperture correction."

- The "sky" will have deviations with a bias toward positive values (i.e., sources). The obvious ones can be removed and replaced with mean surrounding values. In general, it may be desirable to estimate sky as the mode, rather than median or mean, since the mode (most often occurring value) is most immune to bias.

Point spread function (PSF) fitting

The PSF is determined on a bright star and this profile is fitted to the images in the frame by χ^2 minimization. In crowded fields, the fitting can include many stars at once (e.g., DAOPHOT) to extract a best fit to the photometry of overlapping sources.

- PSF fitting is more immune to anomalies than aperture photometry (e.g., a deviant pixel due to a poorly removed cosmic ray hit will be counted at its measured value in aperture photometry, but will generally have a modest effect on the PSF fit to the image).

- Calibration is preserved either with a PSF fit to a standard, or more commonly by comparing aperture photometry on an isolated bright source in the field with the PSF fit result.

- Obviously, PSF fitting is not appropriate for extended sources, for which one needs to use aperture photometry or an automatic program designed for extended sources.[8]

- Programs that do automated PSF fitting over large fields such as DAO-PHOT (Stetson 1987) or DOPHOT (Mateo and Schechter 1989) are essential for many data reduction projects.

That all sounds easy. However, there can be complications, discussion of which we postpone to Section 5.3.

Assuming we have adjusted our optical system to put multiple pixels across the diameter of the image, simple aperture photometry is unlikely to extract the source signal in an optimum way. It amounts to convolving the image of the source with a "top hat" – in this case cylindrical – function, giving equal weight to all the pixels inside the aperture. As a result, the spatial frequencies for the signal in well-sampled images are filtered by the telescope, but no filtering is applied to the noise, which can vary freely from pixel to pixel. Each pixel will contribute roughly an equal amount of noise, but the signal will be strongly weighted toward the brightest parts of the image. In some situations the aperture placed on the source is made relatively large as a way to capture most of the signal and improve the accuracy of the measurement, but this approach further increases the noise (from the sky) in the interest of capturing relatively small increments in the signal.

We should be able to smooth the image to suppress the noise while having little effect on the signal. As one might guess, the way to optimize the signal-to-noise ratio is to convolve the signal with a filter that reproduces the expected shape of the image, that is, the point spread function. Recalling the MTF of the PSF (Figure 2.13), this convolution suppresses the high spatial frequencies in the noise just enough to match the system response to a real signal. Thus, PSF fitting is an optimal way to extract a signal, particularly one detected at limited signal-to-noise ratio. This result extends to

---

[8] Computer programs are available that allow definition of arbitrarily shaped apertures to fit the profile of a source (e.g., IDP3 developed by the NICMOS project). Another approach is to conduct surface photometry within isophotes (lines of constant surface brightness). Sometimes it is appropriate to fit the source with an ellipse (or series of nested ellipses) (e.g, with GALFIT, Peng *et al.* (2002); see also Peng *et al.* (2010)) and define the photometry within them. Sextractor (Bertin and Arnouts 1996) has been developed to identify and obtain photometry of extended sources in large fields. It is not a substitute for surface photometry of well-resolved sources but is useful for analyzing survey observations.

identifying sources in the face of noise; it is often advantageous first to filter the entire image with a kernel given by the point spread function and only then to identify possible faint sources.

5.2.2 Photometric systems and terminology

In discussing photometric systems, we will continue to assume that the measurements have been obtained from images. For many years, the standard photometric instrument was a single detector that measured objects through a fixed aperture at the telescope focal plane. The details of such instruments and their measurements can be found in older books on astronomical instrumentation (e.g., Young 1974a,b,c). The methods for extracting signals differ from those for imaging data. However, once the measurements have been obtained, the following considerations are virtually identical.

If the goal is to measure stellar colors in a relative sense, then we assume the data were obtained as follows:

1. Select a detector and a suite of filters.
2. Define some combination of reference stars that define "zero color."
3. Measure a network of "standard stars" of varying colors relative to these reference stars.
4. Measure the science targets relative to these standard stars.

Assuming we do not want to fight twenty centuries of tradition, we will express the results in the magnitude system. In general, the magnitude difference, $m_1 - m_2$, between objects 1 and 2 at the same photometric band is

$$m_1 - m_2 = -2.5 \log \left(f_1/f_2 \right) \tag{5.2}$$

where f_1 and f_2 are respectively the fluxes of objects 1 and 2 in the band. Thus, the system is logarithmic, useful given the huge range of brightness of astronomical sources, and is a somewhat arcane way of describing flux ratios. To provide a universal system, the *apparent magnitude, m,* is defined as in equation (5.2) but with f_2 a "zero point" flux, that is, a flux that defines the magnitude scale and is based on a star or suite of stars to provide consistent results in multiple photometric bands:

$$m = -2.5 \log \left(f_{\text{star}}/f_{\text{ZP}} \right) \tag{5.3}$$

In general, the zero-point-defining star may not be accessible for direct observation (it may be below the horizon, too bright, or may not exist – if the zero point is defined as a combination of fluxes from more than one star, for example). Therefore, the observations should have used a well-measured "standard star" to allow transfer to the zero point. Thus, we have:

$$m = -2.5 \log (f_{star}/f_{standard}) + m_{standard} \qquad (5.4)$$

Here $m_{standard}$ is relative to the zero point.

Typically the reduced data will initially be quoted in "instrumental magnitudes," that is, in magnitudes relevant for comparing stars in a given dataset but without reference to any zero point. It is necessary to find what signals standards give in instrumental magnitudes to put the data on a standard scale. The magnitude of the standard star in equation (5.4) is then subsumed into a "zero point term."

The absolute magnitude, M, is the magnitude a star would have at a distance of 10 pc. The distance modulus is the difference between apparent and absolute magnitudes, $m - M$:

$$m - M = 5 \log \left(\frac{d}{10\,pc} \right) \qquad (5.5)$$

Here d is the distance of the star with magnitude m, and the effects on m of interstellar extinction are not included (or equivalently the formulation is only useful under the assumption that any necessary extinction corrections have been applied separately).

To go further, we need to introduce a specific photometric system as an example. We select the "Arizona" UBVRIJHKLMNQ system introduced by Harold Johnson with assistance by Frank Low (and H added by Eric Becklin). Other systems are discussed by Bessell (2005). The Johnson UBVR system was originally based on combinations of colored glass and the spectral responses of photomultipliers. The I band was defined by an interference filter (Section 6.6). The bands are shown in Figure 5.1; because the atmospheric transmission is good and relatively featureless across the BVRI bands (Figure 1.5), they could be defined for convenience (U defines the shortest band with useful transmission). This system had the appearance of being cheaply and easily reproduced, but the detector-dependence resulted in a single "system-defining" photometer.

The system was extended into the infrared with interference filters defining the photometric bands. Historically, these bands have varied from observatory to observatory depending on the availability of specific filters; the 2MASS filter set has helped define a more uniform system. The bands are tied to the terrestrial atmospheric windows as shown in Figure 5.2.

With a photometric system, we can continue our discussion of definitions. We extend the definitions of apparent magnitude to the bands of the system:

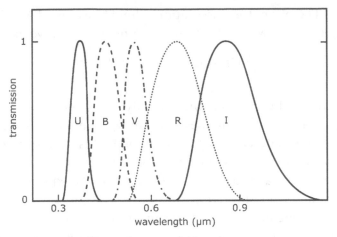

Figure 5.1. Johnson UBVRI photometric bands.

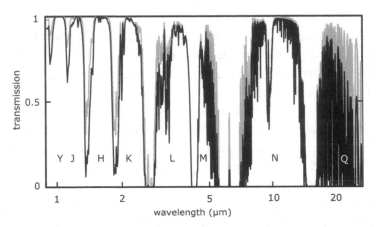

Figure 5.2. Atmospheric transmission in the infrared, with the Johnson/Low/Becklin photometric bands labeled. The Y band (Hillenbrand *et al.* 2002) has been added. The black lines are for 3 mm of water vapor and the gray for 1 mm. Based on Lord (1992) via Gemini Observatory.

for example, the magnitude in the B band is m_B. The color index, CI, is the difference between the magnitudes of a star in two different bands: for example, $CI = m_B - m_V = B - V$, where the last version is frequently used. The color excess, E, is the difference between the observed CI and the standard CI for the stellar type, for example:

$$E_{BV} = CI_{star} - CI_{stellar\ type} \qquad (5.6)$$

The bolometric magnitude, M_{bol}, is the stellar luminosity expressed as an absolute magnitude. The zero point for bolometric magnitudes has been defined in more than one way; the preferred definition is recommended by the International Astronomical Union (IAU): $M_{bol} = 0$ at 3.055×10^{28} W. However, variations at the level of a few percent will be found (generally without warning) in the literature. The bolometric correction, BC, is the correction in magnitudes that must be applied to the absolute magnitude in some band to obtain M_{bol}.

To define the zero point of the UBVRI part of the system, Johnson selected six A0 stars and averaged their colors, setting Vega close to zero magnitude (for arcane reasons related to the five other stars, Vega was not set exactly to zero). This philosophy was extended into the infrared, with the complication that not all the defining stars were detectable at the time in all the infrared bands; thus, the dependence on the brightest, Vega, became much greater. Extensive measurements were made in this system to establish the basic behavior of normal stars. This system is called the Arizona/Vega system of photometry, and magnitudes under it are called Vega magnitudes.

Following astronomical tradition that any prototypical source will prove to be atypical, Vega was a very bad choice for a star to define photometric systems. It has a debris disk that contributes a strong infrared excess above the photosphere, already detected with MSX (the Midcourse Space Experiment) at the ~3% level at 14 μm (Price *et al.* 2004) and rising to an order of magnitude in the far-infrared. Interferometric measurements at 2 μm show a small, compact disk that contributes ~1% to the total flux (Absil *et al.* 2006). Vega is a pole-on rapid rotating star with a 2000 K temperature differential from pole to equator (Aufdenberg *et al.* 2006). This joke nature has played on Harold Johnson and the rest of us accounts for some of the remaining discrepancies in photometric calibration.

5.3 Photometric refinements

Like almost anything else in life, photometry can be made as complicated as you wish. We now describe some of the reasons for complications, although there are others. At the same time, one should approach this issue by thinking carefully about the requirements of the particular science program and only pursue the refinements to the point that is needed.

5.3.1 Corrections for atmospheric absorption

Photometry of multiple sources obtained from the ground will in general be measured along different paths through the atmosphere. Although on occasion it is possible to measure the sources of interest at nearly the same elevation and over a short time interval, in general the measurements need to be corrected for the differing amounts of atmospheric absorption. By convention, the measurements are all corrected to the equivalent values as if they had been made directly overhead, through one "air mass." The different slant paths are described in terms of the factor (>1) in units of air mass by which the atmospheric path (and consequent absorption) is increased relative to one air mass.

For a plane-parallel atmosphere, the air mass is proportional to the secant of the zenith angle, z:

$$\sec z = \{\cos \varphi \cos HA \cos \delta + \sin \varphi \sin \delta\}^{-1} \tag{5.7}$$

where φ is the latitude, HA is the hour angle at which the source is observed, and δ is its declination (compare with the transformation to altitude in Section 4.6.3). This simple formula works remarkably well, with errors $<10\%$ up to four air masses and $<2\%$ up to two! However, the atmosphere is spherical, and so at large air mass a more accurate fit is needed, such as (McLean 2008).

$$\text{air mass} = \sec z - 0.0018167 \, (\sec z - 1) - 0.002875 \, (\sec z - 1)^2$$
$$- 0.0008083 \, (\sec z - 1)^3 \tag{5.8}$$

Assume the extinction is exponential,

$$\frac{dI_\lambda}{ds} = -\kappa_\lambda I_\lambda \tag{5.9}$$

where I_λ is the intensity, κ_λ the absorption coefficient, and s the path length. Then

$$m_\lambda(am) - m_\lambda(am = 1) = 2.5 \, \kappa_\lambda(am - 1)/\ln (10) \tag{5.10}$$

Thus, observations of standard stars at a range of air masses (zenith angles) can be fitted to estimate the absorption coefficient and to correct a suite of measurements to unity air mass. The standard stars selected should have a broad range of color since the differences in the various bands may be a function of atmospheric conditions.

5.3.2 Aperture corrections

Both aperture and PSF-fitting photometry require a good understanding of the point spread function. This is obvious for PSF fitting. In the former case, it is never possible to use a sufficiently large aperture to capture all the signal associated with a source, and optimizing for signal-to-noise ratio or in crowded fields generally leads to apertures that exclude a significant fraction of the signal. It is therefore necessary to determine an "aperture correction" to compensate for the lost signal (except in the special case where one can be sure that the correction is identical for all the measurements). In general, such a correction is also needed for PSF fitting because the "sky" level used in the fit may still contain faint wings of the PSF.

5.3.3 PSF variations

So far, we have assumed that the PSF is a single function for a given instrument and set of images. PSF fitting and determining aperture corrections are relatively easy if this assumption is correct, as it might well be for an instrument in space (above the atmosphere) and with well-corrected optics. However, for optical and infrared imaging from the ground, the seeing can change with time. Even from space, if the photometric band is relatively wide the PSF will vary with the spectrum of the source, and the image quality may vary over the field of view. A solution is to use a PSF-matching filter to adjust the differing PSFs. Calculating this filter by brute force can be clumsy. Alard and Lupton (1998) suggest that the process can be streamlined by calculating in advance a set of basis functions that can be used in linear combination to produce the PSF-matching filter. A variable PSF also makes noise estimation more difficult. A reliable procedure is to place artificial sources with the shape of the local PSF into the image and then extract and measure them to understand the real dependence of photometric accuracy on source strength and position.

5.3.4 Maintaining the photometric system

With traditional stellar photometry, the intercomparison of measurements depends on the ability to reproduce the spectral properties of the system and to maintain them. The original Johnson UBVRI system depended for its definition on a specific photometer – and, whenever a detector failed, there was a crisis to find a replacement with similar spectral properties and then to re-establish the system with it. Now, people have abandoned the idea of a defining photometer; instead, systems are used that are thought to be equivalent from the spectral properties of their optical components.

But be aware that not all "equivalent" systems are equivalent! There are a number of systems advertised as UBVRI that differ dramatically from

Johnson's original. The "I" band is particularly problematic, with center wavelengths ranging by over 0.1 μm! This problem arises with many other photometric systems also. Even identical detectors and filters will have differing spectral properties if operated at different temperatures or in optical systems with different f-numbers at the position of a filter. Furthermore, the atmospheric transmission in some bands depends on conditions. For examples, look at the "M" and "Q" bands for which Figure 5.2 gives transmissions for differing amounts of water vapor. The U band is similarly subject to atmospheric effects.

To deal with the problems caused by differences in photometric systems, two systems can be reconciled through observation of many stars in both and comparing the results. A fit is generated (as a function typically of the color of the source and perhaps the air mass) that allows transformation of the actual values observed with one system to the values that would have been observed with the other. An example is shown in equations (5.11) that convert JHK measurements in the older CIT (California Institute of Technology) system (Elias *et al.* 1982) to the 2MASS system (Carpenter 2001).

$$
\begin{aligned}
(\mathrm{K_S})_{2\mathrm{MASS}} &= \mathrm{K_{CIT}} + (0.000 \pm 0.005)(\mathrm{J-K})_{\mathrm{CIT}} + (-0.024 \pm 0.003) \\
(\mathrm{J - H})_{2\mathrm{MASS}} &= (1.076 \pm 0.010)(\mathrm{J - H})_{\mathrm{CIT}} + (-0.043 \pm 0.006) \\
(\mathrm{J - K_S})_{2\mathrm{MASS}} &= (1.056 \pm 0.006)(\mathrm{J - K})_{\mathrm{CIT}} + (-0.013 \pm 0.005) \\
(\mathrm{H - K_S})_{2\mathrm{MASS}} &= (1.026 \pm 0.020)(\mathrm{H - K})_{\mathrm{CIT}} + (0.028 \pm 0.005)
\end{aligned}
\tag{5.11}
$$

Here, $\mathrm{K_S}$ is the specific 2MASS band in the 2 μm atmospheric window, which differs significantly from previous definitions of the K band. The equations are in terms of the stellar colors because the differing spectral responses of the two systems affect the colors directly. Application of the transformations can homogenize the photometric systems so long as they are being applied to normal stars, since the transformations refer to such stars. However, for demanding work, be suspicious. For example, the number of measurements suitable for computing transformations between some systems is quite small and the transformations are correspondingly uncertain. Also, photometric transformations are generally less reliable in the visible and ultraviolet, where stars have many strong absorption features, than in the infrared where stellar spectra are relatively similar and feature-free.

In the past, another source of error was that standard stars at widely different positions on the sky might not be intercompared to high accuracy. All-sky databases (2MASS for JHK, Hipparcos for BV) make the situation much better, since they provide many stellar observations that are homogeneous and remove a number of possible causes of uncertainty in computing transformations.

5.4 Physical photometry

5.4.1 General approach

Classical stellar photometry is of limited use when studying something other than a relatively normal star, since it assumes a related family of pseudo-blackbody spectra. In addition, the underlying precepts of classical photometry break down at wavelengths where stars are not readily measured. A physical system converts the measurements into flux densities at specific wavelengths.

The procedure begins with a photometric system similar to those used in classical stellar photometry. To each photometric band we assign a wavelength, λ_0 (Figure 5.3), The mean wavelength is the simplest, although we will discuss alternatives later:

$$\lambda_0 = \frac{\int \lambda T(\lambda)\, d\lambda}{\int T(\lambda)\, d\lambda} \tag{5.12}$$

$T(\lambda)$ is the wavelength-dependent system response, called the relative spectral response function. It includes the full spectral transmittance (including filter, telescope, and atmosphere) for the photons reaching the detector, convolved with the detector response. In general for the latter the detector spectral responsivity should be used, the current out of the detector as a function of wavelength and the photon power into it (Figure 3.6), since this current is the signal that we work with.

5.4.2 Absolute calibration

One might expect that physical photometry would utilize a carefully calibrated detector and thorough understanding of the telescope and

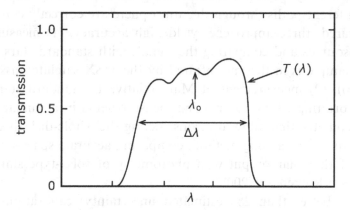

Figure 5.3. System transmission characteristics.

instrumental efficiency, to obtain source measurements directly in physical units. However, it turns out to be very difficult to achieve high accuracy in this way; the telescope and instrumental efficiencies are difficult to measure and are subject to change. Therefore, physical photometry is carried out relative to standard stars or other sources, just as in classical photometry, including all of the corrections to unity air mass (but not to transform to a standard photometric system). Only after the measurements are fully reduced do we determine the flux densities of sources at the fiducial wavelengths of our photometric bands.

The challenge is then to know the flux densities of our standard stars accurately. There are a number of possibilities to get this information.

Direct calibrations: One compares the signal from a calibrated blackbody reference source to those from one or more members of the standard star network. Ideally, one would use the same telescope and detector system to view both, but often the required dynamic range is too large and it is necessary to make an intermediate transfer.

Indirect calibrations: One can use physical arguments to estimate the calibration, such as the diameter and temperature of a source. A more sophisticated approach is to use atmospheric models for calibration stars to interpolate and extrapolate from accurate direct calibrations to other wavelengths.

Hybrids: The solar analog method is the most relevant hybrid approach. It uses absolute measurements of the sun, assumes other G2V stars have identical spectral energy distributions, and normalizes the solar measurements to other G2V stars at some wavelength where both have been measured accurately.

The calibrations in the visible are largely based upon comparisons of a standard source (carefully controlled temperature and emissivity) with a bright star, usually Vega itself. Painstaking work is needed to be sure that the very different paths through the atmosphere are correctly compensated. In the infrared, three approaches yield high accuracy: (1) measurement of calibrated sources and comparing the signals with standard stars (the most accurate example is in the mid-infrared by the MSX satellite mission (Price *et al.* 2004)); (2) measurement of Mars relative to standard stars while a spacecraft orbiting Mars was making measurements in a similar passband and in a geometry that allowed reconstructing the whole-disk flux from the planet; (3) the Solar analog method, comparing accurate space-borne measurements of the solar output with photometry of solar-type stars (for the latter two, see Rieke *et al.* 2008).

Accurate (better than 2% estimated uncertainty) calibrations are now available for the range 0.3–25 µm. They are usually expressed as the flux density corresponding to zero magnitude at each effective wavelength, termed the "zero point." Thus, accurate physical measurements can be deduced over this range of wavelengths by measuring signals relative to standard stars that are traced to the absolutely calibrated network, correcting for atmospheric extinction, and then applying the calibration directly to the measured magnitudes.

5.4.3 Bandpass corrections

Although we want to quote a flux density at a specific wavelength, the signal is proportional to the convolution of the source spectrum with the instrument relative spectral response function (see Figure 5.4). Therefore, sources with the same flux densities at λ_0 but different spectral shapes may give different signals – and, conversely, sources giving the same signals may not have the same flux densities at λ_0. We need to compensate for this effect, or the accurate absolute calibrations now available will be undermined. With knowledge of the source spectrum, one can calculate the necessary corrections by integrating it and the standard star spectra over the passband and comparing the results.

For example, any source so cold that the peak of its blackbody falls at longer wavelengths than the system passband has a spectrum rising very steeply (Wien side of blackbody) across the passband. Most of the signal will

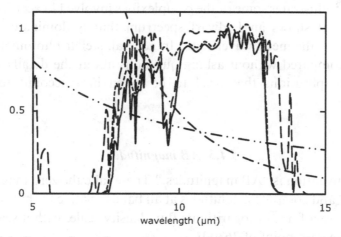

Figure 5.4. Comparison of measurements of a hot star (A0: dash–dot line) and a power law spectrum (λ^{-1}: dash–double dot line) in the N filter. The short-dashed line shows the relative spectral response of the instrument (bandpass-defining filter and detector), the long-dashed line is the transmission of the atmosphere, and the solid line is the convolution of the two to give the net relative spectral response.

be derived from wavelengths longer than λ_0, and the flux density at the center of the passband can be small even with a significant net signal. A correction in the opposite direction is required when the blackbody peaks near λ_0. The size of the corrections scales roughly as $(\Delta\lambda/\lambda_0)^2$, where $\Delta\lambda$ is the filter transmission width (Figure 5.3). The correction is moderate in size for $\Delta\lambda/\lambda_0$ < 0.25, but for broader filters the uncertainties in the correction can dominate other uncertainties in the photometry.

5.4.4 Minimizing bandpass corrections

To reduce the bandpass corrections, some astronomers use alternates to λ_0, for example the effective wavelength (used by IRAS and IRAC on Spitzer among others):

$$\lambda_{\text{eff}} = \frac{\int \lambda^2 \, T(\lambda) \, d\lambda}{\int \lambda T(\lambda) \, d\lambda} \tag{5.13}$$

Yet another approach is embodied in the *isophotal wavelength*, in which the fiducial wavelength of the measurement is adjusted to match the measured flux density. This process is mathematically equivalent to corrections that adjust the flux density. In fact, if one does not want every source to have its own characteristic measurement wavelength, flux-based bandpass corrections need to be used anyway.

   Figure 5.5 illustrates some of the complexities involved in accurate physical photometry. It shows an L dwarf spectrum that is dominated by strong features just at the mean wavelength. In general, stellar photometry – where colors are compared without asking questions about the details – can bury a lot of complications that need to be taken into account for physical photometry.

5.4.5 AB magnitudes

Another convention is "AB magnitudes." These take the zero magnitude flux density at V and compute magnitudes at all bands relative to that flux density. Thus, they are a form of logarithmic flux density scale, with a scaling factor of –2.5 and a zero point of 3630 Jy.

$$m_{\text{AB}} = -2.5 \log(f_\nu(\text{Wm}^{-2}\text{Hz}^{-1})) - 56.085 \tag{5.14}$$

They have the virtue that conventional corrections, such as for interstellar extinction (tabulated in magnitudes), can be applied directly to them. Be

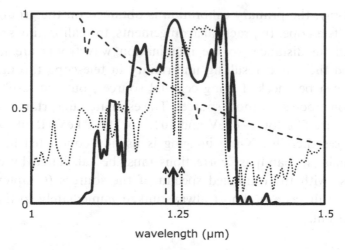

wavelength (μm)

Figure 5.5. J-band photometry of an early L dwarf. The dashed line is the spectrum of an A0V calibrator star, the dotted line is that of the L dwarf, and the solid line is the transmission profile of the J filter. The A0 and L dwarf spectra have been adjusted to give identical signals in the J band. The solid arrow is the mean wavelength of the filter, while the dashed arrow is the wavelength dividing the A-star signal equally within the band, while the dotted arrow divides the L dwarf signal equally.

careful, though, because away from the visible range AB magnitudes can differ substantially from Vega ones: for example, $m_K(AB) - m_K(Johnson) \sim 2$.

5.4.6 Calibration in other spectral ranges

Other spectral ranges lack celestial calibrators that are basically modified blackbody radiators and can be modeled and understood to the level necessary for accurate calibration from first principles. Thus, they rely more on direct measurements with specialized equipment to establish an absolute calibration. For example, in the radio, the primary standards are measured with horn antennas, which have cleaner beams and are easier to model than a paraboloid with a feed antenna (as discussed in Chapter 8), or with dipole antennas, which also can be modeled accurately. The calibrators are then tied in with other standards with conventional radio telescopes. The accuracies are generally in the 3–5% range (e.g., Baars *et al.* 1977; Mason *et al.* 1999). At high frequencies, WMAP (Wilkinson Microwave Anisotropy Probe) observations of bright sources can provide an accurate absolute calibration (Weiland *et al.* 2011). In general a practical limit to the applied calibration arises because of the need to model the variations of the reference sources, either due to secular changes in their outputs, or to rotation in the case of planet calibrators (Weiland *et al.* 2011).

In the X-ray, the primary calibration is obtained on the ground prior to launching a telescope. In practical arrangements, the calibrating source is not quite at infinite distance, so the illumination wavefronts are not exactly parallel, and there is a resulting uncertainty in telescope throughput. The calibration can be checked using celestial sources, but the results are very dependent on models for those sources. The estimated uncertainties for ACIS on Chandra are 5% for 2–7 keV and 10% for 0.5–2 keV (Bautz *et al.* 2000 and references therein). X-ray imaging is usually conducted in very wide spectral bands, so bandpass corrections must be calculated by convolving the response with the estimated spectra of the sources (Chapter 10). The quoted X-ray fluxes are almost always linked immediately to the assumed source spectrum.

5.4.7 *An example*

Suppose you obtained the following measurements of two stars in digital units:

Band	J ($1.25\,\mu$m)	N ($10.66\,\mu$m)
Star Y	10 010 DU	10 016.1 DU
Star Z	983.1 DU	10 097.3 DU
sky	10 DU	10 000 DU

The J and N filter passbands are from Figures 5.5 and 5.4, for which the mean wavelengths are 1.24 and 10.66 µm, respectively. Assume that star Y is a standard star with $m_J = m_N = 5.00$. All of the measurements used aperture photometry and the sky signal was measured with the same aperture as the stars. Star Y was measured at 1.1 air masses and Star Z at 1.3; the corrections for both bands are 0.2 magnitudes per air mass.

Compute m_J, m_N, $m_J - m_N$ for star Z.

Compute the flux densities corresponding to the m_Js and m_Ks. You can take:

$$m_J = 0 \quad \text{at} \quad 1623\,\text{Jy} = 1.623 \times 10^{-23}\,\text{Wm}^{-2}\,\text{Hz}^{-1}$$
$$m_N = 0 \quad \text{at} \quad 35\,\text{Jy} = 3.5 \times 10^{-25}\,\text{Wm}^{-2}\,\text{Hz}^{-1}$$

1. We compute sky- and air mass-corrected signals. The correction to unity air mass for star Y is $10^{(0.4*.2*.1)} = 1.0186$, while for star Z it is 1.0568. We then have for one air mass the equivalent signals:

Band	J (1.24 µm)	N (10.66 µm)
Star Y	1.0186(10 010 − 10) DU = 10 186 DU	1.0186(10016.1 − 10 000) DU = 16.4 DU
Star Z	1.0568(983.1 − 10) DU = 1028.4 DU	1.0568(10 974 − 10 000) DU = 1029.3 DU

2. Assume the output DUs are linearly related to the detected flux density:

$$DU_X = C F_X$$

The reduction program assumes that zero magnitude gives 500 000 DU in J and 3000 DU in N. Then the instrumental J magnitudes are

$$m_{inst} = -2.5 \log (DU/500\,000) = -2.5 \log (DU) + 14.247$$

or, in our case,

$$m_{inst}(Y, J) = -2.5 \log (10\,186) + 14.247 = 4.227 \text{ mag}$$
$$m_{inst}(Z, J) = 6.717 \text{ mag}$$

From the results for star Y, the correction to true magnitudes is $5 - 4.227 = 0.773$ magnitudes at J, so the J magnitude of star Z is $6.717 + 0.773 = 7.49$.

Similarly, the instrumental N magnitude for star Y is 5.674, making the correction to true magnitudes −0.674. The magnitude of star Z is then 0.506 and $m_J - m_N = 6.98$ for this star.
3. To compute flux densities, start with m_J. By definition, the J flux density of star Z is fainter than that of a 0 magnitude star by

$$10^{7.49/2.5} = 990.8$$

Thus, its nominal flux density is 1623 Jy/990.8 = 1.638 Jy.
At N, the flux density is fainter than zero magnitude by $10^{0.506/2.5} = 1.5937$, so the nominal flux density is 35 Jy/1.5937 = 21.96 Jy.
4. However, the two stars have very different spectral slopes, so there may be a significant bandpass correction to get the best possible flux density estimates. We have little information about the slopes across the filter bands, so we need to guess. Suppose we fit a power law to the spectrum of star Z:

$$S_Z(v) = C_Z v^\alpha$$

Then we can determine the index, α, as

$$\alpha = \frac{\log (S(v_2)/S(v_1))}{\log (v_2/v_1)} = \frac{1.127}{-0.934} = -1.207$$

whereas star Y can be taken from its J − N color of zero to be nearly Rayleigh–Jeans,

$$S_Y(v) = C_Y v^2$$

We can simulate the detection process to predict the signals we should have received:

$$\text{Star Z signal} = C_Z \int_0^\infty \text{RRF}(N) v^{-1.207} dv$$

where RRF(N) is the relative response function for the N filter. Similarly,

$$\text{Star Y signal} = C_Y \int_0^\infty \text{RRF}(N) v^2 dv$$

If we adjust C_Z and C_Y to give identical flux densities at the mean wavelength, 10.66 μm, we find that the Z signal is only 0.826 times the Y signal. We conclude that the differing shapes of the spectra in the filter bandpass have biased our assignment of the flux density of star Z relative to that of star Y slightly. To put them on the same basis, we should multiply the derived flux density of star Z by $1/0.826 = 1.21$, while retaining the same effective wavelength. That way, the integrals will correspond to the observed ratio. If we do a similar calculation for the J band, we find that the signal from star Z needs to be multiplied by 1.028 to equalize the signals. Therefore, our best prediction for the flux densities of star Z are 1.638 Jy × 1.028 = 1.684 Jy at J and 21.96 Jy × 1.21 = 26.6 Jy at N. The bandpass correction is much smaller at J than at N because the fractional bandwidth $\Delta\lambda/\lambda$ at J (~0.2) is much smaller than it is at N (~0.5).

5.5 Other types of photometry

5.5.1 Differential photometry

Photometry can usually be made more accurate by using local references, either reference stars nearby on the sky to a science target, or by comparing a spectral region with a nearby one for the same source. "Spatial differential" is the usual approach to monitoring a source for variability by observing nearby stars simultaneously with the measurement target. This gain comes "for free" with imagers. Atmospheric effects and time variability of the system can all be made to cancel to first order. "Spectral differential" is useful when narrow spectral features can be measured by comparing signals from a narrow and broad filter, both centered on the feature (there are other strategies too). An example is a narrow filter centered on the Hβ line and a broader continuum filter centered at the same wavelength. The ratio of the signals from these two filters can be calibrated on stars with spectroscopically measured Hβ equivalent widths. The system can then be used for a quick spectral type classification of hot stars based on the equivalent width of their Hβ absorption.

5.5.2 High speed

There are a number of types of astronomical source that vary more quickly than can be measured with standard procedures. With arrays, the readout time is fixed by the speed of the transfer of signals (and the resulting charge transfer efficiency for a CCD), plus the speed of the output amplifier and the receiving electronics. However, with nearly all arrays, one can get signals from just a subsection. If only a small fraction of the pixels in the array are being read out in this subarray, then they can be read much faster than the entire array while still maintaining the read rate per pixel required for good performance. For example, a CCD can be run fast to shift the charge out over the "unwanted" area, dumping it by resetting. At the subarray, the clocking is slowed down and the pixels are read out at the standard rate. Past the subarray, it is usually desirable to clock through the entire array so that charge does not build up and bleed into regions where it is not wanted. In theory infrared arrays allow full random access – one could address just the subarray. However, since usually electronic circuits are built in to shift the entire array out serially, typically one has to use a fast-advance strategy similar to that with the CCD to get to a subarray. A few arrays have extra logic to allow direct addressing of a subarray (e.g., Teledyne HgCdTe arrays). One can also use a custom-manufactured small array if one wants fast readout without the advancing over pixels.

Higher speeds can be obtained by single-photon-counting using detectors with gain (e.g., photomultiplier, avalanche photodiode, solid-state photomultiplier). Such devices are discussed in Chapter 3. The detectors used in the X-ray (pulse-counting CCDs or microcalorimeters) and radio (heterodyne receivers) are intrinsically fast in response and can be adapted readily to high-speed photometry.

5.5.3 High accuracy

It can be difficult to reach accuracies better than the 1% level with imaging detectors used in standard ways. However, under special circumstances, imagers can do considerably better than they do for routine photometry. Very high photometric repeatability has come to the fore in the past decade, for observing both the transits of planets (Figure 5.6) and also the oscillatory modes of stars (a field called astroseismology). An array detector is virtually

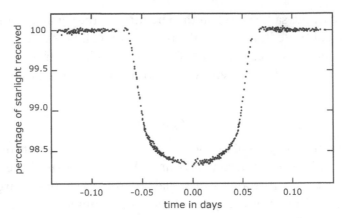

Figure 5.6. Measurement with HST of the reduction in stellar brightness due to the transit of a giant planet (Brown *et al.* 2001). Reproduced by permission of the AAS.

required because an aperture photometer would modulate the signal with tiny telescope motions. It is possible to mitigate the pixel non-uniformity by putting many pixels across the point spread function. The sampling can be enhanced by putting the images slightly out of focus or by moving them systematically over the array (or by other strategies). Accurate placement of the star on the array also improves the relative photometry, since it prevents many of the noise sources due to intra-pixel sensitivity variations and similar causes. Analysis of the images to minimize systematic effects is critical. One effective technique is to convolve a template image of the field with the seeing at the time of observation and then to scale the template to match the image and to subtract the two. The photometry can then be performed on the residual image (Alard 2000). Other methods have also been used – simple aperture photometry works reasonably well if done very carefully.

Because they are free from atmospheric effects, space telescopes perform superbly for transits and astroseismology. Although the many-pixel rule cannot be imposed for instruments optimized for other applications, high-quality data are being obtained by careful modeling of the results, taking advantage of the steady pointing and lack of effects from the atmosphere. HST and Spitzer have turned out to be extremely powerful in studying transits. MOST, CoROT and Kepler are three dedicated satellites for high-accuracy photometry.

5.6 Polarimetry

5.6.1 Background

Electromagnetic radiation consists of transverse vibrations of the electromagnetic field. If they are *incoherent*, such vibrations can separately propagate through space without change and without interfering with each other. If the phase and amplitude of these separately propagating vibrations have no fixed relationship to one another, it is said that the radiation is unpolarized. If there are lasting relationships in phase and amplitude, then the light is at least partially polarized.

Processes that lead to significant polarization include:

- Reflection from solid surfaces, e.g., moon, terrestrial planets, asteroids
- Scattering of light by small dust grains, e.g., interstellar polarization
- Scattering by molecules, e.g., in the atmospheres of the planets
- Scattering by free electrons, e.g., envelopes of early-type stars
- Zeeman effect, e.g., in radio-frequency HI and molecular emission lines
- Strongly magnetized plasma, e.g., white dwarfs
- Synchrotron emission, e.g., supernova remnants, AGN

Polarization is described in terms of the electric vectors of the photons in a beam of light. Two cases are generally distinguished: (1) linear polarization, when the vectors are all parallel to each other and constant in direction; and (2) circular polarization, where the vectors rotate at the frequency of the light. In fact, these examples are the two limiting cases. The general case is elliptical polarization, where the electric vector rotates at the frequency of the light but its amplitude varies at two times the frequency. In this case, the tip of the electric field vector traces an ellipse on a plane that is perpendicular to the wave propagation direction (see Figure 5.7).

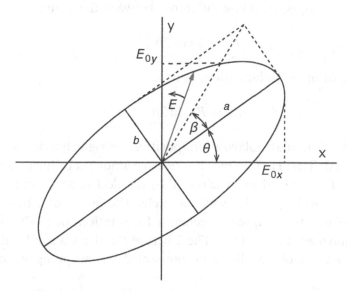

Figure 5.7. Electric vector ellipse.

The electric vector of a totally polarized beam of light can be described as

$$\vec{E} = E_x\,\hat{\imath} + E_y\hat{\jmath} = E_{0x}\cos(\omega t + \varphi_x)\,\hat{\imath} + E_{0y}\cos(\omega t + \varphi_y)\hat{\jmath} \qquad (5.15)$$

where ω is the angular frequency of the radiation and the φ's are the phase. The four Stokes parameters are used to describe this behavior. They are total intensity I, linear polarization given by U and Q, and circular polarization V. The general case, elliptical polarization, should be viewed as the combination of all four, with an intensity of

$$IP_E = \sqrt{Q^2 + U^2 + V^2} \qquad (5.16)$$

where P_E is the degree of polarization (<1).

The Stokes parameters are defined in terms of the electric vector from equation (5.15) as:

$$I = \langle E_x^2\rangle + \langle E_y^2\rangle \qquad (5.17a)$$

$$Q = \langle E_x^2\rangle - \langle E_y^2\rangle = IP_E\cos 2\beta \cos 2\theta = IP\cos 2\theta \qquad (5.17b)$$

$$U = \langle 2E_xE_y\cos\varphi\rangle = IP_E\cos 2\beta \sin 2\theta = IP\sin 2\theta \qquad (5.17c)$$

$$V = \langle 2E_xE_y\sin\varphi\rangle = IP_E\sin 2\beta = IP_V \qquad (5.17d)$$

where $\varphi = \varphi_x - \varphi_y$ is the phase difference between the x and y vibrations,

$$P = P_E\cos 2\beta \qquad (5.18)$$

is the degree of linear polarization, and

$$P_V = P_E\sin 2\beta \qquad (5.19)$$

is the degree of circular polarization, positive for right-handed and negative for left-handed. That is, β defines the relative amounts of linear and circular polarization: for $\beta = 0°$ the ellipse traced by the electric vector collapses into a line at angle θ, while $\beta = 45°$ yields a circle. The factor of 2 before θ arises because a polarization ellipse is degenerate for rotations of 180°; for example, $\theta = 0°$ is equivalent to $\theta = 180°$. The 2 before β reflects a similar degeneracy for a 90° rotation of an ellipse accompanied by swapping its major and minor axes.

Conversely, if we can measure the Stokes parameters, we can solve for the polarization vector as

$$I \qquad (5.20a)$$

$$P = \frac{\sqrt{Q^2 + U^2}}{I} \qquad (5.20b)$$

$$2\theta = \arctan\left(\frac{U}{Q}\right) \qquad (5.20c)$$

The Stokes parameters are a convenient way to describe polarization because, for incoherent light, the Stokes parameters of a combination of several beams of light are the sums of the respective Stokes parameters for each beam.

As applied in astronomy, the angles and intensity determine a polarization vector in spherical coordinates. θ is defined as the angle between the long axis of the polarization ellipse and the direction toward the north celestial pole (NCP), measured counterclockwise from the direction toward the NCP. θ is called the position angle of the polarization vector.

Now imagine that we have a polarization analyzer, a device that subdivides incident unpolarized light in half, with one beam linearly polarized in a direction we define as the principal plane (PP) of the analyzer and the other beam polarized in the orthogonal plane (OP). We designate the angle between the north celestial pole and the principal plane as ξ. When light characterized by the Stokes parameters I, Q, U, and V falls on the analyzer, the intensities of the two beams emerging from it are (Serkowski 1974)

$$I_{PP} = \left(\frac{T_l + T_r}{2}\right)^{1/2} I + \left(\frac{T_l - T_r}{2}\right)^{1/2} (Q \cos 2\xi + U \sin 2\xi) \qquad (5.21a)$$

$$I_{OP} = \left(\frac{T_l + T_r}{2}\right)^{1/2} I - \left(\frac{T_l - T_r}{2}\right)^{1/2} (Q \cos 2\xi + U \sin 2\xi) \qquad (5.21b)$$

where T_l is the transmittance of unpolarized light (e.g., measured with two identical analyzers oriented with their principal planes parallel) and T_r is the transmittance of two analyzers with their principal planes perpendicular. For simplicity, assume a perfect analyzer, $T_l = 0.5$ and $T_r = 0$. Then

$$I_{PP} = \tfrac{1}{2}(I + Q \cos 2\xi + U \sin 2\xi) \qquad (5.22a)$$

$$I_{OP} = \tfrac{1}{2}(I - Q \cos 2\xi - U \sin 2\xi) \qquad (5.22b)$$

Thus, measuring the two beams emergent from the analyzer at a number of angles relative to the direction toward the north celestial pole, ξ, gives a set of values of I_{PP} and I_{OP} that can be solved for Q and U. I can be measured as the total flux from the source. Circular polarization is conventionally measured by inserting into the beam ahead of the analyzer an optical component that converts circular to linear polarization and then measuring the strength of this linear polarization.

5.6.2 Optical elements

Analyzers

These principles underlie all astronomical polarimeters. Conceptually, perhaps the simplest form of analyzer is a fine grid of parallel conductors, placed perpendicular to the beam of light. If the spacing between wires is five times their diameters, then the grid acts as an efficient analyzer for wavelengths longer than about five times the grid spacing. Waves with electric fields perpendicular to the grid lines cannot transfer much energy to the grid because the grid diameter is so small compared with the wavelength, so they are able to travel through the grid. However, if the electric fields are parallel to the grid, the photons interact strongly with the grid and are reflected (or absorbed) efficiently. Given current capabilities to produce fine grids of conductors by photo-lithography, such devices work well throughout the infrared spectrum and can be made to work reasonably well in the visible. A similar concept explains the operation of sheet polarizing material, used in the visible. Sheet analyzers can be fabricated in polyvinyl alcohol plastic doped with iodine. The sheet is stretched during its manufacturing so that the molecular chains are aligned, and these chains are rendered conductive by electrons freed from the iodine dopant.

Another class of analyzer is based on the birefringence of certain crystals – that is, in certain orientations these crystals have different indices of refraction for light polarized in different directions. Examples include magnesium fluoride, calcite, sapphire, and lithium niobate. A simple analyzer can be made with a plane-parallel calcite plate in front of the telescope focal plane, producing images in orthogonal polarization directions separated by 0.109 times the thickness of the plate (at $0.55\,\mu m$). However, because of the differing indices of refraction, the two images come to foci at different distances behind the plate. This issue can be avoided by using two (or more) plates of material. An example is the Wollaston prism (Figure 5.8). It consists of two triangular right prisms cemented (or contacted) along their bases and with their polarization axes oriented orthogonally. When it encounters the

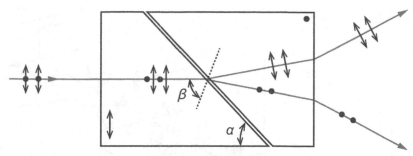

Figure 5.8. A Wollaston prism. The directions of polarization and the birefringent axes are indicated by dots (perpendicular to the page) and arrows (in the plane of the page).

interface between the prisms, a beam of light finds that one polarization direction is transitioning from a medium with a relatively high refractive index to one with a relatively low one, and the other direction is transitioning from low to high. The result is that the two polarizations are deflected in opposite directions. Wollaston prisms can divide the two polarizations by 15° to about 45° in this fashion. There are a large number of alternative arrangements that perform functions similar to that of the Wollaston prism; for example, the Glan-Thompson prism transmits one polarization undeviated but absorbs the other.

Retarders/waveplates

If one direction of polarization is retarded in phase by 180° relative to the other, the direction of the polarization is rotated, as shown in Figure 5.9 (see also equations 5.17 and 5.20). A simple retarder can be made of a plane-parallel plate of birefringent material, suitably cut and oriented. The specific case that gives a 180° phase shift is called a half-waveplate. In general more complex realizations of half-waveplates use multiple materials to compensate for color differences and give a relatively achromatic phase shift.

By symmetry, a half-waveplate with its optical axis oriented in the direction of polarization does not rotate the plane of polarization of the wave. If the optical axis is 45° from the plane of polarization, the half-waveplate rotates the polarization 90° (see Figure 5.9); if the optical axis is 90° off the plane the rotation is 180° (i.e., by symmetry, we are back to the beginning – no effective rotation). Another special case applies when there is a 90° phase shift – then a perfectly plane-polarized wave becomes perfectly circularly polarized (see Figures 5.9 and 5.10, and equation 5.17). Thus, a quarter-waveplate converts linear polarized light to circular polarization, or circular to linear.

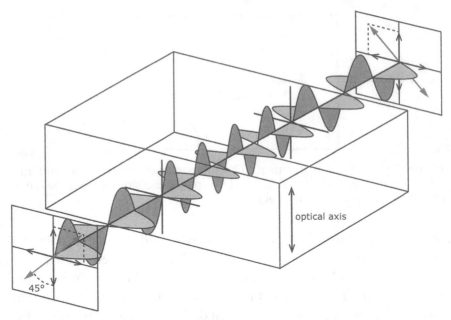

Figure 5.9. Action of a half-wave retardation to rotate the plane of polarization. The input polarization vector (lower left, at 45° clockwise of the optical
axis) is rotated by 90° upon exiting (upper right, 45° counterclockwise of the
optical axis). Redrawn with modifications from Bob Mellish (n.d.),
Wikipedia.

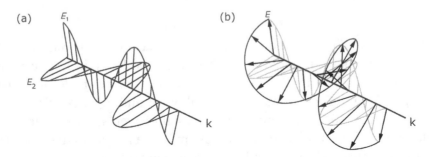

Figure 5.10. (a) Two electric waves $\pi/2 = 90°$ out of phase and (b) their
superposition, producing a circularly polarized wave.

There are other types of retarder besides passive plates of birefringent materials. One example is the Pockels cell. This device is based on materials such as
lithium niobate or ammonium dihydrate phosphate (AD*P) in which the
birefringence can be altered by an electric field. Birefringence is also induced in
normal optical materials by mechanical strain. This effect is used in photo-elastic
retarders, which vibrate a piece of material at its natural resonant frequency to
impose a periodically varying stress and thus a rapidly periodically varying
retardation (e.g., at 20 kHz).

5.6.3 Polarimeters

A very simple polarimeter could be based on placing an analyzer in the beam and making measurements at different rotation angles of the analyzer. However, such an arrangement would have limited accuracy. A fundamental problem is that the measured effect appears synchronously with the rotation of the analyzer, making the instrument susceptible to false indications of a polarized input. One example is wandering of the beam due to optical imperfections in the analyzer. In addition, the optical train behind it may transmit different polarizations with different efficiencies – for example, reflections off tilted mirrors induce polarization in the beam. Such unwanted signals generated within the instrument are described as "instrumental polarization." Although it is possible to calibrate such false signals and remove them to some degree from the measured values, it is much better to use a polarimeter design that is less susceptible to them and has low instrumental polarization.

One such design is to rotate the entire instrument, rather than just the analyzer. In that case, the path through the instrument remains identical for the different angles and the effects of polarization-sensitive optics should not matter. However, the instrument cannot generally be rotated very fast; any changes in conditions (e.g., seeing or atmospheric transmission) can have a large influence on the measurements made at different times. Still, rotating the instrument can be a useful technique to probe possible instrumental errors.

More adaptable designs are based on placing a waveplate or other retarder in the beam before it encounters any tilted mirrors that might induce instrumental polarization. For a normal Cassegrain or Gregorian telescope, the two first reflections (primary and secondary) are nearly normal to the mirror surfaces and thus, by symmetry, also induce very little polarization. Good performance can be obtained if the retarder is placed directly in the beam formed by these two mirrors. An analyzer is placed behind the retarder, and can in fact be behind any number of tilted mirrors or other polarizing optical elements without significantly compromising the performance; any polarimetric activity by these other optical elements combines optically with the activity of the analyzer. For example, if the analyzer is immediately behind the retarder, light fully polarized in a fixed direction will pass through the rest of the instrument; polarizing optical elements behind the analyzer might result in some of this light being removed, but will not change the net modulation of the signal because the loss of efficiency will be in equal proportion for all signals. The spectropolarimeter in Figure 5.11 illustrates this approach. Although its diffraction grating is strongly polarizing, it will have low instrumental polarization.

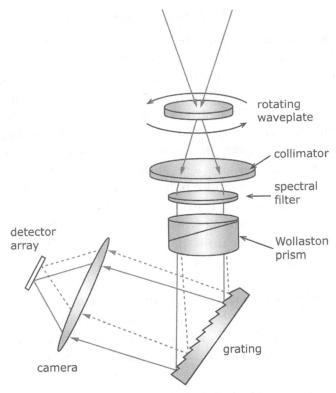

Figure 5.11. A spectropolarimeter. The Wollaston prism divides the input light into two polarizations, shown by the solid and dashed lines.

Polarization measurements are obtained with an instrument such as the one in Figure 5.11 by measuring the signals at different rotation angles of the half-waveplate. The rotation of the angle of polarization will be 2θ for a rotation of the waveplate by θ, that is, changes in signal due to optical imperfections in the waveplate will have a different rotational dependence from changes due to a polarized input beam. These two improvements, (1) immunity to polarization effects and other flaws in the instrument optics, and (2) separating the signal from one-to-one correspondence with the motion of the analyzer, together yield a big gain in the accuracy with which polarization can be measured. A further gain arises because the waveplate can be moved to a new angle quickly, so different measurements can be obtained over time intervals that are short compared with changes in atmospheric transmission, seeing, and other variable effects.

Alternative types of retarder are based on Pockels cells or photo-elastic devices. With them, there is no bulk mechanical motion associated with the rotation of the plane of polarization, and any polarization artifacts associated with the rotation of a conventional waveplate are avoided.

5.6.4 Measurement interpretation and error analysis

If we make measurements of the polarization (1) in the principal plane and (2) in the orthogonal plane of our analyzer when it is at angle ξ relative to the NCP, then (applying equation (5.22)

$$R = \frac{I_{PP} - I_{OP}}{I_{PP} + I_{OP}} = \frac{Q \cos 2\xi + U \sin 2\xi}{I} \tag{5.23}$$

Although this expression assumes a perfect analyzer, it can be generalized to a more realistic case in a straightforward way. We need to make at least two such measurements; assume for simplicity that one is at $\xi = 0$, from which we get $R_0 = Q/I = q$, while the other is at $\xi = 45°$, from which we get $R_{45} = U/I = u$. We can adapt equations (5.20) for the amount and angle of polarization.

$$P = \sqrt{q^2 + u^2} \tag{5.24a}$$

$$\theta = \frac{1}{2} \arctan\left(\frac{u}{q}\right) \tag{5.24b}$$

It is convenient to use a diagram of q vs. u, with angles in 2θ, to represent polarization measurements (Figure 5.12). For example, different measurements can be combined vectorially on this diagram. To see how this works, suppose we made two measurements of a source, the first giving 2% polarization at an angle of 20° and the second 2% polarization at an angle of 110°. These measurements are consistent with the source being unpolarized. If we put them on the q–u diagram and combine them vectorially, we indeed get a result of zero.[9]

Error analysis for polarimetry is generally straightforward, except when it comes to the position angle for measurements at low signal-to-noise ratio. Assume that the standard deviations of q, u, and P are all about the same. Then the uncertainty in the polarization angle is nominally

$$\sigma(\theta) \approx 28.65° \, \frac{\sigma(P)}{P} \tag{5.25}$$

Thus, one might conclude that a measurement at only one standard deviation level of significance (i.e., a non-detection) achieves a polarization measurement within 28.65°. This high accuracy is non-physical – the probability

---

[9] Linear algebra is often used to analyze the behavior of a series of polarimetrically active optical elements, each of which is described by a "Mueller matrix." The basics of polarimetry can be explained without this formalism, but anyone planning to get more familiar with the field should expect to encounter it.

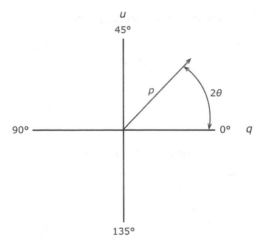

Figure 5.12. The *q–u* diagram.

distribution for θ at low signal-to-noise ratio does not have the Gaussian distribution assumed in most error analyses (see, e.g., Wardle and Kronberg 1974). Similarly, P is always positive and hence does not have the Gaussian distribution around zero assumed in normal error analysis.

5.7 Exercises

5.1 The list on page 151 gives Johnson system photometry of a number of stars at J and K. Kidger and Martin-Luis (2003) report JHK observations of bright stars, Including the ones listed below. Look up these stars in the paper and compute the transformations to put the Johnson J, K photometry on the same basis as the Kidger and the Martin-Luis J, K photometry.

5.2 (a) From the list on page 151 (and your solution to exercise 5.1), how accurate would you estimate the Johnson photometry is?

(b) Johnson took the magnitude of Vega to be + 0.02 at J and K (and 0.03 at V). Other photometric systems would claim they put Vega at zero magnitude at all wavelengths. What would it be in the Kidger and Martin-Luis system? Can you explain why it has this surprising value?

(c) What are the slope terms in the transformations due to?

5.3 Assume you have a somewhat idealized J-band system with spectral response of 0.8 from 1.15 through 1.35 mm and zero outside this range. What is the effective (mean) wavelength? What is the nominal (IRAS definition) wavelength?

Name	Type	J	K	Name	Type	J	K
HR 33	F5V	3.94	3.65	HR 5107	A3V	3.20	3.11
HR 509	G8V	2.16	1.68	HR 5854	K2IIIb	0.76	0.06
HR 1256	K0III	2.63	1.97	HR 5947	K2III	2.09	1.30
HR 1286	K1II-III	3.36	2.49	HR 6623	G5IV	2.18	1.77
HR 1791	B7III	1.97	2.05	HR 6698	G9III	1.68	1.12
HR 1907	K0IIIb	2.35	1.69	HR 6705	K5III	−0.39	−1.34
HR 1963	K1III	2.89	2.11	HR 6707	F2II	3.55	3.23
HR 2077	K0III	2.09	1.46	HR 7236	B9Vn	3.64	3.67
HR 2427	K3Iab	2.79	2.05	HR 7525	K3II	0.30	−0.59
HR 2560	G5III-IV	2.89	2.33	HR 7557	A7V	0.39	0.26
HR 3003	K4III	2.28	1.33	HR 7615	K0III	2.28	1.67
HR 4335	K1III	1.16	0.44	HR 7949	K0III	0.77	0.11
HR 4377	K3III	1.18	0.31	HR 8143	B9Iab	3.95	3.79
HR 4608	G8IIIa	2.48	1.90	HR 8632	K2III	2.36	1.51
HR 4737	K1III	2.55	1.90	HR 8905	F8IV	3.37	3.02
HR 4983	F9.5V	3.24	2.90				

Assume your standard star has a Rayleigh–Jeans spectrum across the J filter. Using the mean wavelength, compute the bandpass correction relative to the standard for the ultraluminous galaxy spectrum of

$$f(\lambda)d\lambda = 0.005\lambda^{1.3}\, d\lambda$$

with λ in µm.

5.4 Consider a bandpass filter with a square transmission function: zero outside the band and a constant value within it. Compare the estimates of the signal passing through this filter (a) when it is determined by integrating the source spectrum over the filter passband; and (b) when only the mean wavelength and width of the passband are used to characterize the filter. Assume a Rayleigh–Jeans source spectrum (in wavelength units, flux density $\propto \lambda^{-4}\, d\lambda$). Show that the error introduced by the simple mean wavelength approximation is a factor of

$$1 + \frac{5}{6}\left(\frac{\Delta\lambda}{\lambda}\right)^2$$

plus terms of order $(\Delta\lambda/\lambda)^4$.

5.5 A main sequence star is observed to have a B magnitude of 7.70 and a V magnitude of 7.02. Using the information for average stellar photometry in the table below, estimate its spectral type, temperature, distance, and luminosity.

Spectral type	M_V	B – V	T_{eff}	BC
F5	3.5	0.44	6650	−0.14
F8	4.0	0.52	6250	−0.16
G0	4.4	0.58	5940	−0.18
G2	4.7	0.63	5790	−0.2
G5	5.1	0.68	5560	−0.21

5.6 Joe Nerd is designing a new polarimeter in which he plans to get better performance by using *two* half-waveplates in series, so the rotation of position angle with plate position will be greater than with a single plate. Is he going to be pleased with the results?

5.7 The point spread function for a telescope with perfect optics, round primary mirror, and round obscuration in the center of the beam (e.g., a secondary mirror) whose diameter is ε times the diameter of the primary is

$$I(\theta) = \frac{I_0}{(1 - \varepsilon^2)^2} \left(\frac{2J_1(x)}{x} - \frac{2\varepsilon J_1(\varepsilon x)}{x} \right)^2$$

where J_1 is the Bessel function, $x \approx \pi r/(\lambda(f/\#))$, with r the radial distance in the focal plane from the optical axis. Consider a 90-inch telescope with a central obscuration 40 inches in diameter; assume it is diffraction limited and that there are no other significant obscurations of the aperture to complicate the diffraction pattern. Compute the aperture correction if photometry is done with an aperture set to the size of the second dark ring.

5.8 You have the following set of measurements, obtained from a telescope at latitude 32 degrees north:

Determine m_V for star 3. Hint: if you use the appropriate units, you can determine the air mass correction from a linear fit.

Star	m_V	Dec.	HA	DU
1	5.73	+55°	−3.5 hours	1554
			1.3 hours	1605
			5.1 hours	1333
2	7.86	+14°	−2.1 hours	220
			3.7 hours	195
3		−3°	1.4 hours	487

Further reading

Bessell, M. S. (2005). Standard photometric systems. *ARAA*, **43**, 293–336.

Budding, E. and Demircan, O. (2007). *An Introduction to Astronomical Photometry*, 2nd edn. Cambridge: Cambridge University Press.

Howell, S. B. (2006). *Handbook of CCD Astronomy*, 2nd edn. Cambridge: Cambridge University Press.

Glass, I. S. (1999). *Handbook of Infrared Astronomy*. Cambridge: Cambridge University Press.

Johnson, H. L. (1966). Astronomical measurements in the infrared. *ARAA*, **4**, 193.

McLean, I. S. (2008). *Electronic Imaging in Astronomy*, 2nd edn. Berlin, New York: Springer.

Serkowski, K. (1974). Polarization techniques. In *Methods of Experimental Physics*, Vol. 12, Part A, Astrophysics, ed. N. P. Carleton. New York: Academic Press, pp. 361–414.

Sterken, C. and Manfroid, J. (2008). *Astronomical Photometry: A Guide*. Berlin, New York: Springer.

Starck, J.-L. and Murtagh, F. (2006). *Astronomical Image and Data Analysis*. Berlin, New York: Springer.

Tinbergen, J. (2005). *Astronomical Polarimetry*. Cambridge: Cambridge University Press.

Young, A. T. (1974a). Photomultipliers, their cause and cure. In *Methods of Experimental Physics*, Vol. 12, Part A, *Astrophysics*, ed. N. P. Carleton. New York: Academic Press, 1–94.

Young, A. T. (1974b). Other components in photometric systems. In *Methods of Experimental Physics*, Vol. 12, Part A, *Astrophysics*, ed. N. P. Carleton. New York: Academic Press, 95–122.

Young, A. T. (1974c). Observational technique and data reduction. In *Methods of Experimental Physics*, Vol. 12, Part A, *Astrophysics*, ed. N. P. Carleton. New York: Academic Press, 123–192.

6

Spectroscopy

6.1 Introduction

Spectrometers divide the light centered at wavelength λ into narrow spectral ranges, $\Delta\lambda$. If the resolution $R = \lambda/\Delta\lambda > 10$, the goals of the observation are generally different from those in photometry, including both measuring spectral lines and characterizing broad features.

There are three basic ways of measuring light spectroscopically:

1. Differential-refraction-based, in which the variation of refractive index with wavelength of an optical material is used to separate the wavelengths, as in a prism spectrometer.
2. Interference-based, in which the light is divided so a phase-delay can be imposed on a portion of it. When the light is re-combined, interference among components is at different phases depending on the wavelength, allowing extraction of spectral information. The most widely used examples are diffraction grating, Fabry–Perot, and Fourier spectrometers. Heterodyne spectroscopy also falls into this category, but we will delay discussing it until we reach the submillimeter and radio regimes in Chapter 8.
3. Bolometrically, in which the signal is based on the energy of the absorbed photon. This method is applied in the X-ray, for example, using CCDs or bolometers, and will be discussed in Chapter 10.

In addition to spectral information, most spectrometers also provide at least some information about the distribution of the light on the sky. Where a full image can be obtained along with the spectral data for each point in the image, we imagine a three-dimensional space called a data cube. It has spectra running in the z-direction and any slice in an x, y-plane produces an image of the source at a specific color. Different spectrometer types differ in their ability to produce data cubes efficiently. Simple grating or prism spectrometers usually use a slit to avoid overlap of the dispersed beam sections and therefore only provide spatial information in one direction, that is, they produce only the x, z-part of the data cube directly. This shortcoming can be mitigated with an integral field unit (IFU) that rearranges the image of the source so that part of it in the direction perpendicular to the slit can enter the spectrometer and be dispersed into a spectrum. For a Fabry–Perot spectrometer, the x, y-slices of the cube are produced directly, that is, the instrument images in a narrow spectral band. The Fourier transform spectrometer, if used with a detector array, yields a value of the Fourier transform of the z-axis and the x- and y-axes simultaneously; by scanning the path difference the full transform is obtained over the field of view. This output can be inverted to provide the full spectral data cube.

6.2 Prism spectrometers

Light passing through a prism is deflected by Snell's Law, in accordance with the refractive index. An empirical formula due to Sellmeir for the wavelength dependence of the refractive index (defined as the dispersion) is

$$n^2(\lambda) = A + \frac{B_1 \lambda^2}{\lambda^2 - C_1} + \frac{B_2 \lambda^2}{\lambda^2 - C_2} \tag{6.1}$$

where A, B_i, and C_i are constants. As a result of the wavelength dependence, the net deflection of the beam depends on the wavelength – the output beam separates into the colors of the light.

This behavior allows construction of perhaps the simplest form of spectrometer as in Figure 6.1, which illustrates many of the essential features of dispersive spectrometers in general. We can think of the optical functions going from the slit to the detector. Then the collimator and camera lenses just form an image of the slit on the detector array. However, the prism spreads the angles into the camera lens in order of wavelength, so we get a series of images of the slit in wavelength order. The prism must be put in a collimated beam; otherwise the range of emergent positions from the prism will cause different wavelengths to overlap in position on the detector. If the prism and slit were removed the instrument would map a two-dimensional set of positions on the sky onto the detector. The slit restricts this mapping so positions parallel to the dispersion direction of the prism do not get mapped onto other positions but at different wavelengths. There is no problem, however, in mapping sky perpendicular to the dispersion onto the detector. Thus, a slit running perpendicular to the dispersion direction is used to restrict the amount of the sky that is imaged through the instrument.

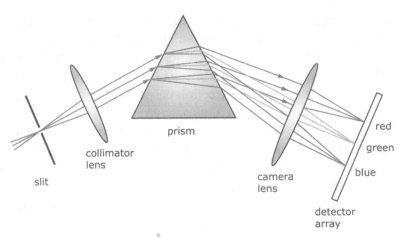

Figure 6.1. Prism spectrometer.

Another approach to collimating the light entering the prism is to place it over the entrance to the telescope, creating an objective prism spectrometer. It is not possible to interpose a slit into this arrangement, so it has the disadvantage that the sky is imaged on top of the spectrum of a source. For example, if there are 100 spectral elements in the spectrum of a source then 100 sky positions at different wavelengths will be imaged to the same place on the detector, raising the background by a factor of 100. Nevertheless, this approach is useful for obtaining spectra of a huge number of sources simultaneously. It is a specific example of slitless spectroscopy in which the extra background normally blocked by the slit is accepted in exchange for observing spectra over a full two-dimensional field of view. However, for more demanding observations this approach has largely been supplanted by multi-object spectrometers that replace the conventional slit with a large number of individual slitlets or apertures each of which can be aligned with a source; the light from each aperture is introduced into the spectrometer in a way that yields a spectrum of its source.

6.3 Grating spectrometers

6.3.1 Diffraction gratings

Prisms have limited design flexibility, since the spectrometer must utilize the spectral dispersion properties of optical materials that have good transmission, can be manufactured in large sizes and to high quality, and so forth. A more flexible family of spectrometers that dominates in astronomical applications is based on diffraction gratings. The simplest form of these instruments is similar to the design of Figure 6.1, with the substitution of a grating for the prism. Therefore, we begin the discussion by concentrating on the properties of a diffraction grating.

The principle behind a grating is to slice the collimated beam in the spectrometer spatially, in narrow strips parallel to the slit. The behavior is analogous to Young's experiment (diffraction and interference through two slits) or, as in Figure 6.2, an expanded version with more slits. The incoming beam propagates through slits cut in a plate with separation d, emerging from each as a pattern of Huygens wavelets. These wavelets are a way to visualize that the light emerging from each slit is spread over a range of angles due to diffraction. The lens relays the light onto the diffraction plane, which is the optimal surface to record the results of the diffraction by the slits. When the light from different slits arrives at the same position on the diffraction plane, it interferes. For the light that proceeds to the plane in a direction centered symmetrically on the direction of the incident beam, all the wavelets for a given wavelength have the same path from slit to slit and interfere constructively. This gives the order $m = 0$, and no spectral information, so it is sometimes called the white light peak. As we move off the white light peak along the diffraction plane, first we encounter a region where virtually all the interference is destructive, and then the first-order peak, $m = 1$ (the sign of m is immaterial), where the path difference from slit to slit is one wavelength and constructive interference occurs. However, since the relevant measure of path lengths from the slits is in units of wavelength, we reach the constructive interference peak for λ_1 (shorter wavelength) before the one for λ_2 (longer wavelength). As we proceed further down the diffraction plane, progressively encountering peaks for a slit-to-slit path difference of two and three wavelengths ($m = 2$ and $m = 3$), the spread between λ_1 and λ_2 grows.

We can understand the behavior mathematically by observing that the path difference, p, for light passing through successive slits, as shown in the inset in Figure 6.2, is

$$p = d\sin i + d\sin \theta \qquad (6.2)$$

where d is the distance between slits, i is the angle of incidence onto the slit plate, and θ is the angle at which the light emerges from the slit (i and θ are measured relative to a line perpendicular to the slit plate). For convenience, we set $i = 0$ in the main part of Figure 6.2 and in the following derivations. We then have a phase difference of

$$\delta = \frac{2\pi p}{\lambda} = \frac{2\pi d\sin \theta}{\lambda} \qquad (6.3)$$

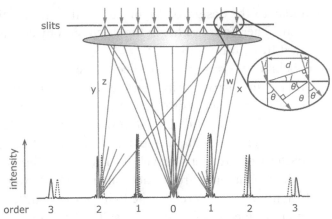

Figure 6.2. Diffraction and interference through multiple slits, indicated for two wavelengths (λ_1 dashed and λ_2 solid, $\lambda_1 < \lambda_2$). The inset shows the general case for the geometry around two slits, for arbitrary angle of incidence, i. The optical path to zeroth order is the same from all the slits; to first order, there is one wavelength difference for adjacent slits (w and x), whereas for second order there are two wavelengths difference (y and z).

The intensity peaks occur when $p = m\lambda$, that is, the path difference between successive slits is an integral number of wavelengths. If a is the amplitude diffracted by a single slit, then the amplitude, A, for N slits is the sum:

$$Ae^{j\phi} = a(1 + e^{j\delta} + e^{j2\delta} + \ldots + e^{j(N-1)\delta}) = \frac{a(1 - e^{jN\delta})}{(1 - e^{j\delta})} \tag{6.4}$$

where ϕ is the phase. The intensity is

$$I(\theta) = AA^* = \frac{a^2 \sin^2(N\delta/2)}{\sin^2(\delta/2)} = a^2 \left[\frac{\sin^2(N\pi d \sin\theta/\lambda)}{\sin^2(\pi d \sin(\theta/\lambda))} \right] \tag{6.5}$$

where A^* is the complex conjugate of A. The term in square brackets gives the profile of the intensity peak – the instrumental profile, $P(\lambda)$. The intensity of the light diffracted by a single slit, a^2, is calculated as the diffraction from the box function, which we found in equation (2.27) to be given by the square of the sinc function,

$$I_{\text{slit}}(\theta) = a^2 = I_0 \frac{\sin^2(\pi w \sin\theta/\lambda)}{(\pi w \sin(\theta/\lambda))^2} \tag{6.6}$$

where w is the slit width. The envelope of the peaks at all the orders has this form (see Figure 6.2).

Although we have a way to distribute the wavelengths in the incoming beam along the diffraction plane, progressively less light gets through as we go to higher orders and larger θ, as shown by equation (6.6). In addition, a field of slits throws away most of the incident light before we even get to discuss it.

Therefore, we need to modify the approach if we are to get good efficiency. To do so, we replace the slits with long thin mirrors – other than the nuisance of keeping the input and output beams separate, this change should make no difference to the physics of the multiple-slit device. If in addition we tilt the mirrors so that the specularly reflected direction is toward a specific wavelength in one of the orders $m > 0$, then we peak the single-slit response in equation (6.6) on that order. We have just built ourselves a diffraction grating blazed for the wavelength and order we have selected for the mirror tilt.

Figure 6.3 shows some of our little tilted mirrors. We can adapt equation (6.2) to this figure to obtain (making the substitution of $m\lambda$ for p)

$$m\lambda = nd(\sin\alpha + \sin\beta) \tag{6.7}$$

where α is the angle of the incoming beam to the grating normal and β is the angle of the outgoing beam (see Figure 6.3). The angles α and β are defined to have the same sign if they are on the same side of the grating normal (as in Figure 6.3); m can be either positive or negative with no difference in meaning. We have added the refractive index, n, because, as shown in the inset to Figure 6.2, the interference depends on path lengths measured in units of wavelengths. A general result includes the effect of the refractive index on the effective slit spacing. For simplicity, however, we have been setting $n = 1$ and will continue to do so. Equation (6.7) is known as the grating equation and will be used for a number of derivations below.[10] In Figure 6.3, the step between adjacent mirrors, δ_B, provides the path difference between mirrors to put the center of I_{slit} (equation (6.6)) on the desired order: for example, $\delta_B \approx \lambda_B/2$ for $m = 1$, where λ_B is the blaze wavelength. Fortunately, the sinc function in equation (6.6) falls off slowly with θ, so the grating can have good efficiency for a substantial range of wavelengths around the blaze wavelength. The useful spectral range in first order is roughly a total of λ_B. From equation (6.7), when operated at higher orders and proportionately shorter wavelengths, the fractional regions of high efficiency decrease inversely with the order. As might be expected from our discussion of wire grid polarizers in the preceding chapter, gratings reflect different polarizations differently, so their efficiency is polarization-dependent.

---

[10] Equation (6.7) assumes that the beam is in the normal plane of incidence, that is, in the plane of the paper on which Figure 6.3 is printed (as is usually the case). If the beam makes an angle γ to this plane, the right-hand side of equation (6.7) should be multiplied by $\cos\gamma$.

Spectroscopy

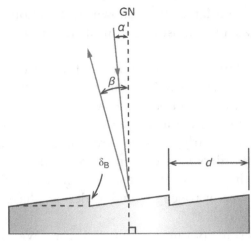

Figure 6.3. A blazed grating, cross-section. GN is the normal to the grating face.

We now compute λ_B in the context of the grating equation and spectrometer design. The blaze angle of the grating is (from Figure 6.3)

$$\theta_B = \frac{\alpha + \beta}{2} \tag{6.8}$$

(since the blaze angle is defined for the purely geometric reflection case). Substituting in equation (6.7),

$$\lambda_B = \frac{d}{m}[\sin \alpha + \sin (2\theta_B - \alpha)] \tag{6.9}$$

Using the trigonometric identity

$$\sin x + \sin y = 2 \sin \left(\frac{x+y}{2}\right) \cos \left(\frac{x-y}{2}\right) \tag{6.10}$$

equation (6.9) can be written

$$\lambda_B = \frac{2d}{m} \sin \theta_B \cos \frac{\Psi}{2} \tag{6.11}$$

where $\Psi = |\beta - \alpha|/2$ is the angle between the collimator and camera beams in a spectrometer.

Gratings are usually specified in terms of the ruling density, $1/d$, specified as grooves per millimeter; the blaze wavelength, λ_B; and the blaze angle, θ_B. Equation (6.11) shows that there is no unique blaze wavelength for a given blaze angle; the catalog values assume the Littrow condition, $\Psi = 0$, as do the other catalog specifications (e.g., efficiency).

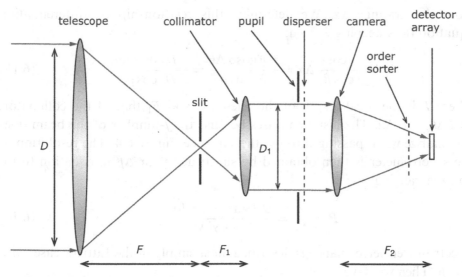

Figure 6.4. Schematic of a spectrometer.

We now consider the resolution of a grating in more detail. We place the grating within the spectrometer illustrated in Figure 6.4, which is qualitatively similar to Figure 6.1 with one addition. From equation (6.11), the series of orders reflected off the grating will all be at the same angle. Leaving all these wavelength ranges piled on top of each other would be disastrous scientifically. The simplest cure is to superpose an "order blocking filter" in the spectrometer optical train to isolate the wavelength range associated with a single order. In Figure 6.4, we leave the possibilities for something more complex open and call it an order sorter.

We obtain the angular dispersion of the grating by differentiating equation (6.7) with regard to β:

$$m\Delta\lambda = d\cos\beta\Delta\beta \qquad (6.12)$$

or

$$\frac{\Delta\lambda}{\Delta\beta} = \frac{d\cos\beta}{m} = \frac{\lambda\cos\beta}{\sin\alpha + \sin\beta} \qquad (6.13)$$

Since $d\beta/dx = 1/F_2$ in Figure 6.4, where x is the linear dimension in the dispersion direction at the camera focus, the linear dispersion at the detector is

$$\frac{d\lambda}{dx} = \frac{d\lambda}{d\beta}\frac{d\beta}{dx} = \frac{d\cos\beta}{m\,F_2} \qquad (6.14)$$

We will now do a similar calculation for the effect of the slit width on the sky, $\Delta\theta$. We start with the resolution, $\Delta\beta$, as a function of the slit width as viewed from the grating, $\Delta\alpha$. We determine this relationship by differentiating equation (6.7) keeping λ fixed:

$$\Delta\beta = \frac{-\cos\alpha}{\cos\beta}\Delta\alpha = \frac{-F\cos\alpha\,\Delta\theta}{F_1\cos\beta} = \frac{-D\cos\alpha\,\Delta\theta}{D_1\cos\beta} \tag{6.15}$$

where F is the focal length of the telescope and F_1 that of the collimator, so $F\Delta\theta = F_1\Delta\alpha$. The last step arises because the f-number of the beam does not change when passing through the slit – see Figure 6.4. The resolution of the spectrometer is then obtained by substituting for $\Delta\beta$ in equation (6.13) and rearranging:

$$R = \frac{\lambda}{\Delta\lambda} = \frac{D_1(\sin\alpha + \sin\beta)}{D\cos\alpha\,\Delta\theta} \tag{6.16}$$

Spectrometer performance is described most simply in the Littrow case, with $\alpha = \beta$. Then we have

$$R = \frac{2D_1\tan\alpha}{D\,\Delta\theta} \tag{6.17}$$

Equation (6.17) makes spectrometer characterization very simple. It says that the resolution of a spectrometer on a telescope of diameter D depends only on the diameter of the collimated beam, D_1, the width of the slit projected onto the sky, $\Delta\theta$, and the tilt of the grating.

If the grating is illuminated so $\alpha \neq \beta$, the behavior is correctly represented by equations (6.7) and (6.16) (and those accompanying them), but is sometimes described in different terms. Figure 6.5 shows an extreme example.[11]

At large angles, a given input collimated beam can spread over a larger area of the grating than the simple cross-section of the beam, a situation called anamorphic magnification. From preservation of etendue, we expect that the increase in beam diameter will be accompanied by a decrease in the range of output angle $\Delta\beta$ for a given range of input angles $\Delta\alpha$. We can confirm this prediction by differentiating equation (6.7) with respect to α while keeping λ fixed, to obtain the anamorphic magnification factor, r:

$$r \equiv \left|\frac{d\beta}{d\alpha}\right| = \frac{\cos\alpha}{\cos\beta} \tag{6.18}$$

The result is that the slit image at the detector is narrower than it would be without this effect, allowing use of a wider slit, for example, without com-

---

[11] To achieve reasonable efficiency, the illumination needs to be from the direction where the cuts between grating facets are not seen, "no shadowing" in the figure.

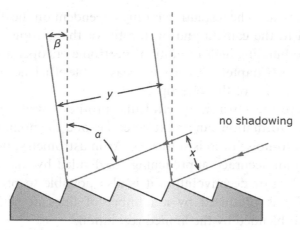

Figure 6.5. Anamorphic magnification at large incidence. The gain is clear from comparing how much larger y is than x.

promising the spectral resolution. Alternatively, it may be possible to increase the spectral resolution for a given slit width, as is seen in equation (6.16), where large values of α can increase the resolution through the cos α term in the denominator (so long as an appropriate value is possible for β).

Does this make spectrometers complex? No, they remain simple if we say that the resolution: (1) is inversely proportional to the size of the telescope; (2) increases with increasing tilt of the grating; (3) is inversely proportional to the field projected onto the sky, that is, the slit width; and (4) is in proportion to the width of the beam delivered to the camera.

This behavior arises for the following reasons. The larger D_1, the larger F_1 must be (Figure 6.4). The scale of the image of the slit on the detector array is proportional to F_2/F_1, so to preserve this scale F_2 must be increased in proportion to the increase in F_1. However, as we calculated for telescopes, the translation from angular range into the camera lens to physical range on the detector is proportional to F_2; the larger F_2, the larger is the image of the spectrum on the array. Thus, if we make D_1 larger and adjust to keep the same size image of the slit on the detector array, we have a proportionately smaller range of wavelengths per slit image, that is, the resolution has increased. Finally, if we keep the same slit width projected onto the sky, the physical width of the slit needs to increase in proportion to increases in the telescope aperture, D; thus the resolution varies inversely as D.

A corollary of these arguments is that the inverse dependence of D need not occur if the projected angular slit width can be decreased with increasing telescope aperture. A limiting case is when point sources are observed with a diffraction-limited telescope, in which case $\Delta\theta$ in equation (6.16) can be replaced with λ/D, and the resolution is only dependent on the diameter of the beam delivered to the camera and on the tilt of the grating. This situation gives impetus to building high-resolution spectrometers operating with adaptive optics systems (Chapter 7) to provide reasonable efficiency with slits sized to the diffraction limit of the telescope.

In cases where the intrinsic spectral line profile is well determined and stable, spectral information can be extracted beyond the nominal resolution, $\Delta\lambda$, if the signal-to-noise ratio is sufficient. As in astrometry, line centers can be measured to an accuracy approaching $\Delta\lambda$ divided by the signal-to-noise ratio. By modeling or deconvolution it is also possible to probe whether a measured profile is produced by a number of lines at slightly different wavelengths and blended by the limited resolution.

6.3.2 Grating types

Ruled, holographic and volume phase gratings

Many astronomical gratings are ruled. This term refers to a process in which a fine tool (diamond, so it does not wear) controlled by a screw drive is drawn over a softer substrate to carve the grooves. The initial master grating is very expensive to manufacture, so one normally uses a grating replicated from it as an epoxy resin casting. Master gratings can also be manufactured holographically. In this case, an interference pattern is projected onto a light-sensitive coating on the grating substrate. After exposure, this coating is developed to remove the unexposed regions, leaving a series of grooves. Because there is no mechanical removal of material, holographic gratings are very smooth and have low levels of scattering. However, it is difficult to blaze them as effectively as for ruled gratings and consequently they generally have lower efficiency. In this paragraph, we have implicitly assumed gratings are used in reflection; similar results can be obtained with gratings used in transmission. In this case the rulings are made in a transparent substrate and the light is dispersed as it exits (or enters) the optical element (see Section 6.3.2.2).

If a ruled grating is configured so that the diffraction occurs in a material with $n > 1$, it is described as an immersion grating. When we introduced the refractive index into equation (6.7), we justified it in terms of the effect of n on the optical slit or groove spacing in wavelength units. From the form of the equation, it also has the effect for a given groove spacing of increasing the order and therefore the resolution. This gain can be substantial if the grating is immersed in a high-index medium; there are interesting possible optical materials for near-infrared gratings, such as silicon ($n = 3.4$).

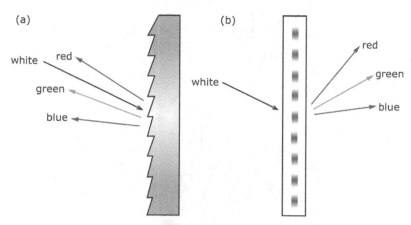

Figure 6.6 Comparison of operation of (a) a conventional reflection grating and (b) a volume phase holographic grating.

Volume phase holographic (VPH) gratings are another approach. They can be made for the optical and near-infrared in thin layers of gelatin (3–30 μm thick) that are sandwiched between plates of glass for protection. Holographic techniques and processing of the gelatin are used to impose a plane of fringes of variable density, resulting in a sinusoidal variation of the refractive index. Fringe periods can range from 300 to 6000 lines/mm. The refractive index variations result in a phase shift across the wavefront of an incident collimated beam, similarly to the operation of a reflection grating (see Figure 6.6). Their behavior in transmission is described by the grating equation (6.7) with the substitution of $n \Lambda/\cos \varphi$ for d, where Λ is the grating period and φ is the angle between the grating normal and the plane of the fringes. The refractive index, n, enters because the grating is immersed in a medium with $n > 1$. There are a number of possibilities for VPH grating configuration: for example, whether the grating operates in transmission or reflection depends on the orientation of the diffracting features. Because the refractive index structure can be realized in three dimensions, constructive interference can be limited to particular orders, improving the efficiency toward 90%. Since the gratings are made holographically, the variations in refractive index can be made very regular and smooth, reducing scattering and avoiding some of the spectral artifacts associated with flaws in ruled gratings.

Grisms

A convenient disperser might be placed in a filter wheel of an imager and could be rotated into the beam to convert from imaging to spectroscopy. However, this application requires that the dispersion occur more or less without deviating the beam, a condition not satisfied by a prism or a

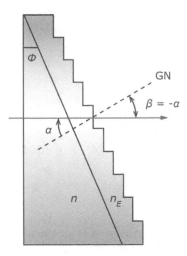

Figure 6.7. A grism.

transmission grating. A solution is to replicate a transmission grating on the hypotenuse of a right-angle prism to make a grism as shown in Figure 6.7.

The deflection of the grating is canceled by that of the prism for a specific wavelength, which can be set to the center of the desired spectral range. The performance of the grism is described by the equivalent to the grating equation:

$$m\lambda = d(n-1)\sin\phi \qquad (6.19)$$

where $\alpha = -\beta = \phi$ (the angle the hypotenuse of the prism makes to the side normal to the beam), n is the refractive index of the prism and grating material (the grism is assumed to be immersed in a medium with a refractive index close to 1), m is the order the grating operates at, and d is the groove spacing.

Echelle gratings

High spectral resolution generally requires large grating tilts. Operating a low-order grating at a large tilt requires a very high groove density. Large gratings can only be ruled in high quality up to about 1500 grooves per millimeter (Richardson Grating catalog). Although holographic gratings can be made with very high groove densities, in general they fall short of ruled gratings in both size and efficiency. A solution is to use an echelle grating, with coarse groove spacing and operated at high order (Figure 6.8). To avoid overlap of the many orders delivered superimposed by this type of grating, another disperser (e.g., a grating or prism) is usually employed to spread the spectrum orthogonally to the echelle grating dispersion.

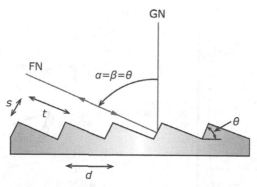

Figure 6.8. Echelle grating. GN grating normal; FN facet normal.

6.3.3 Grating spectrometers

Standard spectrometers

Figure 6.9 illustrates a number of classic approaches to spectrometer design. In the Ebert–Fastie design (a), the entrance slit is placed to the side of the focus of a spherical mirror. The mirror collimates the light that passes through the slit and directs it to the grating, which is flat, so it disperses the light and deflects it still collimated (except for the spread of angles due to the dispersion) back to the spherical mirror. In this reflection, the mirror images the light onto the detector array in the form of the spectrum of the light that passed through the slit. For good optical performance, the two critical focal areas (slit and detector array) should be as close as possible to the optical axis of the mirror. This restriction is removed and the performance improved with the Czerny–Turner design (b). The light passes through the slit to an off-axis paraboloidal mirror, where it is collimated and directed to the grating. Another off-axis paraboloid focuses the dispersed light into a spectrum on the detector array. A Littrow configuration (c) uses the same mirror and optical path for collimator and camera. In theory, this arrangement allows good optical performance by compensating many of the aberrations in the double pass – however, aberrations are introduced by the necessity to offset the slit from the detector array. The Rowland circle spectrometer (d) uses a grating ruled on a concave surface with radius of curvature R. If the grating, slit, and detector are all placed on a circle of diameter R, then the spectrum delivered to the detector is in focus along its entire length, and is also free of coma and with small spherical aberration, but with substantial astigmatism. The astigmatism limits the use of this design in the optical, unless an aberration-reduced concave grating is employed. However, the Rowland circle is used in dispersive X-ray spectrometers because it requires only a single reflection.

(a)   (b)

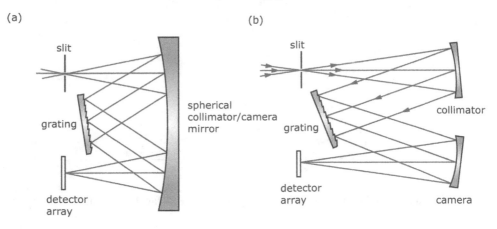

(c)   (d)

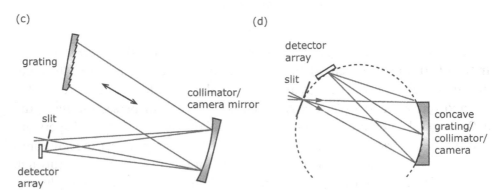

Figure 6.9. Four general types of spectrometer: (a) Ebert–Fastie; (b) Czerny–Turner; (c) Littrow; and (d) Rowland circle.

Spectrometer design: a simple example

We will illustrate some principles of spectrometer design based on the layout in Figure 6.10, which can be considered to be of Czerny–Turner design with lenses substituted for the paraboloidal mirrors. We will determine:

(a) the collimator focal length
(b) the minimum diameter of the collimator
(c) an optimum placement for the grating to be sure it is uniformly illuminated
(d) the width of the entrance slit, assuming the telescope primary has a diameter of 8 m and the seeing is 0.5 arcsec FWHM
(e) assuming a grating with 800 grooves/mm operating in first order, the blaze angle that is optimum for a 60 nm (600 angstrom) wide region of the spectrum centered at Hα

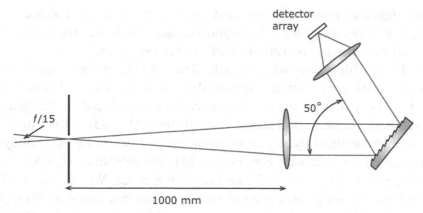

Figure 6.10. A simple spectrometer.

(f) the angles of incidence (α) and diffraction (β) at Hα - are both incident and diffracted rays on the same side of the grating normal?
(g) the focal length of the camera for Nyquist sampling of the image on a detector with 15 μm pixels
(h) the spectral resolution at Hα.

(a) By inspection, the collimator focal length is 1000 mm.
(b) Since the beam from the slit to the collimator is $f/15$, the minimum collimator diameter is $1000\,\text{mm}/15 = 66.7\,\text{mm}$.
(c) The grating will be uniformly illuminated if it is placed at a pupil, which will fall roughly one focal length behind the collimator, or 1000 mm behind.
(d) The equivalent focal length of the telescope is $15 \times 8\,\text{m} = 120\,\text{m} = 120\,000\,\text{mm}$. The scale at the focal plane is then $1/120\,000$ radians per mm, or $1.72''/\text{mm}$. The narrowest plausible slit would be just the FWHM of the seeing disk, or 0.29 mm. In most applications, a better tradeoff is to make the slit about twice this width, so slit losses are smaller and holding the image on the slit easier. For the following we assume a slit that is twice the seeing FWHM, or 0.58 mm.
(e) The wavelength of Hα is 0.6563 μm. From Figure 6.10, the collimator-camera angle is 50°. Equation (6.11) expresses the blaze condition for a grating; we substitute $\lambda_B = 0.6563$ μm for the blaze wavelength; $d = 1/800\,\text{mm}^{-1}$ for the groove spacing; $m = 1$ for the order; θ_B is the blaze angle, to be found; and $\Psi = 50°$ is the collimator-camera angle. The solution is $\theta_B = 16.84°$.
(f) From Figure 6.5, to avoid light hitting the wrong facet of the grating "shadowing," we need to illuminate so $\alpha > \beta$. Using equation (6.9), but now letting the incidence angle be $\alpha - \theta_B$, and solving, one finds $\alpha = 41.84°$.

Substituting this result into equation (6.7) – the grating equation – and solving for β, one gets $-8.2°$. The opposite signs show that the incident and diffracted rays are on opposite sides of the grating normal.

(g) We begin with a first-order estimate. The FWHM of the image is 290 μm at the slit. If we are to Nyquist sample this image, we want a 15 μm pixel to project to 145 μm (ignoring any complications introduced by the grating). That requires a camera focal length of 1000 mm $\times$ (15/145) = 103.5 mm. This value will be modified by any anamorphic magnification by the grating. To see its effects, we start with equation (6.15). We substitute $D = 8$ m, $D_1 = 66.7$ mm, $\alpha = 41.84°$, $\beta = -8.2°$, and $\Delta\theta = 0.5$ arcsec. We get $\Delta\beta = 2.19 \times 10^{-4}$ radians. We want this angle to be imaged onto two pixels, or 30 μm, by a camera of focal length f_{cam}. That is, 30 μm $= \Delta\beta f_{cam}$, or $f_{cam} = 137$ mm. In this example, the anamorphic magnification is about 32%.

(h) The spectral resolution is given by equation (6.16). We have determined all the quantities that go into it. Assuming a 1 arcsec slit, $R \sim 1200$. The corresponding resolution at Hα is about 0.54 nm (5.4 Angstroms).

Some examples of astronomical spectrometers

Real spectrometers follow closely the design concepts we have discussed. As an example, Figure 6.11 shows the optical layout of the Gemini Multi-Object Spectrometer (GMOS). Although the instrument uses all-refractive optics, it is well corrected over the 0.35–1.5 μm range. The first set of elements in GMOS corrects for atmospheric dispersion. The collimated beam is 9.8 cm in diameter; flexure is compensated with an open loop correction system. Spectral resolutions of a few thousand are typical, as well as an imaging mode. An echelle spectrometer would have a similar layout, with the cross-dispersing grating or prism between the grating and the camera and with a higher priority in illuminating the grating close to the Littrow condition (see exercise 6.3).

As another example, the High Accuracy Radial Velocity Planet Searcher (HARPS) at ESO is optimized for measurements of very accurate radial velocities. It is fed by two optical fibers that convey the light from the telescope focus to the input of the spectrometer optics, so that those optics can be fixed and there is no flexure associated with different viewing angles. The spectrometer is inside a vacuum vessel that allows very accurate control of its temperature and also avoids calibration changes associated with variations in atmospheric pressure. It has a 20.8 cm diameter collimated beam that illuminates an echelle grating blazed at 75° (spectral orders near 100), with a grism cross-disperser to achieve a final spectral resolution of 120 000 over the 378–691 nm spectral range.

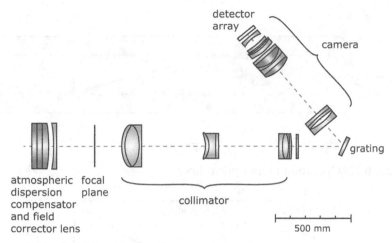

Figure 6.11. GMOS optical train.

The use of fibers to convey the light to the spectrometer is a key feature of HARPS (and many other optical spectrometers). Figure 6.12 shows the operation of a fiber. The fiber has two zones, the cladding and the core, with the refractive index in the cladding lower than that in the core. Over a range in input angles (within the acceptance angle, θ_{max}), the light is held within the fiber by total internal reflection, while light entering outside θ_{max} is not reflected and escapes. The maximum acceptance angle is given by

$$\text{NA} = n \sin \theta_{max} = \sqrt{n_1^2 - n_2^2} \tag{6.20}$$

where NA stands for "numerical aperture," and n is the refractive index of the medium the fiber is within; usually $n = 1$. Typical fibers have a NA of 0.22, corresponding to an input beam of $f/2$.

Fibers can carry light virtually losslessly for many meters, particularly in the visible to about 1.5 μm. However, as the light is conveyed down the fiber, the spread of angles is increased, a process called focal ratio degradation. Thus, a price to be paid for using fibers is an increase in the etendue. By making the input solid angle relatively large, the additional degradation can be kept small, as shown in Figure 6.13 (Ramsey 1988); f-ratios between 3 and 6 allow conveying the light with relatively little growth in etendue for the system.

A difficult challenge for the optics of a large spectrometer is to match the scale of the spectrum to the detector array for optimum performance. Too coarse a scale undersamples the spectrum and makes it difficult or impossible to extract all the information from it. With an infrared array, since on-chip

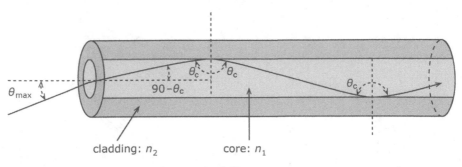

Figure 6.12. Operation of an optical fiber.

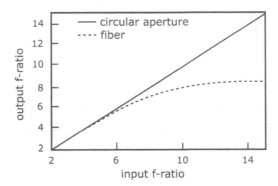

Figure 6.13. Degradation of the focal ratio in optical fibers.

binning is not possible, too fine a scale reduces the signal-to-noise ratio achievable in the spectrum, since more pixels need to be read out and combined to make a measurement. With a CCD detector, the degradation in signal-to-noise ratio due to an overly fine scale can be overcome if the pixels are combined before readout. However, information is lost that would potentially be available with the full array format.

In general, matching the pixel scale is a job that falls on the camera optics. GMOS uses a complex chain of refractive optics for its camera. This option is attractive because a small amount of chromaticity in the camera optics is not a serious problem – if it is in the dispersion direction, it can just be subsumed in the corrections required to calibrate the spectra, while most spectrometers do not have large fields in cross-dispersion for individual spectra. An alternative shown in Figure 6.14 is a Schmidt design – with a corrector plate to compensate aberrations. An on-axis beam is delivered to the CCD by folding the light off a flat mirror and imaging through a hole in the center of the flat (a layout named after its inventor, A. H. Pfund). This hole lies where the obscuration due to the secondary mirror of the telescope removes light anyway, so little is lost.

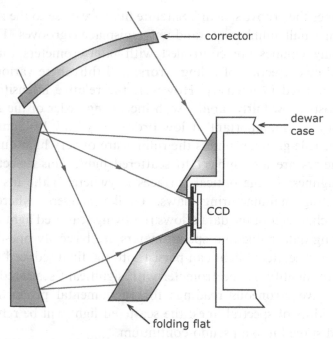

Figure 6.14. A Schmidt–Pfund camera.

Spectrometer optical behavior

Grating spectrometers are, of course, subject to all the standard aberrations: spherical, coma, astigmatism. Slight degradations of the images may be hidden because of the relatively large pixels and the effects of the slit and spectral dispersion. However, the extreme f-numbers required for good matching of the pixels of the detector to the projected slit can result in a substantial level of distortion. This aberration must be corrected very carefully in data reduction, since it would otherwise result in shifts of the apparent wavelengths of spectral features and would undermine many of the applications for spectrometers. As a result, a key step in the analysis of spectroscopic data is to conduct a fit to the apparent wavelengths of lines from a calibration source and to correct the spectra for the indicated errors in the wavelength scale.

There can also be optical issues associated directly with the grating. Periodic errors in the groove spacing produce spurious lines offset from the real one that are called ghosts. Rowland ghosts result from large-scale periodic errors, on the scales of millimeters. They are located symmetrically around the real line, spaced from it according to the period of the error and with intensity that increases with the amplitude of the error. Lyman ghosts are farther from the real line and result from periodic errors on smaller scales, just a few times the groove spacing. Satellite lines are close to the real one and arise from a small number of randomly misplaced grooves. Fortunately, modern ruling engines are controlled with interferometers that generate signals to allow correction of ruling errors, and thus these various spurious lines are minimized in intensity. However, the relative intensities of some forms of ghosts grow fairly rapidly with increasing order of the real line, so although they are unimportant at low order, they may be significant with high-order echelle gratings, unless the rulings are of very high quality.

Spectrometers are also subject to scattered light. This defect may arise through roughness in the optical surfaces anywhere in the instrument, or through grating manufacturing flaws. Unlike imagers, where the two-dimensional character of the data allows removing scattered light as a natural process during data reduction, spectrometers are basically one-dimensional instruments and scattered light can persist into the final reduced spectra and be difficult to identify. A spectrometer with significant scattered light in its spectra can give erroneous readings for fundamental properties such as equivalent widths of spectral lines; the scattered light will be removed from the lines and spread into a pseudo-continuum.

6.3.4 Imaging spectrometers

As described so far, grating (and prism) spectrometers are best suited for measurements of point sources that fit within their slits, or of long thin sections of a source, for example to determine a velocity gradient. There are two approaches to improve the spatial measurement capabilities of these instruments: (1) multi-object front ends that allow simultaneous spectra of a number of sources at various positions within the field of the instrument; and (2) integral field units that provide full imaging information simultaneously with spectroscopy.

Multi-object spectroscopy

There are two basic forms of multi-object spectrometer. In the first, a mask is substituted for the simple slit at the telescope focal plane with holes (or short slits) at the positions of sources. If this mask is properly laid out, it allows for the dispersed spectrum of each source at the detector array, without overlaps with the spectra of other sources. The spectrometer then operates in a normal fashion, except that rather than one long slit with a series of spectra for each position along this slit, it obtains individual spectra for each of the small slits in the mask.

Usually, the masks discussed above need to be prepared well in advance, and they also require the spectrometer to be mounted in the usual way directly on the telescope. A second approach relaxes both of these requirements – feeding the spectrometer with optical fibers. The input end of each fiber is located in the telescope focal plane by motor-driven "fishing rods" or some equivalent procedure, and can be reprogrammed readily. The fiber outputs are at the input to the spectrometer and are arranged to provide a well-organized pattern of spectra. Fibers can convey the light to a spectrometer either on or off the telescope, allowing a much broader selection of mounting arrangements. Because high-performance fibers are only available for the optical and near-infrared, fiber-fed spectrometers are limited to these wavelength regimes.

Integral field units

Integral field units (IFUs) are devices that take the two-dimensional image at the focus of a telescope and rearrange it into a format that can be received by a traditional grating spectrometer: for example, by organizing it to fall along a line (emulating a slit). Figure 6.15 shows three basic types of IFU: (a) using a field of lenslets; (b) using a bundle of optical fibers to sample the image and rearrange it, possibly with an assist by feeding the fibers with lenslets; and (c) dividing the image using mirrors arranged as an image slicer.

An example of the second kind has been built for GMOS on Gemini. It has a total of 1500 fibers, organized into a 5×7 arcsec object field and a 3.5×7 arcsec sky field, with each fiber sampling a 0.2 arcsec diameter part of the image and covering the 0.4–1.1 μm spectral range. To avoid losses due to focal ratio degradation, a fiber is terminated at each end with a tiny lens, speeding up the beam to the range that fibers can transmit without significant $A\Omega$ degradation.

Image slicers (Figure 6.15(c)) were first demonstrated by Bowen in 1938. The primary challenge is to fabricate and align the tiny mirrors with the necessary accuracy. One solution to this problem is to direct the light from the telescope to an optical element that substantially increases the f-number of the beam and enlarges the telescope focal plane scale. At the large scale it is easier to make the necessary array of mirrors that slice the image and deflect it for each one in the direction required for the IFU output (see, e.g., Wells *et al.* 2000). These image slices pass through an array of baffles to eliminate stray light from one slice to another and then to the spectrometer input.

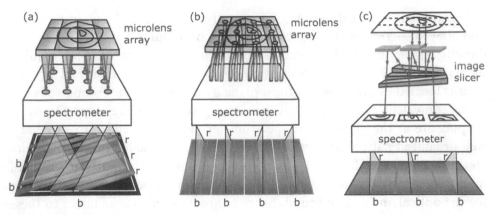

Figure 6.15. Basic IFU types: (a) microlens array; (b) fibers fed by micro-lenses; and (c) image slicer. The blue ends of the spectra are labeled "b" and the red ends "r."

6.3.5 *Data reduction*

In general, there should be detailed instructions on reducing spectra for any spectrometer, including a lot of important details individual to that instrument. Here we can only give an overview of typical procedures, assuming a simple slit spectrometer.

The reductions start with the identical steps described in Chapter 4 for imagers, and the product will be a clean two-dimensional image from which the spectrum must be extracted by additional steps. Thus, it is necessary to obtain data, response, and dark-current frames, subtract the dark current from the other two, and divide the resulting data frame by the resulting response (flat field) frame. A difficulty is that the sky may not be suitable to obtain a response frame, because its spectrum may contain emission lines or the signal level may be too low. Therefore, the response frame is usually measured with lamps shining off a diffuse scattering screen. Such a response frame will have its own spectral character. To compensate for the spectrum of the lamps, one should observe a star of known well-behaved type and reduce it in the same way. We will call the result the spectrum of the reference star.

At this point, the reduction should yield an artifact-free two-dimensional image of the spectrum. There are a number of steps to extract from it a one-dimensional spectrum suitable for measuring line strengths and other parameters.

1. The distortions introduced by the spectrometer optics will result in the spectrum not being perfectly straight – trace it to determine its shape (if it is very faint, it may be better to determine the shape from a similar spectrum of a brighter star).

2. Locate the spectrum and suitable regions for measuring the sky and extract the measured values from each. A number of specific methods may be used for the extraction. For example, one can just take the signal in a zone centered on the spectrum and following its shape (analogous to aperture photometry for photometry). Alternatively, there may be a computer program that fits a profile of the appropriate cross-dispersion shape to the spectrum, analogous to PSF-fitting photometry.

3. Perform a similar extraction on an adjacent region of blank sky and subtract the result from the object spectrum.

4. Put the reference star spectrum through the same steps.

5. Divide the target object spectrum by the one of the reference star and multiply the result by a standard template spectrum of the reference star (i.e., a library spectrum that is free of all telluric atmospheric absorptions). This step removes the spectrum of the flat field lamp and also removes residuals from features in the spectrum of the reference star.

6. Then, assuming a calibration lamp exposure or equivalent data are available, do a similar extraction of these wavelength-calibration data.[12]

7. Determine the observed wavelengths for the calibration lines by fitting line profiles to them and comparing the fitted values with the laboratory measured wavelengths. The wavelength errors can then be removed from the spectrum of the object, for example through a polynomial fit to their run with wavelength.

8. If data on standard stars were obtained, subject them to the identical steps and determine the flux level of the new spectrum by comparison with the signals from the standards.

9. Identify and extract spectral lines by fitting them with an appropriate line profile. If the spectrum is not flux calibrated, the information can be used in a relative sense, for example as line equivalent widths, or by measuring the strength of one line compared with that of another.

To support these reduction steps, a number of precautions should be observed while taking the data. First, if the detector array is subject to fringing, accurate strengths of narrow spectral lines can only be obtained if the spectral resolution element is significantly narrower than the fringe period – otherwise, the apparent strength of the line will depend on where it falls relative to the fringes, something that the standard reduction steps cannot determine. This caution applies not only to fringes but to any form of intra-pixel sensitivity variations (e.g., Figure 4.5); if the spectral feature

---

[12] In the near-infrared, the OH airglow lines can be used for wavelength calibration without a specific lamp exposure.

is smaller than the pixels, errors can result that are difficult to correct in the conventional calibration approach. Second, it is desirable to measure the spectrum on more than one set of pixels on the detector array – just as in imaging, doing so allows substitution of real data for data lost due to bad pixels. This goal can be achieved with a slit spectrometer by taking a series of spectra moving the source along the slit between integrations. This procedure also gives a set of sky measurements taken with the same pixels as the source spectrum that may prove to be a useful way to subtract the sky signal (as opposed to the subtraction based on neighboring regions in a single integration assumed above). If the spectrometer is subject to flexure, calibration lamp exposures and other calibration frames may need to be taken at the same elevation as the object being observed. If the spectrometer does not have atmospheric dispersion correction, one must be aware of the position angle of the slit when spectra are measured. If the atmospheric dispersion is along the slit, then the result will be a distortion of the spectrum, and the full information can often be recovered by fitting this effect. However, if the atmospheric dispersion is perpendicular to the slit, light will be lost at one end of the spectrum or the other (or both!). Additional issues arise with fiber-fed spectrometers. There is no long slit to obtain sky measurements automatically, and for very deep observations using some of the fibers to do so can be limited by mismatches between the performance of the sky fibers and those dedicated to the source. The safest procedure is to use pairs of fibers, one for source and the other for sky, and periodically to move the telescope to switch their roles. This approach allows the sky to be measured in exactly the same configuration as is used for the source.

6.4 Fabry–Pérot spectrometers

6.4.1 Principles of operation

Another approach to obtaining spectral information through interference is to divide the photons into two beams using partially reflecting mirrors, delaying one of these beams relative to the other, and then bringing them together to interfere. An implementation of this approach is the Fabry–Pérot spectrometer, based on two parallel plane plates with reflecting surfaces.

Figure 6.16 concentrates on the region between these surfaces. We assume illumination at a single wavelength. At the surface to the left, a portion of the input beam is transmitted and a portion reflected, and the reflected portion is again partially transmitted and reflected at the right surface. However, interference modifies this simple picture. If the spacing, l, causes a 180° phase shift

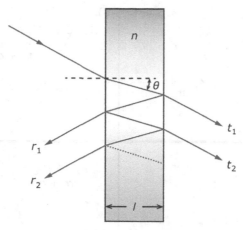

Figure 6.16. Operation of a solid Fabry–Pérot etalon.

between the incoming ray and r_1, then there will be no light escaping to the left. Under this condition, the phase shift at t_2 will be 360° and the interference at t_2 will be constructive, so the light will escape there. As the spacing between the surfaces is changed, the proportions of light escaping to the right and left will change as varying amounts of constructive and destructive interference occur at the reflective surfaces.

The functioning of this device as a spectrometer, where it is called a Fabry–Pérot etalon, can only be understood through these interference effects. Without them, since the two surfaces are typically 90% reflective, the transmission would be only 1%. Instead, the interior of the etalon acts like a resonant cavity for the photons. This behavior makes it possible for a large portion of the incident beam to be transmitted through the device if the wavelength just matches the resonant wavelength of the cavity.

We now treat this process more quantitatively. First, consider a single reflective surface as in Figure 6.17. To the left we show a wave of amplitude a incident on a surface that reflects r of the amplitude and transmits t of it. The wave is partially reflected, ar, and partially transmitted, at. By the principle of reversibility, if we reverse the directions of all the rays, there should be no change in the amplitudes. We show this situation to the right, where the reflection and transmission in the reverse direction are r' and t'. We then require

$$att' + arr = a$$
$$art + atr' = 0 \tag{6.21}$$

from which we conclude $r' = -r$ and $tt' = 1 - r^2$. Now consider the reflection and transmission of a wave of amplitude a between two such surfaces

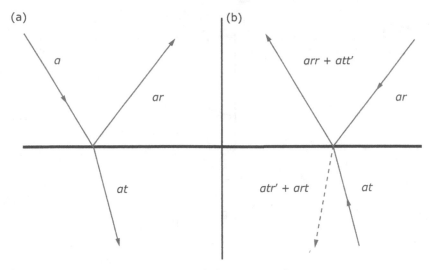

Figure 6.17. Reflection and transmission at a surface.

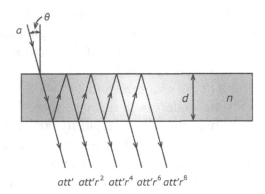

$$att' \quad att'r^2 \quad att'r^4 \quad att'r^6 \quad att'r^8$$

Figure 6.18. Transmission of a wave through two parallel reflective surfaces.

(Figure 6.18). The amplitude for the emerging light is the sum of the contributions of each emerging ray. Thus:

$$Ae^{j\phi} = att' + att'r^2e^{j\delta} + att'r^4e^{j2\delta} + att'r^6e^{j4\delta} + \ldots \tag{6.22}$$

where δ is the phase difference imposed by the reflections,

$$\delta = \frac{4\pi\,d\,n\cos\theta}{\lambda} \tag{6.23}$$

and n is the refractive index of the material between the reflecting surfaces. Applying the results derived from equations (6.21), we obtain

$$Ae^{j\phi} = a(1 - r^2)\left(1 + r^2 e^{j2\delta} + r^4 e^{j4\delta} + r^6 e^{j6\delta} + \cdots\right)$$

$$= \frac{a(1 - r^2)}{1 - r^2 e^{j\delta}} \tag{6.24}$$

We get the emerging intensity by multiplying the complex amplitude by its complex conjugate, yielding

$$I_{\text{out}} = \frac{a^2(1 - r^2)^2}{1 - r^2(e^{j\delta} + e^{-j\delta}) + r^4} = \frac{I_0(1 - r^2)^2}{1 - 2r^2 \cos\delta + r^4}$$

$$= \frac{I_0}{1 + \frac{4r^2}{(1-r^2)^2}\sin^2\left(\frac{\delta}{2}\right)} = \frac{I_0}{1 + \frac{4\mathfrak{R}}{(1-\mathfrak{R})^2}\sin^2\left(\frac{2\pi dn \cos\theta}{\lambda}\right)} \tag{6.25}$$

where the last step depends on $\cos(2\delta) = 1 - 2\sin^2(\delta)$ and $\mathfrak{R} = r^2$ is the reflected intensity.

Equation (6.25) shows that I_{out} has maxima when

$$m\lambda = 2d\,n \cos\theta \tag{6.26}$$

where m is the order. It is also apparent that these maxima are narrower in spectral range the closer $\mathfrak{R}$ is to 1 (the closer the reflectivity of the surfaces is to being complete). The sharpness of the spectral transmission bands is determined by a parameter called the finesse of the device. It is

$$\mathfrak{J} = \frac{\pi\sqrt{\mathfrak{R}}}{1 - \mathfrak{R}} \tag{6.27}$$

(valid for $\mathfrak{R} > 0.5$). The spectral resolution to full width at half power of the transmission profile is

$$R_{\text{res}} = \frac{\lambda}{\Delta\lambda} = m\mathfrak{J} \tag{6.28}$$

and the free spectral range between transmission orders is the finesse times $\Delta\lambda$. Figure 6.19 shows the changes in spectral performance with increasing values of $\mathfrak{R}$ and hence of finesse.

Because of the angle dependence of the response (the θ dependence above), an image of the output has a ring-like appearance as different orders appear with increasing distance off-axis; see Figure 6.20. The performance description we have provided (illustrated in Figure 6.19) applies on-axis.

6.4.2 Practical spectrometers

In the optical and infrared, Fabry–Pérot etalons are manufactured by evaporating suitable reflective coatings on substrates of optically transmitting

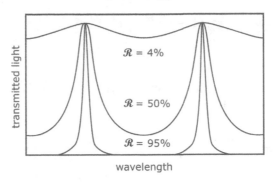

Figure 6.19. Performance of a Fabry–Pérot spectrometer as a function of $\mathcal{R}$.

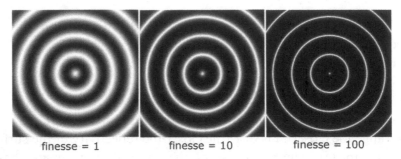

Figure 6.20. Field properties of a Fabry–Pérot spectrometer as a function of its finesse.

material. Etalons can also be made by evaporating the reflective layers on opposite sides of a plane-parallel transmissive plate, producing a solid etalon. In the far-infrared, they can be built using free-standing metal meshes as reflectors. The etalon needs to operate in a collimated beam, so the basic optical layout of a full Fabry–Pérot spectrometer is similar to that for a dispersive spectrometer. Thus, a typical optical arrangement is similar to Figure 6.4 with the interferometer placed at the disperser position (and with no slit).

Air-spaced Fabry–Pérot etalons are scanned and set in wavelength either by controlling the separation of the reflecting plates, or by varying the gas pressure so the refractive index of the medium between the plates changes. Solid etalons are scanned in wavelength by tilting them. In either case, it is necessary to isolate a single spectral order so that the observations refer to a single wavelength range. This isolation can be provided by narrowband filters, or by a second etalon of low spectral resolution.

The natural advantage of Fabry–Pérot spectrometers is that they provide images in a very narrow spectral band, the appropriate data format if the science objective is to trace the distribution of an emission line on the sky. Another advantage is that they can be operated at very high order, for example $m \sim 500$, significantly higher than is easily achieved with an echelle grating. As a result, a Fabry–Pérot spectrometer can provide high spectral resolution in a compact package. They are less optimum for other types of measurement. For example, since different wavelengths must be measured in a time sequence, any change in the measurement conditions can affect the calibration of the measurements. As a result, these spectrometers are not ideal for measuring spectral features of very small equivalent widths; such features are better measured with a dispersive spectrometer where all the surrounding wavelengths are measured simultaneously. Fabry–Pérot spectrometers are also relatively finicky devices, requiring a high degree of temperature control and a vibration-free environment to operate correctly. Errors in control and departures from perfect flatness of the reflecting plates appear as degradation in the finesse and hence a reduction in the resolution.

6.5 Fourier transform spectrometers

6.5.1 Principles of operation

The Fourier transform spectrometer (FTS) is yet another way to obtain spectra by dividing a beam of light, introducing a phase difference, and recombining it. The principle behind it is shown in Figure 6.21. The beam of light is divided, ideally 50/50, at the beamsplitter. One of the resulting beams is reflected off a stationary mirror and back through the beamsplitter to the detector. The second beam goes to a moving mirror and from there to the beamsplitter, where the two beams are combined. Half of the combined beam escapes through the input (although more complex designs can recover it) and the other half goes to the detector. As the moving mirror is scanned, the phase between the two components of the combined beam changes, causing the interference between them to change. Thus, for a monochromatic input, the signal varies sinusoidally.

To analyze this behavior we start with the amplitude of the two beams after they have been divided and re-combined:

$$E = h_1 e^{j\varphi_1} + h_2 e^{j\varphi_2} \tag{6.29}$$

Assuming that the beam splitting is perfectly 50/50 ($h_1 = h_2 = h$), the resulting intensity is

$$I = EE^* = 2h^2(1 + \cos(\varphi_1 - \varphi_2)) = 2h^2\left(1 + \cos\left(\tfrac{2\pi\Delta}{\lambda}\right)\right) \tag{6.30}$$

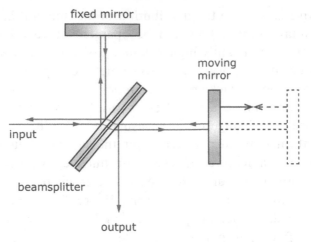

Figure 6.21. Principle of operation of a Fourier transform spectrometer.

where the phase difference has been converted to an expression involving the path- length difference, $\Delta = \lambda(\varphi_1 - \varphi_2)/2\pi$. Now we scan the moving mirror at velocity v, $\Delta = vt$. We then have that the phase difference $\varphi_1 - \varphi_2 = 2\pi f t$, where $f = 2v/\lambda = 2v \, (v/c)$, and v is the frequency of the photon. We can set the frequency f by controlling the speed of the moving mirror. Thus, we can down convert the photon frequency to a range where the detector can track it. From equation (6.30), the modulation sits on top of a constant signal, $2h^2$, which contains no useful information other than the intensity of the source and can be discarded.

Each wavelength in the input beam will produce a similar sinusoidal response in units of wavelength as the mirror is moved. Thus, if the moving mirror is scanned at a constant rate, there will be a unique frequency at the detector for each wavelength in the input beam. The signal for n wavelengths becomes

$$I(t) = \sum_n 2h_n^2 \cos\left(2\pi f_n t\right) \tag{6.31}$$

This has a form that can be inverted by discrete Fourier transformation to convert it to a spectrum in frequency units.

6.5.2 Practical applications

Fourier transform spectrometers were popular in the era of low-performance infrared detectors because of the "Fellgett advantage" (Fellgett 1971). They collect information on all the wavelengths simultaneously. If the system is limited by the detector noise, then the net noise is not increased by broadening the spectral range to measure many wavelengths, and the resulting spectrum can have a significantly greater ratio of signal to noise than would have been obtained with the same detector measuring one wavelength at a time. Thus, in the detector-limited situation, the signal-to-noise ratio is improved in principle in proportion to the square root of the number of spectral elements using the FTS (Saptari 2003). Also, like the Fabry–Pérot spectrometer, the FTS is capable of imposing very large phase differences in a relatively compact package and hence is useful for very high spectral resolution. Another advantage is that a FTS provides images simultaneously with spectra if it is used with an array of detectors. This can be described as a throughput advantage, since a grating spectrometer must use one dimension of its detector array for spectral information while a FTS can use both for imaging and encodes the spectral information in the time dependence of the signal. This capability is exploited, for example, in the SPIRE instrument on Herschel.

There are some cautions about the performance of an FTS. The throughput and Fellgett advantages may be difficult to realize in practice (Saptari 2003). The Fellgett advantage vanishes if the system is limited by background noise. In principle, the FTS performance is then similar to that of a dispersive spectrometer with a single detector that measures one wavelength at a time. However, it may operate at the "Fellgett disadvantage" if the noise varies significantly across the free spectral range – for example, if there is a large change in the background level with wavelength. Then the higher level of noise for part of the spectrum will degrade the entire measurement. The operation of the FTS also requires that the detector be read out frequently, which can degrade the noise if it is an integrating device. In addition, the encoding of the spectrum in a FTS results in a modest degradation in signal-to-noise ratio compared with an equivalent multi-channel spectrometer (Treffers 1977).

There is a sharp limit on the phase difference that can be obtained with a FTS, set by the constraints on the distance the moving mirror can go. As a result, the measured profile of an unresolved spectral line is a sinc function, see Figure 6.22. The ringing arises because the limit on the path-length difference (imposed by the limit on the travel of the moving mirror) means that the Fourier series for the spectrum is truncated, and there are no terms to cancel the oscillations associated with the signal that is produced just below this truncation. The ringing can be reduced by reducing the weight of the measurements just below the truncation point, at the expense of spectral resolution. This process is called apodization. A triangle weighting is perhaps the most popular but there are many other options.

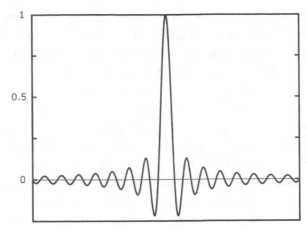

Figure 6.22. Line profile for an unapodized FTS output.

6.6 Interference filters

Colored glasses, crystal absorptions, or interference filters are used to define spectral bands for photometry, order blocking in spectrometers, and other applications. Colored glasses were the basis for Johnson's UBVR bands, and in the far-infrared a similar approach can be taken using the transmission bands of crystals. However, interference filters are more widely used because they have better-defined transmission characteristics (sharp onset and conclusion of transmission) and can be designed to have central wavelengths and wavelength ranges as desired.

Figure 6.23 shows the basic building block of an interference filter. It is a low-order (in this case $m = 1$) solid Fabry–Pérot interferometer. The layers are built up by evaporation onto a transparent substrate that provides the mechanical strength for the filter. The gap is determined by a spacer layer, which is made $m/2$ waves thick, with m generally small. The reflective surfaces are provided by ¼ wave layers alternating between low and high refractive index and designed to provide the desired finesse. Each Fabry–Pérot-like unit is called a cavity. The performance of the filter can be improved by using a series of cavities, as in Figure 6.24. Additional blocking of wavelengths far from the filter passband is usually provided by other filter elements based on the same principles. Often absorption by the substrate is used to supplement that provided by interference to assist the blocking.

As would be expected from the behavior of Fabry–Pérot spectrometers in general, the spectral properties of an interference filter depend on the angle it makes with a beam of light.[13]

---

[13] In addition, the spectral characteristics of interference filters shift slightly toward shorter wavelengths as they are cooled.

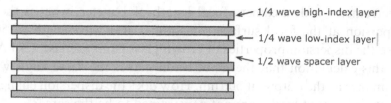

Figure 6.23. First-order interference filter cavity. Based on image by Russell Scaduto (n.d.), with permission.

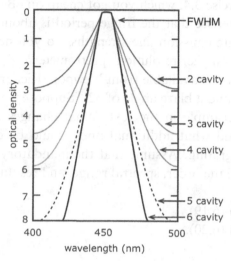

Figure 6.24. Improvement of filter performance with multiple cavities. Based on image by Russell Scaduto (n.d.), with permission.

The central wavelength moves from λ_0 to shorter wavelengths, λ_c, as the filter is tilted by θ according to

$$\lambda_c = \lambda_0 \left(1 - \frac{\theta^2}{2n^{*2}} \right) \qquad (6.32)$$

where n^* is the effective refractive index of the filter. At large angles the transmission tends to be reduced and the passband broadened. However, at small angles the behavior can provide a convenient way to tune the wavelength of the passband. The behavior also results in a degradation of the purity of the passband when a filter is placed in a beam with a large angle of convergence (small f-number).

6.7 Exercises

6.1 In a neighborhood yard sale, you bought an equilateral prism of crown glass, designed with three 60° angles (equal sides). If you put it into a spectrometer with a camera focal length of 1 meter, what is the linear dispersion at the focal surface ($nm\,mm^{-1}$) at a wavelength of 500 nm? (Use the dispersion properties of crown glass from exercise 2.5.) Assume for this calculation that the input beam makes the same angle with the prism face as the output at 500 nm. How does the dispersion change if the input and output angles are not constrained to be the same?

6.2 To stay within your budget, you are building a spectrometer with a spectral resolution of only 100 to work near 0.9 μm. You will be using the CCD described in exercise 3.4, which you got cheap on eBay. You should have found in that exercise that the fringe period is about 0.0076 μm. If your goal is to measure emission line strengths, do you need to increase your budget to build a higher-resolution spectrometer?

6.3 An echelle grating is available from Newport/Richardson Gratings with 79 grooves/mm and a blaze angle of 62°. Suppose we substitute it for the grating in the example in the text. What changes are required in the spectrometer, and what additional ones would be desirable, if it is to work with this grating? Assume that the mandatory changes have been made; determine the order, spectral range, and resolution at Hα with this grating.

6.4 Derive equation (6.19).

6.5 Derive equation (6.20).

Further reading

Hearnshaw, J. (2009). *Astronomical Spectrometers and Their History*. Cambridge: Cambridge University Press.

Palmer, C. and Loewen, E. (2005). *The Diffraction Grating Handbook*, 6th edn. Newport Corporation, http://gratings.newport.com/information/handbook/handbook.asp

Mediavilla, E., Arribas, S., Roth, M., and Cepa, J. (2010). *Three-D Spectroscopy in Astronomy*. Canary Islands Winter School on Astrophysics, Vol. 17. Cambridge: Cambridge University Press.

7

Adaptive optics and high-contrast imaging

7.1 Overview

The images obtained by groundbased optical and infrared telescopes are degraded by the effects of turbulent cells in the atmosphere. Each cell is characterized by a slightly different temperature and, because the refractive index of air is temperature-dependent, by a slightly different index of refraction (from Cox 2000):

$$n_{air}(P[\text{mbar}], T[\text{K}], \lambda[\mu\text{m}]) = 1 + 7.76 \times 10^{-5}\left(1 + 7.52 \times 10^{-3}\tfrac{1}{\lambda^2}\right)\tfrac{P}{T} \quad (7.1)$$

The result is that the wavefronts of the light from a point-like astronomical source, which are plane-parallel when they initially strike the atmosphere, become distorted and the images are correspondingly degraded. A rough approximation of the behavior is that there are atmospheric bubbles of projected diameter $r_0 = 5$–15 cm with temperature variations of a few hundredths up to $1\,°\text{C}$, moving at wind speeds of 10 to 50 m/s; r_0 is defined by the typical size effective at a wavelength of $0.5\,\mu\text{m}$ and called the Fried length, defined to be the length over which the rms phase variation is one radian. The argument in the exponential of the extended Maréchal relation (equation (2.2)) is $(-2\pi\sigma/\lambda)^2$, so a one radian $(\lambda/2\pi)$ phase error yields a Strehl ratio of $1/e$. The timescale for variations over a typical size of r_0 at the telescope is of order 10 ms, the time for the air to move a distance of r_0 (e.g., for $r_0 = 10$ cm and a wind velocity of 10 m/s). For a telescope with aperture smaller than r_0, the effect is to cause the images formed by the telescope to move rapidly as the wavefronts are tilted to various angles by the passage of warmer and cooler air bubbles. If the telescope aperture is much larger than r_0, many different r_0-sized columns are sampled at once. Images taken over significantly longer than 10 ms are called seeing-limited, and have typical sizes of λ/r_0, the familiar expression for the FWHM for a diffraction-limited telescope

189

of aperture r_0. This result arises because the wavefront is preserved accurately only over a patch of diameter $\sim r_0$. For example, with $r_0 = 10\,\mathrm{cm} = 0.10\,\mathrm{m}$ at $\lambda = 0.5\,\mu\mathrm{m} = 5 \times 10^{-7}\,\mathrm{m}$, the image diameter will be about 5×10^{-6} radians, or 1 arcsec, independent of the telescope aperture (so long as it is significantly larger than r_0). With conventional instruments, the success of an observing night can be critically dependent on the seeing – how large r_0 happens to be and therefore how small the images delivered to the instrument are. The same atmospheric effects discussed above are also responsible for fluctuations in the flux arriving at the detector, termed atmospheric scintillation.

7.2 Speckle imaging

The phase of the light varies quickly over each r_0-diameter patch. A fast exposure (e.g., $\sim 10\,\mathrm{ms}$) freezes this pattern and the image appears speckled, within the overall envelope of the seeing limit (Figure 7.1). The speckles result from interference among coherent patches separated by distances up to the full aperture of the telescope, D, and hence can have a range of diameters including some close to the traditional diffraction limit, λ/D. One way to recover the intrinsic resolution of a telescope, called speckle imaging, is to obtain many exposures of $\sim 10\,\mathrm{ms}$ length and to analyze them to reconstruct the source structure. To avoid blurring of the speckles by wavelength-dependent diffraction effects, these exposures are usually taken through a relatively narrow filter to restrict the spectral range. A general way to extract information from speckle images is to Fourier transform them, take the squared modulus of the Fourier transform to obtain the energy spectrum, and to compare the energy spectrum with similar information from an unresolved reference object, a procedure called speckle interferometry. The energy spectrum contains spatial frequencies up to the diffraction-limited cutoff of the telescope, so in principle this procedure allows extracting information up to the diffraction limit. However, to make an actual image of an object requires the phase of the Fourier transform, which can be obtained in a number of ways making use of the high-order moments of the complex transform of the speckle image.

A related technique, called "lucky imaging," also begins with many thousands of short exposures. However, rather than analyzing the speckles in these images, the images are sorted to identify the best – those obtained when by a lucky accident the atmospheric turbulence had minimal effect on the wavefronts and the energy was concentrated rather than being spread over a large seeing disk. These few exposures can be combined to yield the final image of the object.

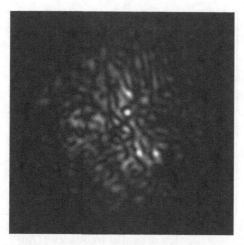

Figure 7.1. Simulated image for r_0 of 5% the telescope diameter, showing complex speckle structure. For an image in the visible, this image might correspond to seeing of 0.5 arcsec on a 4-m telescope. From R. N. Tubbs (n.d.).

7.3 Atmospheric behavior

Because of the very short exposure times, any approach based on speckles is severely limited by detector noise, and it is only useful in the imaging domain (e.g., it is of no help in concentrating light onto a narrow spectrograph slit). A more versatile approach is to compensate for the atmospheric turbulence effects *pre-detection* to deliver a diffraction-limited (or nearly so) image to the instrument focal plane. This procedure, termed adaptive optics (AO), uses a natural or artificial star to measure the atmospheric-induced wavefront distortions and corrects them in real time. Before discussing some of the implementations, we will describe some of the requirements that must be met for AO to succeed.

To derive these requirements, we need to have a mathematical procedure to characterize the atmospheric turbulence. Suppose the refractive index is a function of position, $n(x)$. The fluctuations in n could then conventionally be described by its covariance, which is the average of $n(x) - \bar{n}$ times a similar quantity at a nearby location, where $\bar{n}$ is the mean value. However, the covariance includes all dimensional and temporal scales; relevant parameters such as pressure, temperature, and humidity have slow changes (e.g., weather) that are not of interest for imaging through the atmosphere. Instead, we are interested in only a subset of the variations: refractive index differences between nearby points on our wavefront. Kolmogorov (1941) realized that he could separate long-term drifts in atmospheric properties from the shorter-term turbulent fluctuations by basing his analysis on structure functions.

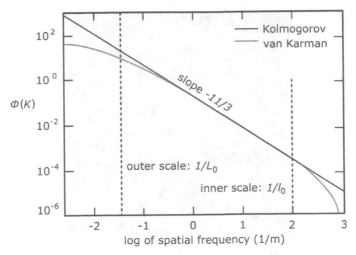

Figure 7.2. Power spectral density (PSD), $\Phi(K)$, in arbitrary units, of wave-front aberrations due to atmospheric turbulence. The simple power law of the Kolmogorov description is contrasted with the von Karman formalism, which allows for the damping of turbulence by viscosity on small scales (l_0) and saturation on large scales (L_0).

Thus, if we want to describe small-scale random fluctuations in the refractive index, we begin with the difference function

$$n(x,r) = n(x+r) - n(x), \tag{7.2}$$

that is, avoiding reference to some long-term mean value. The fluctuations in the refractive index over small scales are characterized by the structure function,

$$D_n(r) = \langle [n(x+r) - n(x)]^2 \rangle \tag{7.3}$$

Since a structure function is based purely on differences, it is not affected by smooth changes over large distances (unless they are very big). The problem for imaging through the atmosphere can be posed almost entirely in terms of various structure functions, for example in temperature and velocity. Using this insight, Kolmogorov was able to develop a first-order description of atmospheric turbulence (Figure 7.2).

The Kolmogorov derivation makes a number of assumptions, which have turned out to be remarkably good. In it, turbulence starts with some large cell size, L_0 (typically 15–40 meters; Martin *et al.* 2000), and transfers energy down through a cascade of smaller cells with no energy loss, until the viscosity regime is reached at size scale l_0 (typically millimeters in scale), where viscosity dissipates the turbulent energy, converting it to heat and damping it out. The turbulence is assumed to be homogeneous and isotropic. The resulting theory states that the optical effects of turbulence can be fitted by a power law in spatial frequency ($= 2\pi/l$ where l is the local spatial scale of the disturbances) between the two size scales l_0 and L_0. Kolmogorov extended the fit beyond these two cutoffs for mathematical convenience. A more complete treatment by von Karman includes the behavior at low and high frequencies explicitly, but the Kolmogorov spectrum remains a useful approximation.

The turbulence can be characterized by the power spectral density (PSD), that is, the turbulent power as a function of the spatial frequency. By the Wiener–Khinchin Theorem, the PSD of the atmospheric behavior is the Fourier transform of the covariance, $\Phi(K)$, where K is the three-dimensional spatial frequency. As shown in Figure 7.2:

$$\Phi(K) = 0.033 C_n^2 K^{-11/3} \tag{7.4}$$

(Noll 1976). $\Phi(K)$ is a measure of the relative contribution of atmospheric effects to the total wavefront distortion as a function of the spatial frequency. Here, C_n^2 is the structure "constant," perhaps the least constant constant you will encounter. It is proportional to the mean square of the difference in refractive index for points separated by 1 meter (compare equation (7.3)). It therefore describes the overall strength of the turbulence in terms directly relevant to imaging, with units of $m^{-2/3}$. It can range from $10^{-15} m^{-2/3}$ to $< 10^{-18} m^{-2/3}$, depending on time, season, elevation in the atmosphere, and other variables.

Figure 7.3 shows C_n^2 as a function of altitude on a typical night at a good observing site. It has the characteristic behavior of one peak close to the ground with a height of about 3 km, and another in the tropopause at a height of about 12 km and roughly 8 km thick. There can be even lower-level contributions, such as turbulence in the telescope dome due to heat sources there, or wind-flow patterns around the telescope. In fact, the seeing in older telescopes was almost always dominated by turbulence within the dome (including turbulence from heat trapped in their massive primary mirrors and mounts), and a major advance in image quality has resulted from efforts to reduce these effects.

The Kolmogorov formalism allows derivation of the key aspects of the influence of this turbulent air on imaging. The Fried length can be calculated as (Tyson 2000)

$$r_0 = \left[0.423 \, k^2 \sec \beta \int C_n^2(z) \, dz \right]^{-3/5} \tag{7.5}$$

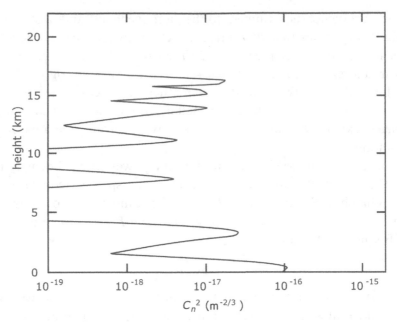

Figure 7.3. Typical behavior of the structure constant with altitude.

where k is the wavenumber, $k = 2\pi/\lambda$; β is the zenith angle; and the integral is taken over the path traversed by photons at the zenith. The exponent of $-3/5$ is a result of assuming the Kolmogorov description of turbulence. This equation shows the dependence of $r_0 \propto \lambda^{6/5}$. It also makes the effect of increasing path due to non-zero zenith angles explicit. The integral term reflects the phase correlation, which describes the phase coherence received at the telescope. The integral is only over C_n^2, with no additional dependence on altitude. This behavior is because the deviation from straight paths for the photons is very small, so the refractive index variations at any altitude affect the phase correlation similarly. The atmospheric layers near the ground, where C_n^2 is the largest, therefore have the largest effect on r_0.

From the definition of r_0, the variance in the wavefront, σ_{wf}^2, for images delivered by a telescope of diameter D is

$$\sigma_{wf}^2 = 1.03 \left(\frac{D}{r_0}\right)^{5/3} \tag{7.6}$$

This value can be used with the Maréchal relation (equation (2.2) with $(\sigma/\lambda)^2$ replaced with σ_{wf}^2) to estimate the Strehl ratio to be achieved in an image.

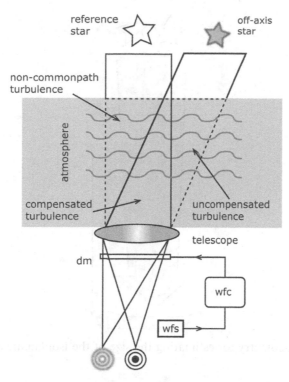

Figure 7.4. Limits of the isoplanatic patch. A schematic AO system is shown with a star on-axis and a science target off-axis. The wavefront distortions measured by the wavefront sensor (wfs) and corrected by wavefront controller (wfc) are not fully appropriate for the science target; dm = deformable mirror.

A critical parameter is the size of the area on the sky where a single set of corrections does a reasonably good job of taking out the atmospheric effects. Figure 7.4 shows the geometry involved. As the angular distance increases between the star used for the corrections and the science target, the corrections become increasingly irrelevant for the air path toward the target. The area with useful corrections is defined as the isoplanatic patch and often is described in terms of its angular radius around a guide star, the isoplanatic angle. It can be understood as shown in Figure 7.5. Above some altitude, the paths of air toward the guide or reference star and the target are separate and thus will impose uncorrelated distortions on the wavefronts from the two objects. The isoplanatic angle is defined to be where the Strehl ratio has degraded by $1/e \sim 0.37$ from the value right on the guide star (Hardy 1998), corresponding to one radian wavefront error (see equation 2.2). As an approximation, if we define the atmospheric distortions to occur at a single layer at altitude h (Figure 7.5), then the isoplanatic angle is:

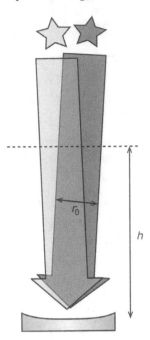

Figure 7.5. Geometry for estimating the size of the isoplanatic angle.

$$\theta_0 \approx 0.31 \frac{r_0}{h} \tag{7.7}$$

In this expression, a typical value for h is 5 km, so with $r_0 = 10$ cm equation (7.7) suggests that θ_0 in the visible is only about 1.3 arcsec and is about 7 arcsec at 2.2 μm. A useful rough approximation is that the radius (in arcsec) of the useful field of view defined by θ_0 is about 3–10 times the wavelength in microns under excellent conditions. The isokinetic angle is that over which the image motions are strongly correlated. Since the motions involve wavefront averages over the telescope aperture, the isokinetic angle is roughly 0.3 D/h, where D is the telescope aperture. For a big telescope, it is therefore much larger than the isoplanatic angle, of order 2 arcmin for 8–10-m telescopes.

In the formalism of the other critical parameters, a more complete description of isoplanatism is (Tyson 2000)

$$\theta_0 = 0.057 \, \lambda^{6/5} \left(\sec \beta \int C_n^2(z) \, z^{5/3} \, dz \right)^{-3/5} \tag{7.8}$$

As one might expect from Figure 7.4, this term has a relatively strong altitude dependence; that is, strong atmospheric effects near the telescope where there is good beam overlap are largely canceled, whereas weak turbulence high above the telescope can have a large effect along the different lines of sight.

The time variation of the atmospheric effects is calculated under the assumption that the turbulent structure changes slowly in the frame of reference moving with the local wind velocity, so the variability in time is just due to the turbulent spatial structure moving through the field of view (Tatarski 1961). Thus, the simplest estimate of the timescale for the wavefront to remain reasonably stable ($\sim$1 radian rms) is the coherence time t_C:

$$t_C \approx 0.3 \frac{r_0}{V} \tag{7.9}$$

where V is the wind speed. The coherence frequency is then

$$f_C \propto \frac{1}{2\pi t_C} \approx \frac{V}{2r_0} \tag{7.10}$$

For $V = 40\,\text{m/s}$ and $r_0 = 100\,\text{cm}$ (values appropriate at $2\,\mu\text{m}$), $f_C \approx 40\,\text{Hz}$. Similarly, substituting D for r_0 in equation (7.10), the frequency for overall image motions with a 10-m telescope can be estimated to be several hertz.

A more detailed derivation also divides the wavefront errors (as we have above) to distinguish timescales for image structure changes from those for image motions. Structure changes depend on the rate of the wavefront variation ($\sim$1 radian rms), as given by the Greenwood frequency (Tyson 2000):

$$f_G = 2.31\,\lambda^{-6/5} \left[\sec \beta \int C_n^2(z)\, V^{\frac{5}{3}}(z)\, dz \right]^{3/5} \tag{7.11}$$

where $V(z)$ is the wind velocity. Depending on conditions, including the quality of the telescope site, f_G can range from tens to hundreds of hertz (in the visible). Equation (7.11) shows how the changes occur more slowly with increasing wavelength, as $f_G \propto \lambda^{-6/5}$ and also illustrates the effects of increasing zenith angle. There is no explicit altitude dependence, only one on the structure constant combined with the wind velocity. That is, the frequency of the fluctuations will be dominated by atmospheric layers with both large values of C_n^2 and of the wind velocity.

The cells that tilt the wavefront appear to be at least as large as the largest telescopes currently available ($\sim$10 m) and the resulting image motion occurs at a characteristic tilt frequency (Tyson 2000):

$$f_T = 0.33\,D^{-1/6}\,\lambda^{-1} \sec^{1/2}\beta \left[\int C_n^2(z)\, V^2(z)\, dz \right]^{1/2} \tag{7.12}$$

where D is the telescope aperture. This frequency decreases with increasingly large telescopes, is inversely proportion to the wavelength, and is typically a few to tens of hertz (in the visible).

As contrasted with the wavefront tilts that dominate seeing and image motion, wavefront curvature induced by atmospheric turbulence causes scintillation – focusing and defocusing of the light (visible as rapidly moving shadows in an out-of-focus star image), resulting in fluctuations in the flux received from a source. Scintillation, as applied to reasonably large telescopes ($D \gg r_0$), is predominantly due to turbulence high in the atmosphere, 7–15 km, described by:

$$\sigma_{\rm I}^2 \propto D^{-7/3} \sec^3 \beta \left[\int C_n^2(z)\, z^2\, dz \right] \tag{7.13}$$

where $\sigma_{\rm I}{}^2$ is the variance in the intensity (Dravins *et al.* 1997). The larger the telescope, the more the shadows are averaged out, causing the rapid decrease in the variance with increasing D. Scintillation also occurs in many other situations where signals propagate through turbulent media (e.g., interplanetary scintillation of radio waves due to the solar wind).

7.4 Natural Guide Star AO systems

These three boundary conditions – the Fried length, speed of fluctuations, and isoplanatic patch size – define the design and operation of a Natural Guide Star (NGS) AO system. Full correction (to ~1 radian rms) requires that each footprint on the primary mirror of diameter r_0 be corrected individually. The necessary adjustments must be derived from a guide star within θ_0 of the scientific target. They also must be fast enough to track the variations. Corrections need to be applied at about ten times $f_{\rm G}$ to keep up with the wavefront changes and impose a correction that tracks them sufficiently closely.

7.4.1 Performance tradeoffs

If all other sources of image degradation were absent, a system correcting on the scale of r_0 and at a frequency well above $f_{\rm G}$ would put significant amounts of energy into the diffraction-limited beam, for example a Strehl ratio ~ 0.3 (or higher in the mid-infrared). However, this requirement can lead to very complex designs – an AO system designed for the visible on an 8-m telescope and with $r_0 = 10$ cm would require of order 5000 corrected footprints. As more footprints are measured and corrected, the light of the NGS must be divided into correspondingly more parts. Good correction requires that each of these parts be measured very quickly, making difficult demands on the detector system to operate with low noise.

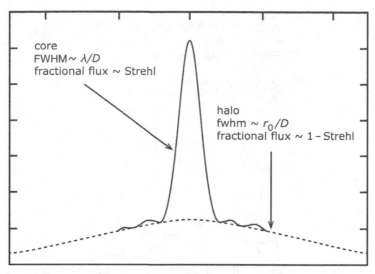

Figure 7.6. Cross-section of a partially corrected image.

Therefore, many systems are built to provide fewer corrected footprints with a goal of achieving partial correction. The images yielded by partial correction have sharp, diffraction-limited cores, surrounded by low surface brightness halos of size similar to that for the traditional seeing-limited case, as in Figure 7.6. The effect of improving the AO correction is to put relatively more light into the core, not to decrease its diameter. Therefore, for observations dependent on maximal resolution without regard to overall efficiency, even modest correction can be very powerful; obviously, for goals such as maximizing the amount of light passing through a narrow spectrometer slit, it is less beneficial.

The small size of the isoplanatic patch requires either that the correction be derived from the target star itself (if the program is based on bright stars, e.g., an exoplanet detection effort), or from a star very near the target. Where the guide star is distinct from the target, relatively faint guide stars must be used if any reasonable fraction of the sky is to be accessible. This behavior is strongly wavelength-dependent since r_0 and θ_0 increase as $\lambda^{6/5}$. From equation (7.5), we would estimate that full correction at $2\,\mu$m would require 25 times fewer corrected footprints compared with the 5000 in our example in the visible. For this reason, most AO systems operate in the infrared and we will use $2\,\mu$m as a fiducial wavelength in the following.

7.4.2 Basic layout

Figure 7.7 shows the basic layout of a NGS AO system. As a result of atmospheric turbulence, a distorted wavefront is delivered to the telescope.

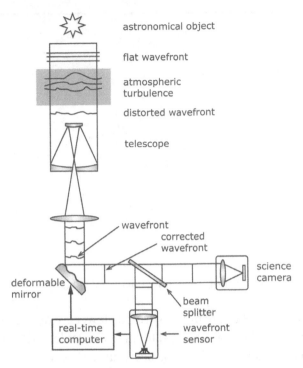

Figure 7.7. A "simple" NGS AO system.

Behind the telescope focus, the beam is divided with a dichroic mirror that reflects the visible component and transmits the infrared one. The reflected beam is brought to an optical device that measures the distortions; because the optical and infrared wavefronts traverse the same column of air, the distortions are the same in both spectral regions. The result is analyzed by a signal processor, or reconstructor, to generate commands to the deformable mirror (DM), placed at a pupil. The surface of this mirror is adjusted to compensate for the wavefront distortions so they have been removed after the wave is reflected. Because the wavefront sensor sees the corrected wavefronts from the preceding cycle, the system works by feedback – the new corrections represent only the relatively minor changes in successive snapshots of the wavefront, and if there were no changes the system would quickly settle to a static solution.

7.4.3 *Wavefront sensors, reconstructors, and deformable mirrors*

The AO system needs three dedicated components: the wavefront sensor, the deformable mirror, and the signal processor (or reconstructor). We discuss each in turn.

A variety of approaches can be used to sense the distortions in the wave-front. For adjustments just in tip and tilt, a simple approach is based on a quadrant sensor (or "quadcell") – four individual detectors that come together at their corners. Such a device can measure at high signal-to-noise ratio the position of an image placed near the junction of the cells because the light is spread over so few individual detectors. By comparing the signals from the four detectors, it is possible to calculate the offset of the image from their junction, if the image size is known (it may not be if the seeing is variable). The use of additional detectors in a quadcell can bring advantages in the accuracy of the offset determination (e.g., by real-time measurements of image size), potentially at the expense of requiring additional light for adequate signal-to-noise ratio.

Independent of image size, the quadcell can identify when the image has been placed to balance the detector outputs to values indicating the source is at a standard position (although possibly not centered on the junction of the detectors if they are not matched well).

A more powerful option is the Shack–Hartmann sensor. This device uses an array of lenses at a pupil to divide the wavefront and focus different sections of it as individual images on a detector. Local tilts in the wavefront shift the positions of the images of the segments, so the deviations from planarity in the wavefront can be measured by centroiding the images (see Section 2.5.2 for more details). The image centroiding can use a quadcell or a CCD.

Another approach (Roddier 1988) measures the curvature of the wavefront directly. Conceptually, it is based on the fact that, where the wavefront has concave curvature, it will tend to produce concentrated light ahead of the true focus of the optical system, whereas where the curvature is convex the concentration will tend to be behind the true focus. Therefore, the curvature of the wavefront can be determined from the structure of the out-of-focus image. Although in principle only one such image is required, determining the image structure both in front of and behind the plane of good focus has a number of advantages: (1) it helps compensate for systematic errors; (2) it allows correction of atmospheric scintillation, which otherwise might add structure to the image, and (3) it allows for a simple control signal. Therefore, the optics feeding the wavefront sensor is arranged so it can alternate rapidly between one position in front of and another behind focus, far enough apart that they provide pupil-like images. The wavefront errors have been removed when the distributions of intensities are equal on both sides of focus; this behavior is the signature of a flat wavefront.

Finally, the pyramid sensor divides the light of the guide star by focusing it on the tip of a transparent pyramid (Ragazzoni 1996). The pupils associated

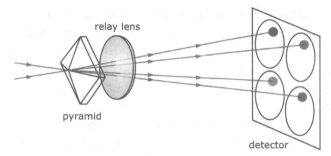

Figure 7.8. Pyramid wavefront sensor. After S. Egner (2006), with permission.

with the resulting four beams are imaged onto a detector (Figure 7.8). If the wavefront were flat and the image were held exactly on the tip of the pyramid, the four beams would have equal amounts of light. Distortions in the wavefront change the shape of the point spread function and thus change the distribution of light over the pupil, for example among the four beams. The relative brightness of corresponding patches in the pupil images therefore is a measure of the local tilt of the wavefront. It would seem that the output of this type of sensor depends critically on holding the image exactly on the tip of the pyramid. A solution is to move the image in a pattern, for example a circle. Then the image will be in each of the beams for some fraction of the time and this fraction can be used like a quadcell to measure the necessary tilt adjustment to keep the image centered (Ragazzoni 1996). Other corrections can be determined after the tilt correction has been made in this way (feasible because f_T is substantially less than f_G, so the image can be centered stably while the wavefront errors are measured). An advantage over other wavefront sensors is that the spatial sampling can be changed easily by changing the binning on the detector, or by changing the reimaging optic. Like the curvature sensor, the pyramid sensor does not subdivide the light finely as is done with a Shack–Hartmann sensor, so it can be used with relatively faint guide stars.

Ultimately the output of the wavefront sensor must control a deformable mirror (DM) that can be shaped to correct the wavefront errors. These mirrors must meet a number of requirements: (1) they must have enough degrees of freedom in their shape to provide the desired degree of wavefront correction; (2) they must have smooth surfaces, particularly on the scales that are smaller than their ability to correct; (3) they should have good control characteristics, such as being fast enough to track the wavefront errors and being free of hysteresis (i.e., when commanded to take a certain shape, they should do that accurately independent of the shape they had assumed

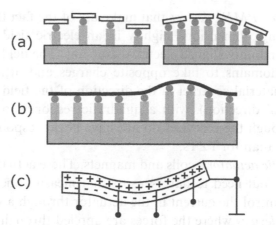

Figure 7.9. Three different types of deformable mirror: (a) segmented; (b) continuous face sheet; and (c) bimorph.

previously); (4) they should have low power dissipation so that they do not add their own turbulent air to the necessary corrections; (5) their dynamic range must be large enough (i.e., they must be capable of deformations up to ~5 μm for 10-m telescopes, three times larger for 30-m ones); and (6) they need to be large enough for the overall optics design (small DMs need to be fed by fast optical beams). There are at least three basic designs currently in use as illustrated in Figure 7.9.

Segmented mirrors are made up of individual facets, each with one to three actuators. In these mirrors there is no connection from one facet to its neighbors. With a single actuator per segment (Figure 7.9 (a) left), there must be a large number of segments to achieve reasonably good wavefront correction. However, with three actuators per segment (Figure 7.9(a) right) so one can adjust tip/tilt as well as piston, accurate correction needs far fewer segments.

Continuous face sheet mirrors are made of thin, flexible mirrors with an array of actuators glued onto their backs (Figure 7.9(b)).

Bimorph mirrors are made by joining two plates of piezoelectric material with a suitable pattern of electrodes distributed over their areas (Figure 7.9(c)).

A variety of actuator types can be used with either of the first two approaches:

1. *Piezoelectric (PZT) devices*, crystals made of molecules with dipole moments that have been aligned across the sample. An applied voltage then exerts a force that stretches or compresses the molecules and the sample changes length in proportion to the voltage. A shortcoming of these actuators is that they have a significant level of hysteresis.

2. *Electrostrictive (PMN) devices* that make use of the fact that all dielectric materials change dimensions slightly in an electric field because of the presence of randomly aligned electrical domains. The field causes opposite sides of the domains to take opposite charges and attract each other, making the material contract in the direction of the field (and expand in the orthogonal direction), with a quadratic response to the size of the voltage. Although these devices do not have linear response, their hysteresis is smaller than for PZTs;

3. *Electromagnetic actuators*, coils and magnets. These actuators are not stiff in themselves, but need to have sensors that measure the position of the mirror and control the current in the actuator through a feedback loop.

4. *Electrostatic devices* where the forces are applied through an electrostatic force between parallel plates.

For bimorph mirrors, voltages put on the electrodes cause the plates to expand or shrink and the device then bends. These mirrors have the advantage over other types in that their electrodes can be laid out to match the subapertures in a curvature wavefront sensor. The translation from wavefront curvature error to required DM curvature is then simplified.

Two special approaches to DMs deserve mention. Deformable secondary mirrors (Salinari and Sandler 1998) have an advantage over other systems because they subject the photons to no additional reflections, nor do they introduce additional thermal emission into the beam (Lloyd-Hart 2000). Thus, for systems operating in the thermal infrared, they have significantly lower background and somewhat higher throughput than other types of DM. A modest price is that the DM is close to but not exactly at the telescope exit pupil (see example in Section 2.3.2). DMs built around micro-electronic mechanical systems (MEMS) may be able to achieve a breakthrough in cost (Cornelissen *et al.* 2010). They are constructed by etching silicon (termed "silicon micromachining") to provide a thin membrane mirror that is bent by electrostatic forces imposed by parallel conductive plates. They can be manufactured efficiently because they do not involve the large number of discrete parts that must be assembled to construct a conventional DM. One limitation is that they are small, for example 25 mm square for 4096 segments. When used with a large telescope, according to the Lagrange Invariant the angles of incidence onto the DM are large, limiting the field over which good corrections to the wavefront can be obtained.

Between the wavefront sensor and the DM comes a substantial amount of computing power. The computer is often called a reconstructor because it must take the output of the wavefront sensor and reconstruct the phase, and then compute a set of appropriate electrical voltages and currents to the actuators on the DM so it assumes the shape required to correct the wavefront. This process must occur in a few milliseconds; fortunately, the computations lend themselves to highly parallel architectures to help achieve the necessary speed.

The AO system must be designed so that this problem is well-determined. For example, if there are fewer measurements over the wavefront than there are actuators to move, the solution is underdetermined – there may be a number of solutions and the reconstructor will not necessarily converge to the right one. This situation can be improved by using the wavefront sensors to determine large-scale modes, such as the Zernike functions. However, most systems are designed with more wavefront measurements than actuators, so they are overdetermined and there is a single best solution for the reconstructed phases. To make the job of the reconstructor tractable the sensor subapertures are aligned with the DM actuators; there are a number of standard layouts. A critical calibration for this process on a continuous faceplate mirror is to determine the influence function for each actuator – that is, to measure exactly how the wavefront sensor reacts to deformations the actuator imposes on the mirror surface. This information can be summarized in a linear algebra matrix. Another matrix relates the errors measured at M positions (e.g., the subapertures in a Shack–Hartmann sensor) to the overall wavefront error. Still another matrix captures the actions that need to be taken by N actuators to apply the necessary corrections. The reconstructor can then use the methods of linear algebra to manipulate these matrices to determine the correction signals.

The performance of an AO system is typically quoted in terms of the Strehl ratio, SR, or just "Strehl." In the chapter on telescopes, we cited the Maréchal condition (equation (2.2)); in AO, units of radians are often used, in which case

$$S = e^{-\sigma_{rad}^2} \tag{7.14}$$

where $\sigma_{rad}^2 = \sigma_1^2 + \sigma_2^2 + \sigma_3^2$ is the quadratic combination of all the sources of wavefront error. (This approximation is only reasonably accurate for $S \gg 0.1$.) The optimization of a system to get a high Strehl ratio then depends on minimizing a set of errors:

1. σ_{fit} is the error in the reconstructed wavefront due to the limited number of subapertures and is given by $\sigma^2 \propto C\,(d/r_0)^{5/3}$, where d is the subaperture diameter and the proportionality factor depends on details of the deformable mirror design but is of order $C \approx 0.2$;

2. σ_{photon} is the error due to counting statistics on the NGS and is given by $\sigma^2 \propto 1/N$, where N is the number of photons collected and the proportionality factor depends on the efficiency of the detector and optical system feeding it;

3. σ_{delay} is the error associated with the non-instantaneous time response of the corrective system and is given by $\sigma^2 = 28.4 \, (t/t_G)^{5/3}$, where t and t_G are respectively the response time and the critical time $= 1/(2\pi f_G)$; and

4. σ_{iso} is the error due to incomplete isoplanatism and is given by $\sigma^2 = (\theta / \theta_0)^{5/3}$ where θ is the angular distance from the guide star (Hardy 1998).

All of these dependencies are strong, so all of them must be taken into full account in optimizing a system.

There may be additional errors due to parts of the optical path that differ between the wavefront sensor and the science instrument (called non-common-path errors, e.g. within the science instrument in Figure 7.7, or as in Figure 7.4). The system must also be able to operate correctly on enough guide stars that it has scientifically useful sky coverage, setting upper limits on the number of subapertures and the frequency of the corrections. Thus, optimizing a NGS AO system involves a complex series of scientific and technical tradeoffs.

7.5 Enhancements to NGS AO

7.5.1 Laser guide stars

Satisfying the constraints on sampling frequency and the number of corrected areas requires use of a relatively bright guide star. The exact magnitude, of course, depends on a lot of details, but it is likely that your favorite target will not be within the isoplanatic patch for any suitable NGS, unless it is one of these stars itself. For example, if the star must be brighter than visible magnitude 12 and the radius of the isoplanatic patch is 20 arcsec (an optimistic value at 2 μm), then the system will have access to less than 1% of the sky. The solution is to make an artificial star wherever it is needed, using a powerful laser.

There are two basic types of laser guide stars. In both, a powerful laser beam is projected up along the direction the telescope is looking. A Rayleigh scattering guide star is created with a pulsed laser; a tiny fraction of the laser light is scattered back into the measurement beam by the atoms and molecules in the atmosphere. The scattering is all along the column of light but a region can be selected by gating the response of the wavefront sensor to just a

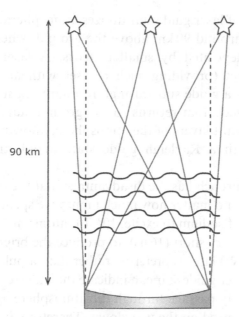

90 km

Figure 7.10. Use of multiple laser guide stars to probe the turbulence over the entire column of air above a telescope.

short period of time with the appropriate delay after the laser pulse has been launched. The return signal is proportional to the laser output times the scattering cross-section times the density of scattering molecules divided by the square of the range, $z \sec\beta$ (Hardy 1998):

$$N_{\text{signal}} \propto \frac{N_{\text{laser}}}{(z \sec \beta)^2} \sigma_B^R n_{\text{mot}} \propto \frac{N_{\text{laser}}}{(z \sec \beta)^2} \left(\frac{P(z)}{T(z)} \lambda^{-4.0117} \, \text{m}^{-1} \, \text{sr}^{-1} \right) \quad (7.15)$$

where $P(z)$ and $T(z)$ are the pressure and temperature at altitude z, respectively. The behavior is basically λ^{-4} Rayleigh scattering with the additional temperature and pressure terms. $P(z)$ falls exponentially with z (with a scale height of about 8 km), so this type of guide star can probe only the lower layers of the atmosphere, say up to 10 km. In addition, the return beam is in the form of a cone with its tip at the position of the guide star and its base at the telescope primary mirror (Figure 7.10). Thus, the cylinder of light from the astronomical source is not covered to outside the atmosphere, and the distortions imposed outside the cone from guide star to telescope are also not sensed, leading to "cone error." In compensation, a very powerful laser (preferably operating in the blue to take advantage of the λ^{-4} dependence of the scattering cross-section) can make the guide "star" bright.

The second type of laser guide star utilizes atmospheric sodium, in a layer in the mesosphere around 90 km above the ground, where the sodium and other metals are deposited by small meteors. A laser tuned to the D_2 transition of sodium (providing such a laser with adequate power is a challenge) produces a guide star through resonant scattering (i.e., by emission when the excited atom returns to the ground state). This beam does traverse all the relevant layers of the atmosphere, allowing a more complete correction than with a Rayleigh guide star. In addition, cone error is reduced.[14]

However, this approach has the disadvantage that the artificial star can be made only so bright, no matter how much money we spend on the laser; there is a limited amount of sodium to excite, 10^3–10^4 atoms cm^{-3}. Since the sodium de-excitation time is very short (16 ns), to produce the brightest possible star a continuous wave (CW) laser is preferred rather than a pulsed one.

Neither type of laser guide star can indicate the tilt errors, because the laser light makes a two-way passage through the atmosphere and the tilt imposed on the way up is reversed on the way down. Therefore, it is still necessary to use a tilt sensor on a natural star to stabilize the image. However, the necessary bandwidth of these corrections is relatively low; even more importantly, with the corrected wavefront the entire telescope aperture is effective in forming the image for measuring the tilt. In addition, the tilt correction is valid over the isokinetic angle – typically a few arcminutes at 2 μm and thus many times larger than the isoplanatic angle. As a result, guide stars that are both faint and relatively far from the science target can be used for the tilt correction, opening up the majority of the sky for access.

However, we have to tolerate some degradation of performance compared with NGS systems. The laser beam is distorted by the atmosphere on the way up, so the artificial guide star is not a point source, but has a typical seeing-limited size of 0.5 to 2 arcsec, potentially increasing the wavefront sensing errors. Also, because the artificial stars are not sufficiently far from the telescope, there is a focus offset and the returning wavefronts are spherical rather than planar. Consequently the turbulence at the edges of the pupil is not sampled well.

7.5.2 *Multiple guide stars*

So far, we have discussed correcting wavefront errors accumulated over a three-dimensional volume of the atmosphere by use of a small number of

---

[14] The wavefront variance due to the cone effect scales as $D^{5/3}$ where D is the telescope aperture (Hardy 1998), so with really large telescopes ($D > 10$ m) cone error is still an issue.

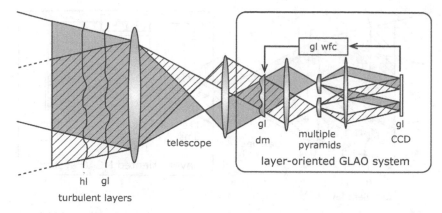

Figure 7.11. A layer-oriented ground layer adaptive optics (GLAO) system. gl ground layer; hl high layer; wfc wavefront control; dm deformable mirror. After S. Egner (2006), with permission.

guide stars and with a planar deformable mirror. These approaches are now in general use at a number of observatories. Current research addresses the improvements that can be made by recognizing the full three-dimensional complexity of the atmosphere-induced wavefront errors. For example, a constellation of lasers would allow a whole range of improvements in the adaptive corrections, not only eliminating cone error (Figure 7.10) but also making corrections appropriate for the individual turbulent atmospheric layers. A separate DM could be placed at an image of each atmospheric layer that is producing wavefront distortions needing correction, that is, at the conjugate image of this layer.[15] In this way, multiple laser guide stars could provide larger fields of view and allow imaging of extended sources. These improvements are not only central to current research; in addition, sharp images obtained in this manner underlie plans for very large future telescopes.

As an example, Figure 7.11 shows an AO system to correct the effects of low-lying layers of the atmosphere. It is assumed that the wavefront sensors are of the pyramid type. A number of them are used, one for each guide star. Their outputs are brought to a single detector focused to the low-lying atmosphere. The multiple guide star signals are used to isolate the turbulence of this ground layer from the effects of the rest of the optical path. The resulting correction for the ground layer is then fed to a single deformable mirror at an optical position that is conjugate with the low-lying atmosphere. Since more than half of the overall wavefront distortion is usually associated

---

[15] According to AO lingo, saying that "a mirror is conjugated to an atmospheric layer" translates into "an image of an atmospheric layer is formed on the surface of a mirror." For ground layer adaptive optics (GLAO), the DM is placed at an image of the layer to be corrected.

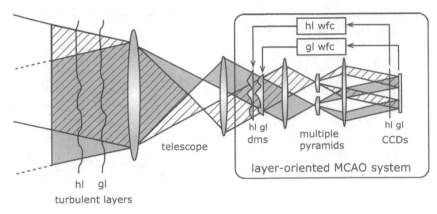

Figure 7.12. A layer-oriented multi-conjugate adaptive optics (MCAO) system. gl ground layer; hl high layer; wfc wavefront control; dm deformable mirror. After S. Egner (2006), with permission.

with these low-lying layers, there is a significant improvement in the image quality by making this type of correction. Since the light over a large field of view still passes through nearly the same path in these layers, the resulting improvement is maintained over fields a number of arcminutes in diameter. However, the Strehl is low for these systems because the upper atmospheric layers are completely uncorrected.

Improving the overall performance further requires correcting additional atmospheric layers. Doing so requires multiple DMs conjugated optically to the appropriate atmospheric levels, plus expanding the wavefront sensor optics to multiple detectors, each focused to a different atmospheric layer and controlling the appropriate deformable mirror (Figure 7.12). For obvious reasons, this approach is termed layer-oriented multiple conjugate adaptive optics (MCAO).

An alternative approach is to employ a complete wavefront sensor train for each of a number of guide stars (Figure 7.13) as in a traditional AO system. This information is applied through tomography, a term that means building up an image in layers. With multiple guide stars, we can obtain different projections of the turbulence that together cover the entire column of air of interest (Figure 7.10). This information can be utilized with the Projection-Slice Theorem: each piece of projection data at some angle is the same as the Fourier transform of the multidimensional object at that angle.[16] Applying this theorem to measurements from a range of angles, one can reconstruct an

---

[16] This powerful approach was invented by Bracewell (1956) for interpretation of strip scans of images in radio astronomy.

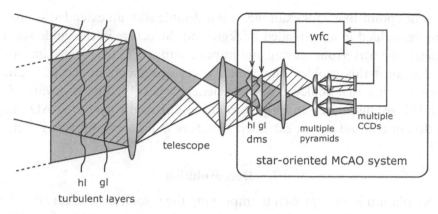

Figure 7.13. Star-oriented MCAO. gl ground layer; hl high layer; wfc wave-front control; dm deformable mirror. After S. Egner (2006), with permission.

image of the atmospheric turbulence. Bello *et al.* (2003a, b) compare these two approaches to MCAO.

MCAO is being developed to expand the full well-corrected field beyond the simple isoplanatic patch. Another approach, multi-object adaptive optics (MOAO), has a different goal of providing well-corrected images at a suite of subfields within a larger field, but without correcting the wavefronts between these points. It is made technically attractive because of the availability of MEMS DMs at modest cost and complication. It can address feeding a set of spectrometers simultaneously with diffraction-limited images to improve their sensitivity. In MOAO, multiple stars are used for wavefront sensing and to build a complete picture of the turbulence over the telescope through tomography. The correction for the specific line of sight for a DM is extracted from this result and used to correct the wavefronts in the subfield.

7.6 Cautions in interpreting AO observations

There are a number of conditions that can undermine the reliability of AO measurements. The first is when only a low Strehl ratio is achieved (say < 10%) in a conventional (not GLAO) system. The ability of the system to concentrate energy into the central part of the image is likely to be variable, and if the amount of concentration achieved is low the relative amount of the variations can be large – that is, photometry obtained with a low Strehl is likely to have large errors. In addition, the point spread function is likely to have a variable shape that may undermine conclusions based on the observed source structure. A second type of systematic problem occurs when the guide star is not point-like – for example, it is a double star, an extended source, or a star embedded in an extended background. Structure in the guide star can influence the wavefront sensing and impose artifacts in the science image. To guard against this problem, independent measurements of the PSF can be compared with that achieved on the science target. Finally, be aware of the extended, seeing-limited envelope around any partially corrected AO image, and do not confuse it with extended structure in an astronomical object.

7.7 Deconvolution

Deconvolution is an approach to improving the resolution of data after it has been obtained rather than in real time as with adaptive optics. A variety of methods can be used to process high-signal-to-noise images or spectra to enhance their resolution. For simplicity, we describe the situation in one dimension, as is appropriate for a spectrum; the arguments can be readily extended to two-dimensional data such as images. We describe the true spectrum as $\Psi(\lambda)$ (a function of wavelength λ), the observed one as $O(\lambda)$, and the instrumental line profile (analogous to the PSF) as $P(\lambda)$. Then $O(\lambda)$ is the convolution of $\Psi(\lambda)$ and $P(\lambda)$:

$$O = \Psi \otimes P \tag{7.16}$$

It is far more convenient to work in Fourier space, where

$$[O(\zeta)] = [\Psi(\zeta)][P(\zeta)] \tag{7.17}$$

from the Convolution Theorem, ζ is the "spatial" frequency characterizing the structure of the spectrum, analogous to the spatial frequency for imaging, and $O(\zeta)$, $\Psi(\zeta)$, and $P(\zeta)$ are the Fourier transforms of $O(\lambda)$, $\Psi(\lambda)$, and $P(\lambda)$. The spectral information has been degraded by the line profile, which in general will be deficient in fine spectral structure, that is, at high values of ζ. The situation is analogous to the attenuation of high spatial frequencies for the telescope PSF as shown by the MTF in Figure 2.13. The true spectrum can be recovered as

$$\Psi(\lambda) = F^{-1} \left[\frac{O(\zeta)}{P(\zeta)} \right] \tag{7.18}$$

Unfortunately, we introduce three problems if we take this step. First, the abrupt termination of spatial frequencies at some maximum of ζ (analogous to the maximum spatial frequency for a telescope, D/λ) may cause ringing in the deconvolved spectrum. Second, the noise spectrum is usually independent of ζ, so when we divide $O(\zeta)$ by the Fourier transform of the line profile we amplify the noise substantially at high frequencies where the values in the Fourier transform of $P(\zeta)$ are small. Third, if there are any errors, even very small, in our determination of $P(\lambda)$ at the fine structure (large ζ) limit where it is small and the correction in equation (7.18) is large, they get amplified into substantial spectral artifacts.

The usefulness of deconvolution depends on the development of approaches that provide improvements in some aspect of an image while avoiding ringing, suppressing noise, and minimizing artifacts. If this goal sounds a little ill-defined, it is because there is a large parameter space of goals for image improvement, acceptable levels of artifacts and ringing, and final signal-to-noise ratio. Consequently, there are many approaches to deconvolution.

The most reliable form of deconvolution is based on fitting parameters, assuming that the parameters are well chosen to describe the aspect of the image that is of interest. We have already seen one example in the discussion of astrometry. If the PSF is well-determined and the image is a field of point sources, so no free parameters other than brightness need to be used to describe each one, then a source position (described by two parameters) can be determined to an accuracy roughly corresponding to the FWHM of the image divided by the signal-to-noise ratio. Such positions can be substantially more accurate than the diffraction limit of the telescope. Simple parametric modeling can be applied in many other circumstances. Examples include estimating the diameter of a uniform brightness disk, or fitting the properties (inclination and diameter) of a disk around a central point source.

However, in many cases the object cannot be described by a limited number of parameters, or, even if we think it can, we want to avoid biases that we might introduce with parametric modeling. We might then build on the example at the beginning of this section by using some kind of filter to control the noise amplification and soften the abrupt cutoff that produces ringing. The Wiener filter, $W(\zeta)$, has been shown to be the optimum choice:

$$\Psi(\lambda) = F^{-1}[W(\zeta)\,O(\zeta)] = F^{-1}\left\{\frac{P^*(\zeta)}{P(\zeta)P^*(\zeta) + 1/\mathrm{SNR}(\zeta)}O(\zeta)\right\} \quad (7.19)$$

where SNR is the ratio of signal to noise. By inspection, where the SNR is large, this procedure becomes identical to equation (7.18). However, at frequencies where the SNR gets smaller, the strength of the deconvolution decreases.

Although the Wiener filter addresses some of the issues with simple Fourier deconvolution, it provides only a one-try approach and is too inflexible to support the improvements possible by generating a series of models of the target and testing which ones are most consistent with the noise properties. A large variety of iterative approaches have been introduced to overcome these shortcomings. They have the further advantage, shared with parametric modeling, that they can utilize information above the spatial frequency cutoff of the optical system. In general, however, they introduce another issue: the definition of the best model is ambiguous – does it mean minimum artifacts, minimum ringing, minimum noise, or some global minimum over all three? All iterative deconvolution procedures try to resolve this dilemma by imposing some a priori conditions on the solution. Convergence is assisted immensely by a few very non-restrictive ones, such as that the image must always be positive, and that fringes should be suppressed. Nonetheless, the selection of a "best" solution retains a degree of arbitrariness.

The approach with the best claim to being non-arbitrary is the Maximum Entropy Method (MEM) (Narayan and Nityananda 1986). It is based on deriving the smoothest deconvolution that is consistent with the input data within the noise. The definition of smooth is derived by analogy with entropy in statistical physics. In this case, if $I(x,y)$ defines an image, its entropy is

$$S = \iint f[I(x,y)] \, dxdy, \tag{7.20}$$

The minimization of S is determined by χ^2 calculation. Two forms have been used for $f(I)$:

$$f_1(I) = \ln(I)$$
$$f_2(I) = I \ln(I) \tag{7.21}$$

The performance of MEM seems not to be affected by which is used. Narayan and Natyanandan (1986) demonstrate that both forms share important characteristics: (1) they do not permit negative values of I; (2) given a single value of flux for the entire image, they force the image to be uniformly illuminated as required by the definition of maximum entropy; and (3) their negative second derivatives suppress fringing.

The MEM can be taken to provide a conservative deconvolution. This advantage comes with the disadvantage that the amount of improvement in the resolution is limited. In addition, MEM can propagate data defects over the entire deconvolved image. The amount of resolution improvement is non-uniform, with maxima on peaks of the image where the signal-to-noise ratio is highest. Moreover, MEM tends to work poorly on point sources, particularly those embedded in extended emission – which can be merged into an extended object.

A number of alternatives have various advantages and disadvantages. For example, pixon deconvolution models images from a library of pseudo-images (Peutter *et al.* 2005). As a result, it can base its deconvolution on a variable number of pixels according to the local information density – using larger pixons where the image is smooth, for example. It avoids noise amplification, but suppresses the noise so strongly that sources that are not readily apparent in the input image tend not to appear in the deconvolved one. That is, bright objects are restored well, but faint ones tend to disappear in the output image. In Chapter 9 we will discuss CLEAN, which deconvolves under the assumption that the image consists of an ensemble of point sources. It has excellent noise performance for such sources, but not surprisingly introduces significant artifacts in extended sources due to its pixel-by-pixel operation without consideration of surrounding structure. The Lucy–Richardson method defines a log-likelihood function from the elements of the model image, M_i, and the corresponding data points, D_i:

$$\Lambda = 2 \sum_i (M_i + D_i \ln(M_i)) \tag{7.22}$$

which it minimizes by making iterative multiplicative corrections (Puetter *et al.* 2005). It conserves flux, does not introduce negative sources, and converges well. The calculation converges to the maximum likelihood solution if the noise is Poisson-distributed (Shepp and Vardi 1982). However, artifacts appear if it is carried through too many iterations (e.g., ringing on sharp features such as point sources), or if the signal-to-noise ratio is low (in which case, it tries too hard to fit the noise).

There are many variations on the approaches listed above, in some cases customized to specific applications. It is often desirable to deconvolve data in more than one way to judge the reliability of any marginal features.

7.8 High-contrast imaging

With the focus on detection of planets orbiting nearby stars, technical means for high-contrast imaging – detecting extremely faint objects very near to bright ones – are under rapid development. These techniques also have other applications, such as studying the environments of bright active galactic nuclei. Their power has grown immensely with the development of adaptive optics that can deliver diffraction-limited images as a starting point for high-contrast imaging. We will discuss two basic approaches: (1) apodization and phase manipulation; and (2) coronagraphy.

7.8.1 Apodization

The prominent rings in the Airy function arise because of the abrupt termination of the spatial frequency spectrum transmitted by the telescope, corresponding to the edge of the primary mirror. These rings are obviously detrimental to high-contrast imaging. They can be reduced in amplitude by reducing the weight of the outermost zone of the primary mirror in forming the image. We could achieve this goal by grading the reflectivity of the mirror so that it gradually became less and less with increasing radius. In this example, the FWHM of the central image increases, since we are suppressing the large baselines that are responsible for the highest-resolution part of the image. However, we can almost completely suppress the diffraction rings. This process is called apodization. Since other users of the telescope would object if we actually reduced the reflectivity of the primary mirror, apodization is carried out by reimaging the primary to a pupil, where a suitable mask can be placed. The design of the pupil apodization mask can produce a variety of beam weightings over the telescope primary.

We will illustrate this behavior with an example. As discussed in Section 2.3.4., the point spread function is the square modulus of the field amplitude, EE^*, where E is the Fourier transform of the distribution of the electric field of the signal, $E(x)$, onto the telescope:

$$E(\phi) = \int E(x)e^{-j2\pi x\phi}\,dx \tag{7.23}$$

We will contrast the case of a one-dimensional uniformly illuminated aperture (which was derived starting from equation (2.23)) with an aperture illuminated with a triangle function, $\mathrm{Tr}(x)$. The Fourier transform of $\mathrm{Tr}(x)$ is $F(u) = \mathrm{sinc}^2(u)$. Therefore, equation (7.23) becomes

$$E(\phi) = \int_{-\infty}^{\infty} \mathrm{Tr}(x)e^{-j2\pi x\phi}\,dx = \mathrm{sinc}^2(\phi) \tag{7.24}$$

The corresponding PSF is proportional to the fourth power of the sinc function (Figure 7.14), with greatly reduced energy into the diffraction rings. The width of the central maximum has been increased as a consequence.

Slepian (1965) demonstrated that a prolate spheroid weighting maximized the concentration of energy into the central response and minimized the diffraction rings. Outside a radius of about $4\lambda/D$, where D is the telescope aperture, apodization with this function can result in intensities of 10^{-10} or less compared with the central intensity of an image (Figure 7.15). However, manufacturing a mask with this performance is challenging.

One approach to ease the mask manufacturing issues is to use a binary mask, in which appropriately shaped holes are cut in an opaque

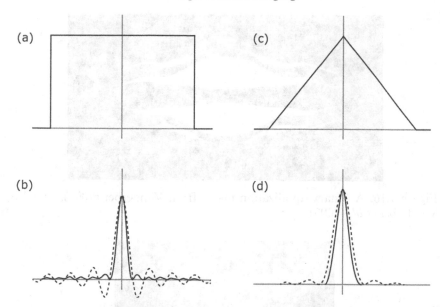

Figure 7.14. Illumination patterns compared with the resulting distribution of electric field and point spread functions. (a) is uniform illumination, and (b) shows the resulting field (dashed) and PSF (solid). (c) shows illumination that has been apodized to have a triangle distribution; (d) shows the resulting field (dashed) and PSF (solid).

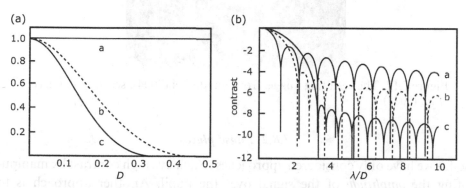

Figure 7.15. Effects of prolate spheroidal apodization on the PSF. Panel (a) shows the weighting over the pupil, in units of radius running from 0 to 0.5 D. Panel (b) shows the resulting PSFs plotted radially as a function of λ/D. Curve a is no apodization; curves b and c show increasingly strong prolate spheroidal apodization. After Aime (2005).

sheet to allow the correct weighting over the pupil. Such a mask produces similar performance in one direction to the prolate spheroid apodization; probing around a source requires rotating the mask (Figures 7.16 and 7.17).

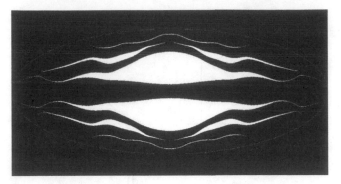

Figure 7.16. A binary apodization mask, from Vanderbei (2004); see also Vanderbei *et al.* (2003).

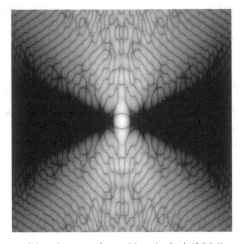

Figure 7.17. The resulting image, from Vanderbei (2004); see also Vanderbei *et al.* (2003).

7.8.2 Phase plates

So far we have only considered approaches to achieve high contrast by manipulating the *amplitude* of the signal over the pupil. Another approach is to manipulate the *phase* (Codona and Angel 2004; Yang and Kostinski 2004), and in many cases doing so can provide high contrast at greater efficiency (since manipulating the amplitude usually involves losing some portion of the signal). A generic term for the most common class of device to work with phases is an apodizing phase plate (APP). An APP can be used to shift the diffraction pattern so that the rings in the Airy pattern are canceled in a zone around a star, while only modestly affecting the Strehl of the images (Figure 7.18). Searches for faint companions are conducted in these dark zones; the APP is then rotated to shift the search region to another zone around the star.

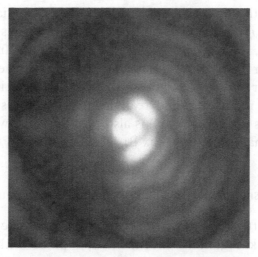

Figure 7.18. Diffraction pattern around a star with the use of an apodizing phase plate (Kenworthy *et al.* 2010; see also Quanz *et al.* 2010). Reproduced by permission of the AAS.

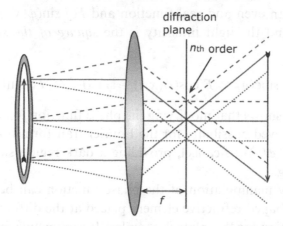

Figure 7.19. Concentration of diffraction artifacts at the diffraction plane.

To describe the function of the APP, we concentrate on the diffraction plane (see Figure 7.19), since the diffraction artifacts in the image are optimally concentrated there. We have illustrated this behavior previously in discussing diffraction gratings (Figure 6.2 and surrounding discussion). In the current case, we will derive the behavior in one dimension (following Yang and Kostinski 2004).

If the transmission function of the system pupil is given by $T(x)$, then the illumination field at the diffraction plane is

$$E(\eta) = F(T(x)e^{j\phi(x)}) = F(T(x)) \otimes F(\cos(\phi(x)) + j\sin(\phi(x))) \qquad (7.25)$$

where F denotes the Fourier transform, $\phi(x)$ is the spatial phase over the pupil, and x and η are the (one-dimensional) coordinates in the pupil and diffraction planes, respectively. The second expression is from the Convolution Theorem. For a uniformly illuminated pupil of diameter D, $T(x) = 1$ for $|x| \leq D/2$ and 0 for $|x| > D/2$. Its transform is sinc η, which is an even and real function. If $\phi(x)$ is an even and real function, then a standard result in Fourier analysis is that $F(\cos(\phi(x))$ is also even and real, while $F(j\sin(\phi(x)))$ is even and imaginary (see Appendix B). The light intensity on the diffraction plane is then

$$I(\eta) = EE^* = [\text{sinc}(\eta) \otimes F(\cos(\phi(x))]^2 + [\text{sinc}(\eta) \otimes F(\sin(\phi(x))]^2 \quad (7.26)$$

The form of $I(\eta)$ (the *sum of the squares* of two real functions) shows that generating a dark region in the diffraction plane with an even phase function requires that each of the Fourier transforms (of $T(x)\cos(\phi(x))$ and of $T(x)\sin(\phi(x))$) must be very small, a difficult requirement to meet. However, this derivation also indicates that when $\phi(x)$ is an odd function, $F(\cos(\phi(x)))$ is an even and real function and $F(j\sin(\phi(x)))$ is an odd and *real* function, and the light intensity is the *square of the sum* of two real functions

$$I(\eta) = EE^* = [\text{sinc}(\eta) \otimes F(\cos(\phi(x))) + \text{sinc}(\eta) \otimes F(j\sin(\phi(x)))]^2 \quad (7.27)$$

Equation (7.27) shows that modifying the phase function to be odd opens the more easily achieved possibility for the Fourier transforms of $T(x)\cos(\phi(x))$ and of $jT(x)\sin(\phi(x))$ to cancel, providing a dark region suitable for high-contrast imaging.

The necessary manipulation of the phase function can be provided by a slightly wedge-shaped refractive element placed at the diffraction plane. The design optimization for this plate is complex. It can include features to cancel artifacts such as the diffraction resulting from the secondary mirror support and flaws in the telescope optics. To understand how image artifacts can be suppressed, imagine that we expand a mirror flaw into Fourier components. Each such component can be considered to be a physical ripple-like structure that diffracts light in specific directions, creating speckles in pairs distributed symmetrically about the optical axis. If an appropriate ripple structure is put at the appropriate position on the APP, it can manipulate the phase so that one of these pairs is canceled, at the expense of doubling the intensity of the other, that is, it can shift the effect of the artifact entirely to one side of the optical axis. Such optimization produces the very clean high-contrast region shown in Figure 7.18.

7.8.3 Coronagraphy

The Lyot coronagraph. A second tool for high-contrast imaging is based on the principle of never letting the bright light from the central source enter the instrument. This light can contaminate the signal either by scattering and diffracting, or by over-stressing the detector array so that bleeding or some other form of charge leakage occurs.

The most direct approach would be to place an occulting mask far in front of the telescope and along its optical axis to block the direct light from the star. To avoid light diffracted by the occulter entering the telescope, the mask must be significantly larger than the telescope aperture. To allow imaging at the telescope diffraction limit, it must be far enough away that it is unresolved. These two constraints lead to concepts like a 50-m diameter occulter placed 50 000 km in front of the telescope. From these parameters, the approach would only be feasible in space. A simple round occulter would create a bright central spot (the spot of Arago) due to diffraction, but this problem can be solved with a complex edge shape. For good performance, the shape of the occulter must be optimized and controlled accurately (to about 1 mm at the edge) and it must be kept accurately in the correct position (to within about a meter), requiring very precise station keeping between the telescope and the occulter satellite. The benefits would include the ability to look at high contrast very close to the star without using a heavily optimized telescope and sophisticated instrument, but there are clearly significant practical engineering difficulties that need to be overcome.

The classic Lyot coronagraph is a more easily implemented (but probably lower performance) way to improve the contrast around a bright source. As shown in Figure 7.20, light enters the telescope (represented by a lens) from the left, uniformly illuminating the telescope aperture. The telescope forms an image, and most of the light from the central object can be blocked from entering the instrument by placing the image on a small occulting spot. This spot takes the place of the large occulter far in front of the telescope. Nonetheless, some extraneous light escapes: (1) as the diffraction pattern associated with the telescope aperture; (2) as scattering and diffraction from structures in the beam entering the telescope, for example diffraction from the supports for the secondary mirror; and (3) due to diffraction at the occulting spot. To remove at least some of this unwanted light, the telescope entrance pupil is reimaged, where a mask is placed. The first of these sources of extraneous light can then be mitigated by appropriate treatment of the pupil mask to apodize the aperture. The mask can also be made to block the view of the secondary supports and other structures within the telescope. Finally, the

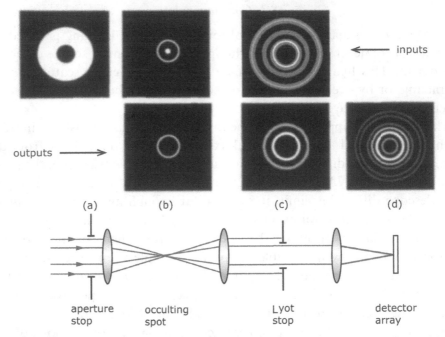

Figure 7.20. Layout of a classic Lyot coronagraph. The input line shows (a) the uniformly illuminated telescope (with central obscuration); (b) the diffraction-limited image and as output the same image with the central peak removed by the focal plane mask; (c) the illumination pattern input to the pupil, and as output the pattern after masking the outer zone; and (d) the output to the detector.

diffraction from the occulting spot appears at the outer zone of the pupil and can be blocked there.

Of course, each of these mitigations loses light. Improvements over the classical Lyot approach center on achieving better compromises between the rejection of unwanted light from the central source and the throughput of the instrument for the signal from faint nearby objects. Some possibilities are discussed below, after some definitions.

Coronograph lingo. To expand on the characterization of coronagraphs, we need to define the following terms.

The *throughput* is the ratio of the light received at the detector to the light into the coronagraph. It generally depends on the radial distance from the center of the coronagraph field and may have more complex behavior.

The *inner working angle* (IWA) is the minimum angular separation between a faint source and the bright one being suppressed by the coronagraph. The IWA is expressed in units of λ/D and is usually defined as the point where the source throughput is 50% of the maximum throughput.

The raw contrast, or just the *contrast*, is the ratio of local surface brightness to peak surface brightness of the point spread function (i.e., the bright source).

The *coronagraphic rejection* is the central brightness of the bright source divided by the brightness of its image through the device.

The *detection contrast* is a similar parameter after all the possible tricks have been employed to remove the residual signal (e.g., taking multiple images under different conditions and subtracting them from each other to remove residual stray light without removing the light from the faint source).

The *null order* (example: fourth-order null coronagraph) describes the coronagraph throughput as a function of angular separation close to the optical axis. In general, the higher the null order the deeper and wider is the region with good suppression of the central source and the more immune the performance is to residual pointing error and stellar angular size, but also the larger the IWA.

The *angular resolution* can be considered to be the full width at half maximum of the image delivered to the coronagraph detector.

The optimization of a coronagraph centers on these terms. We want the highest possible throughput, the smallest inner working angle, the highest contrast, and the greatest immunity to pointing errors (i.e., a high null order) while preserving the basic resolution of the telescope. As usual in life, we can't have it all, and coronagraph designs always involve painful tradeoffs among these parameters. We discuss three modifications of the classic Lyot concept to illustrate some of the improvements that are possible.

Phase mask coronagraphs are designed to reduce the IWA. In the classical Lyot coronagraph, there is a sharp cutoff on the IWA, determined by the radius of the occulting spot. However, the occulting spot can be replaced by a mask that imparts phase differences in different parts of the source wavefront, so when re-combined into an image, the light interferes destructively. A simple implementation is to put an optical element where the image will be formed that retards the phase by π in two opposite quadrants. If a monochromatic source is placed exactly at the center of the resulting four-quadrant phase mask, the rejection is formally complete. These devices are not achromatic, however, and generally operate with spectral bandpasses of about 10%, and with rejection by about two orders of magnitude. There are various implementations besides the four-quadrant phase mask, including round retarding regions and spirally tilting ones called optical vortices (Figure 7.21).

Band-limited coronagraphs use an occulting spot designed to limit the area where the light from the bright source falls at the pupil, so the Lyot stop need block minimal area, therefore improving the throughput. The operating

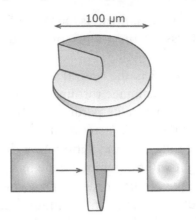

Figure 7.21. A transparent plate formed into an optical vortex. In the lower part of the figure, the input image to the left has had its central component removed by interference, to the right.

principle can be understood by considering the performance of a telescope with its primary mirror masked off except for two small round apertures opposite each other and near the edge of the mirror. The resulting image will be the Airy pattern corresponding to the diameters of the apertures, with interference fringes imposed upon it at the spatial frequency corresponding to the separation of the apertures. This arrangement is a basic interferometer. For our current purpose, however, we observe what would happen if we could reverse the direction of time and the photons at the focal plane flowed to the primary mirror and from there out into space. If we reproduced the exact same spatial distribution and phases of the photons as in the forward-time situation, we would expect only the two small apertures to be illuminated. If we form a pupil, we should find that it is illuminated only at the images of the two apertures.

In the band-limited coronagraph, the occulting spot is designed to impose the basic interferometer pattern (or an equivalent one) on the image of the bright object. The light is then directed at the pupil to specific areas, just as in our time-reversal experiment. The Lyot stop removes this light. The light from other objects in the field but away from the occulting spot is distributed over the entire pupil and can pass through at high efficiency.

Phase-induced amplitude apodization coronagraphs

An ideal approach would combine (1) the removal of the light of the dominant source with (2) apodization to reduce the artifacts outside the central image peak. The phase-induced amplitude apodization (PIAA) coronagraph (Figure 7.22, Guyon 2003) is an example; moreover, it provides the advantages of apodization without the usual loss of light. It uses aspheric reflective

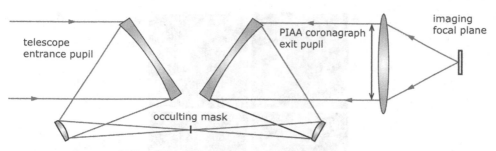

Figure 7.22. Basic layout of the PIAA coronagraph.

optics to form a pupil on the second mirror in Figure 7.22, which remaps the light into roughly a Gaussian distribution. Something has to be compromised in this process, and it is the off-axis image quality, which has strong coma since the remapping cannot be done without seriously violating the Abbe sine condition (equations (1.21) and (1.22)). Nonetheless, the on-axis image is adequate to allow a mask to remove the light from the very center of the field, that is, from a star. Thereafter, an additional set of optics is required behind the focus to correct the wavefronts to give acceptable images over a useful field (Figure 7.23). This approach not only avoids losing light, but also preserves the full resolution of the telescope (delivered image diameters near the field center are $\sim\lambda/D$), and provides a small IWA. The major issue is that the mirror surfaces require very large curvature at the edges, making them difficult to manufacture sufficiently accurately.

7.8.4 When a coronagraph is advantageous

A coronagraph can be useful simply for reducing the light of the bright central source in the image, thus making less demand on the dynamic range of the detector and possibly eliminating artifacts such as bleeding of the signal. However, under some circumstances much greater gains in contrast between the bright source and nearby target can be achieved. We now consider the conditions for such gains.

Assuming a perfect telescope in the seeing limit, the image structure is determined by speckles. When the wavefronts are partially compensated, a diffraction-limited core image will appear. This image is made of the identical photons responsible for the speckles – it can be viewed as a sort of super-speckle. Thus it interferes freely with the ordinary speckles. We can approximate the situation by considering the total image to consist of a portion that we maintain in a static form through wavefront correction, and a portion that varies due to uncompensated variable wavefront distortions. These two

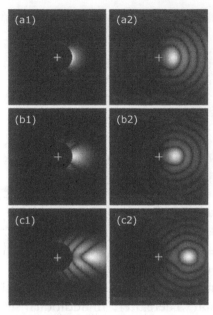

Figure 7.23. Images in the PIAA at the first focal plane (left) with the occulting mask, and at the second focal plane (right) after restoration of the wavefronts. Cases (a) through (c) are for increasing separation of the target object and the bright central source.

components can be represented as the sum of two complex terms, which we will describe as the diffraction and speckle terms respectively. The intensity is the square of the absolute value of the amplitude of this two-component signal. It contains the cross-product of the diffraction and speckle terms, representing the interference between these signal components. Thus, the net image is a complex and variable interference pattern in the regions where the two components are of comparable strength, for example in the zone around the central peak where the bright diffraction rings lie. Where the interference is constructive, there is coherent amplification of the speckles by the diffraction pattern in a phenomenon called speckle pinning. Even where there is little or no signal from the diffraction term, the speckles exhibit correlated behavior and do not average out as the inverse square root of the integration time, as uncorrelated noise (e.g., photon noise) would (see, e.g., Soummer *et al.* 2007).

Therefore, stellar coronagraphs have become of much greater interest with the development of techniques to acquire images approaching the diffraction limit, that is, with adaptive optics used in the infrared. However, the conventional diffraction limit, rms wavefront errors less than $\lambda/14$, is not adequate for very-high-contrast imaging. Any optical imperfection can produce speckles and, since it is not possible to make any system *perfectly* stable, these speckles carry many of the issues discussed for ones due to atmospheric turbulence. High-contrast imaging requires optics at least an order of magnitude more precise than the conventional diffraction limit (see, e.g., Stapelfeldt 2006). Even in space, residual optical errors produce speckles that are not completely stable. Achieving high contrast therefore requires development of extremely high-quality optics, including a degree of active control to reach the performance goals.

7.9 Exercises

7.1 The seeing is observed to be 1.2 arcsec at the zenith. What is the expected seeing at a zenith distance of 60°?

7.2 Do spy satellites (looking down to the surface of the earth) need AO? Why or why not? (Assume the largest spy telescopes have mirror diameters of 3 meters, so they can fit inside rocket shrouds.)

7.3 Suppose you are designing an AO system to work well whenever $r_0 > 15$ cm and with a wind speed at the critical altitude < 20 m/s. You plan to use a Shack–Hartmann wavefront sensor operating at 0.55 μm, bandwidth 0.2 μm, and with each lenslet matched to the minimum design r_0. Zero magnitude at 0.55 μm is ~3900 Jy; assume also that the effective system throughput is 40%. If a ratio of signal to noise of 7 is needed on the total stellar signal for each Shack–Hartmann channel for adequate centroiding, what is the faintest natural guide star that the system can use? For operation at 2.2 μm, what is the probability of finding a suitable star within the isoplanatic patch? You can take the number of stars brighter than m_V per square degree to be

$$N(<m_V) = 0.00138 \ 10^{0.4896 \ m_V + 0.001159 \ m_V^2 - 0.000235 \ m_V^3}$$

7.4 You are thinking of carrying out a lucky imaging observation at 2.15 μm, bandwidth 0.4 μm. On the night you have been assigned, you find that the average number of speckles in an image is 10. Assume that the distribution of the number of speckles can be represented by a Poissonian, and that you want to get images that have collapsed to a single speckle. How many images do you expect to take to get on average the one you want? Suppose you can only integrate for 10 ms to catch the speckles in a stationary state, and that it takes 10 ms to read out the detector array. Suppose the read noise is 10 electrons, your telescope is 4 meters in aperture, and 20% of the photons into the telescope from a star contribute to the central image. Your target is a 12th magnitude star (zero magnitude is 647 Jy) and you need a signal-to-noise ratio of 200:1 to search for a faint companion. For how long do you need to take data?

7.5 Use equation (7.5) to evaluate r_0 and equation (7.8) for the isoplanatic angle at $2\,\mu m$ corresponding to the atmospheric turbulence shown in Figure 7.2 (and assuming one is looking at the zenith). Compare with the estimate we made using equation (7.7). For convenience, take, $C_n^2(h) = 1 \times 10^{-16}\,m^{-2/3}$ from the ground to 1 km altitude, $1 \times 10^{-17}\,m^{-2/3}$ from 1 km to 4 km, $3 \times 10^{-19}\,m^{-2/3}$ from 4 km to 10 km, $1 \times 10^{-17}\,m^{-2/3}$ from 10 km to 17 km, and zero above 17 km. What region(s) of the atmosphere have the dominant effect on r_0 and θ_0?

7.6 Use equations (7.11) and (7.12) to evaluate f_G and f_T for the same night as in 7.5 (and also looking at the zenith). You can assume wind velocities of $6\,m/s$ from the ground to 1 km, $10\,m/s$ from 1 to 2 km, $20\,m/s$ from 2 to 3 km, $50\,m/s$ from 3 to 6 km, $60\,m/s$ from 6 to 12 km, $40\,m/s$ from 12 to 16 km, and $30\,m/s$ from 16 to 20 km. What altitudes have the dominant effect on these frequencies? What is the required update rate for the AO system on this night?

7.7 You are planning AO observations at $2\,\mu m$ on the MMT. The AO system there has equivalent subapertures of 1-m diameter and can update its wavefront corrections at a rate of 500 per second. Assume the values from the preceding two exercises: $r_0 = 1.07\,m$, $\theta_0 = 10.6\,arcsec$, and $f_G = 11\,Hz$. If the measurements require a Strehl of 30%, how much wavefront error (in radians) can be allowed due to counting statistics from a natural guide star? Assume that the number of photons per subaperture required for error σ_{star} is given approximately by $N = 6/\sigma_{star}^2$, that the system throughput is 60%, and that the system works at V ($0.55\,\mu m$, bandwidth $0.2\,\mu m$, zero magnitude = $3900\,Jy$). There is a candidate guide star $5\,arcsec$ from the target. How bright does it have to be?

7.8 Consider a simple quadcell with four detectors producing four outputs, I_i, $i = 1$ to 4. Let the light of a source be put onto the quadcell – assume that the signal is a perfect uniform disk. If the quadcell is aligned along x and y, then show for small offsets of the source from the center position that the offset in the x direction is given by

$$\delta_x \approx \frac{d}{2}\left(\frac{(I_1 + I_2) - (I_3 + I_4)}{I_1 + I_2 + I_3 + I_4}\right)$$

where d is the diameter of the source image and cells 1 and 2 are on the side of the quadcell toward increasing x. Make a graph of the displacement signal with offsets in x for the cases $d = 0.25$, 0.5, and 0.75 of the

width of a quadcell pixel. Evaluate the range and linearity of the displacement signals in terms of the optimum spot size for good quadcell
performance.

7.9 A creative but eccentric colleague wants to build a telescope with an
elliptical mirror, 4 meters for the minor axis of the primary and 8 meters
for the major axis. He claims that this design will improve the throughput
of slit spectrometers, if the slit is oriented parallel to the major axis of the
primary. The plan is to use the design with a high-resolution infrared
spectrometer working near 2 μm. Will this concept work as he thinks? If
you could use high-performance adaptive optics on the telescope, would
it work?

Further reading

Beckers, J. M. (1993). Adaptive optics for astronomy – principles, performance, and
applications. *ARAA*, **31**, 13–62.
Hardy, J. W. (1998). *Adaptive Optics for Astronomical Telescopes*. Oxford,
New York: Oxford University Press.
Narayan, R. and Nityanandan, R. (1986). Maximum entropy image restoration in
astronomy. *ARAA*, **24**, 127–170.
Oppenheimer, B. R. and Hinkley, S. (2009). High-contrast observations in optical
and infrared astronomy. *ARAA*, **47**, 253–289.
Puetter, R. C., Gosnell, T. R., and Yahil, A. (2005). Digital image reconstruction:
Deblurring and denoising. *ARAA*, **43**, 139–194.
Starck, J. L., Pantin, E., and Murtagh, G. (2002). Deconvolution in astronomy:
A review. *PASP*, **114**, 1051–1069.
Tyson, R. K. (2000). *Introduction to Adaptive Optics*. Bellingham, WA: SPIE Press.

8

Submillimeter and radio astronomy

8.1 Introduction

The submillimeter and millimeter-wave regime – roughly $\lambda = 0.2$ mm to 3 mm – represents a transition between infrared and radio ($\lambda > 3$ mm) methods. Because of the infinitesimal energy associated with a photon, photodetectors are no longer effective and we must turn to the alternative two types described in Section 1.4.2. Thermal detectors – bolometers – are useful at low spectral resolution. For high-resolution spectroscopy (and interferometers), coherent detectors are used. Coherent detectors – heterodyne receivers – dominate the radio regime for both low and high spectral resolution (Wilson *et al.* 2009).

As the wavelengths get longer, the requirements for optics also change. The designs of the components surrounding bolometers and submm- and mm-wave mixers must take account of the wave nature of the energy to optimize the absorption efficiency. "Pseudo-optics" are employed, combining standard lenses and mirrors with components that concentrate energy without necessarily bringing it to a traditional focus. In the radio region, non-optical techniques are used to transport and concentrate the photon stream energy. For example, energy can be conveyed long distances in waveguides, hollow conductors designed to carry signals through resonant reflection from their walls. At higher frequencies, strip lines or microstrips can be designed to have some of the characteristics of waveguides; they consist of circuit traces on insulators and between or over ground planes.

We first discuss bolometers, followed by a summary of heterodyne principles of operation; we then derive the general observational regimes in the submm/mm range where each is predominant. We follow with a more detailed description of heterodyne receivers in both the submm-/mm-wave and the radio regimes. For bolometers the basic instrumentation – imagers, polarimeters, spectrometers – usually follows the design principles in Chapters 4–6. We will concentrate in this chapter on describing bolometer operation. Because heterodyne receivers encode source spectra and the signal phases in their outputs, however, instrumentation for them takes fundamentally different forms, to be discussed below.

8.2 Bolometers

8.2.1 Principles of operation

Bolometers absorb the energy of a photon stream, but rather than freeing charge carriers, this energy is thermalized. The deposited energy raises the temperature of an absorber that is connected to a heat sink through a poorly conducting thermal link (Figure 8.1). This change is sensed by a thermometer mounted on the absorber, which produces an electronic signal. The signal is proportional to the total absorbed energy, not to the number of photons absorbed. Because of these differences in operation, the performance of bolometers needs to be described differently from that of the photodetectors that have held center stage up to now.

The smallest photon stream power a bolometer can detect is the basic measure of its performance. This measure is characterized by the noise equivalent power, NEP, defined as follows. If the output of the bolometer is taken to electronics filtered to have an electronic bandwidth of 1 Hz, then the NEP is the power onto the detector that produces a signal that is equal to the root-mean-square (rms) noise from the detector into the same electronics. This definition is based on a "square" electronic frequency passband (infinitely abrupt cuton and cutoff), which cannot be realized in practice. However, equivalent behavior can be achieved with a circuit with exponential response, since an exponential time constant of τ corresponds to a frequency bandwidth of $1/4\tau$.

A number of noise components are contained within the NEP. A fundamental noise source is associated with the storage of energy as a result of the

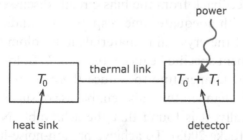

Figure 8.1. Thermal model of a bolometer. Important aspects of the design are the strength of the thermal link, G, the heat capacity of the detector, C, and the heat sink temperature, T_0.

heat capacity of the detector and the fluctuations of entropy across the thermal link; by the laws of thermodynamics, this energy is released (and restored) at a fluctuating rate (van der Ziel 1976, Mather 1982), yielding:

$$\mathrm{NEP}_T = \frac{(4kT^2G)^{1/2}}{\eta} \tag{8.1}$$

where $T \, (= T_0 + T_1$ in Figure 8.1) is the temperature of operation, G is the conductivity of the thermal link in watts per degree (K), and η is the fraction of these photons that are absorbed (the quantum efficiency). In addition, the signal noise still has a component proportional to the square root of the number of absorbed photons:

$$\mathrm{NEP}_{\mathrm{ph}} = \frac{hc}{\lambda} \left(\frac{2\varphi}{\eta}\right)^{1/2} \tag{8.2}$$

where φ is the photons s^{-1}. The noise also includes Johnson noise, the thermal noise current for a resistor R:

$$\langle I_J^2 \rangle = \frac{4kT\,df}{R} \tag{8.3}$$

where df indicates the frequency bandwidth. The corresponding contribution to the NEP depends on construction details of the device (Rieke 2003).

The fundamental speed of response of a bolometer is set by the ratio of its heat capacity, C, to the strength of the thermal link, which determines the thermal time constant τ_T:

$$\tau_T = \frac{C}{G} \tag{8.4}$$

Making G small increases the temperature excursion for a given power and tends to reduce the NEP as in equation (8.1) (which is good). It also makes the detector slower as in equation (8.4) (which is bad); high heat capacity also slows the detector. Even though the realized speed of the detector can be increased through feedback from the bias circuit (discussed in Section 8.2.3), high performance with adequate time response demands low heat capacity. The specific heats of the crystalline materials in a bolometer decrease as T^3; the metallic component specific heats decrease as T. Bolometers are therefore operated at very low temperature to reduce thermal and Johnson noise and heat capacities. These temperatures also enable use of superconducting readout electronics. Empirically, it is found that the achievable NEP scales approximately as $T^{2 \text{ to } 2.5}$ (Rieke 2003). To achieve photon-noise-limited performance requires temperatures of ~0.3 K on the ground and ~0.1 K when using cold optics in space (or moderately high spectral resolution on the ground).

Very-high-performance bolometers have been built into small arrays for some time, but until recently these devices were based on parallel operation of single pixels. The obstacle to true array-type construction was that the very small signals required use of junction field effect transistor (JFET) amplifiers that needed to operate above about 50 K, far above the operating temperature of 0.3 K or below for the bolometers themselves. It is difficult to implement the simple integration of detector and amplifier that is the heart of array construction with this temperature difference.

8.2.2 The PACS bolometer array

With the development of adequately low-noise readouts that can operate near the bolometer temperature, high-performance bolometer arrays for the far-infrared and submillimeter spectral ranges are now available. To illustrate their design, we describe two of them specifically. One channel of the Herschel/PACS (photodetector array camera and spectrometer) instrument uses a 2048-pixel array of bolometers (Billot *et al.* 2006). The architecture of this array is vaguely similar to the direct hybrid arrays for the near- and mid-infrared. One silicon wafer is patterned with bolometers, each in the form of a silicon mesh, as shown in Figure 8.2. A second silicon wafer is used to fabricate the MOSFET-based readouts, and the two are joined by indium bump bonding.

The development of "silicon micromachining" has enabled substantial advances in bolometer construction generally and is central to making large-scale arrays. As an example, the delicate construction of the PACS detectors, as shown in Figure 8.2, depends on the ability to etch exquisitely complex miniature structures in silicon. In this instance, the absorbers are fine grids of silicon (grid spacing ≪ the wavelength of operation) to minimize their mass and heat capacity. The silicon mechanical structure around the grid region provides the heat sink; the grid is connected to but isolated from it with thin and long silicon rods. The rods and grid both need to be designed to achieve appropriate response and time constant characteristics. Each grid is blackened with a thin layer of titanium nitride. Quarter-wave resonant structures surrounding it tune the absorption to higher values over limited spectral bands. For each bolometer, a silicon-based thermometer doped by ion implantation to have appropriate temperature-sensitive resistance lies at the center of the grid with a reference thermometer on the frame. Large resistance values are used so that the fundamental noise is large enough to utilize MOSFET readout amplifiers. To minimize thermal noise and optimize

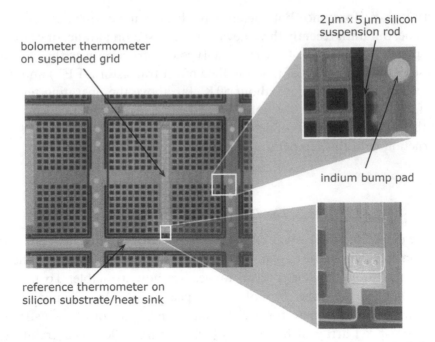

2 μm x 5 μm silicon
suspension rod

bolometer thermometer
on suspended grid

indium bump pad

reference thermometer on
silicon substrate/heat sink

Figure 8.2. A single pixel in the Herschel/PACS bolometer array, pixel size about 750 μm. Image adapted from one supplied by Patrick Agnese, CEA Electronics and Information Technology Laboratory (Leti).

the material properties, the bolometer array is operated at 0.3 K. Further details are in Billot *et al.* (2006).

8.2.3 *The SCUBA-2 bolometer array*

Another approach is taken in transition edge sensor (TES) arrays such as the ones used in the submillimeter camera SCUBA-2 (Holland *et al.* 2006; Craig *et al.* 2010). The name of these devices is derived from their thermometers, which are based on thin superconducting films. In superconductivity, pairs of electrons with opposite spin form weakly (order of a few meV) bound Cooper pairs. The pairs are not subject to the Pauli Exclusion Principle for Fermions and can move through the material without significant interaction; hence, the electrical resistance vanishes. In the transition region between normal conductivity and superconductivity, the films have a stable but very steep dependence of resistance on temperature. A TES is held within this transition region to provide an extremely sensitive thermometer.

The resistance of a TES is so low that it cannot deliver significant power to JFETs and MOSFETs. Instead, the signals are fed into superconducting

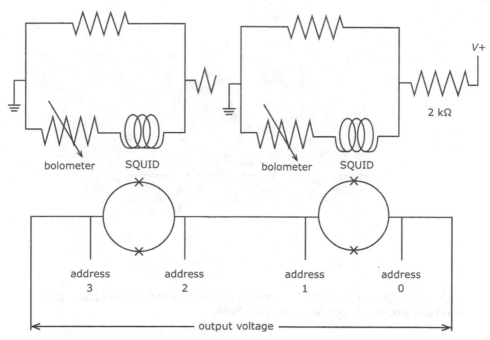

Figure 8.3. Bias circuit for TES bolometer (upper circuit) and SQUID readout (lower circuit). The Josephson junctions are indicated with "X." The circuit is repeated twice with appropriate address lines to operate as a simple SQUID multiplexer.

quantum interference devices (SQUIDs). A SQUID (Figure 8.3) consists of an input coil that is inductively coupled to a superconducting current loop. Two Josephson junctions – junctions of superconductors with an intervening insulator – interrupt the loop. Because of quantum mechanical interference effects, the current in the loop is very strongly affected by the magnetic field produced by the coil. Thus, changes in the bolometer current produce a large modulation of the SQUID current – that is, when its output is made linear by using feedback, the device works as an amplifier. SQUIDs are the basis for a growing family of electronic devices that operate by superconductivity.

TES bolometer arrays use SQUIDs for the same readout functions that we have discussed for photodetector arrays. The operation of two units of a simple SQUID time-domain multiplexer is illustrated in Figure 8.3 (Benford *et al.* 2000). The biases across the SQUIDs are controlled by the address lines. Each SQUID can be switched from a normal operational state to a super-conducting one if it is biased to carry about $100\,\mu A$. The address lines are set so that all the SQUIDs in series are superconducting except one, and then only that one contributes to the output voltage. By a suitable sequence of bias settings, each SQUID amplifier can be read out in turn.

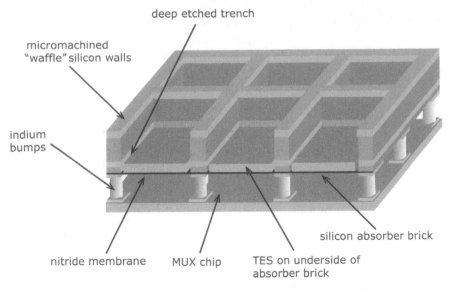

deep etched trench

micromachined
"waffle" silicon walls

indium
bumps

silicon absorber brick

nitride membrane   MUX chip   TES on underside of
                                 absorber brick

Figure 8.4. Design features 2 × 3 pixels of the SCUBA-2 bolometer array, pixel size about 1.1 mm (Walton *et al.* 2004).

When the TES temperature rises due to power from absorbed photons, its resistance rises, the bias current drops, and the electrical power dissipation decreases. These changes partially cancel the effects of the absorbed power and limit the net thermal excursion. This behavior is called electrothermal feedback. It can make the bolometers operate tens or even hundreds of times faster than implied by equation (8.4), because it reduces the physical temperature change in the detector. In fact, if the TES is too fast, the bolometer/ SQUID circuit can be unstable, and measures must be taken to slow the response.

An important feature of these devices is that the superconducting readouts operate with very low power dissipation and at the ultra-low temperature required for the bolometers. Therefore, integration of detectors and readouts is simplified and the architecture can potentially be scaled to very large arrays. The SCUBA-2 bolometer arrays are an example. The design is implemented by silicon micromachining and is illustrated in Figure 8.4. Each of these arrays is made of four subarrays, each with 1280 transition-edge sensor pixels. The detector elements are separated from their heat sinks by a deep etched trench that is bridged by only a thin silicon nitride membrane. The absorbing surface is blackened by implanting it with phosphorus. The dimensions of the array pixels are adjusted to form a resonant cavity at the wavelength of operation, to enhance the absorption efficiency. The superconducting electronics that read out the bolometers are fabricated on separate wafers; the two components are assembled into an array using indium bump bonding.

There are two basic approaches to multiplexing TES signals. We have described the time-domain approach, but multiplexing in the frequency domain is also possible. In this case, each TES is biased with a sinusoidally varying voltage at a unique frequency and the signals from a number of TESs are encoded in amplitude-modulated carrier signals by summing them. They are read out by a single SQUID and then brought to room-temperature electronics that recovers each of the signals by synchronous (frequency-dependent) detection.

8.2.4 Kinetic inductance detectors

The pace of technical progress in submillimeter detection is so high that before one advance has been fully implemented (e.g., SCUBA-2 TES arrays), another is vying to take its place. Kinetic inductance detectors are an example (Mazin 2009). Although these devices have promise for operation in the microwave through the X-ray, the greatest interest has been in their use at the low-frequency end of this range; hence we discuss microwave kinetic inductance detectors (MKIDs).

In a superconductor, electrons are either bound in Cooper pairs or "free;" because their behavior is still strongly influenced by quantum effects, the "free" electrons are termed quasi-particles. The density of thermally generated quasi-particles decreases exponentially with Δ/kT, where Δ is the Cooper pair binding energy per electron. Thus, for a superconductor well below the critical temperature, the great majority of the electrons are bound in Cooper pairs. When photons are absorbed with energy greater than the Cooper pair binding energy, 2Δ (0.4–4 meV depending on the material, corresponding to submm-/mm-wavelengths), they break the pairs and release excess quasi-particles. For incident power, P, the number of excess quasi-particles, n_{qp}, in a device is

$$n_{qp} \propto \frac{\eta P \tau_{qp}}{\Delta} \tag{8.5}$$

where η is the energy transfer efficiency from the incident radiation to quasi-particle production (~ 0.57) and τ_{qp} is the lifetime of the quasi-particles. If the thermally generated quasi-particle density is low (aided if the superconductor is in the form of a thin film), the excess quasi-particles can dominate the behavior of the material.

Free charged particles are accelerated by any high-frequency electric field; because of conservation of momentum, it takes a finite time for them to react.

As a result, they impose a phase lag similar to that created in a conventional electrical circuit by an inductor, an effect described as kinetic inductance. Although kinetic inductance occurs generally, it is prominent in superconductors because of the high mobility of the Cooper pairs. A MKID is based on this process. The breaking of the Cooper pairs by incident photons to create excess quasi-particles affects the reactance of the detector to increase the inductance (and the resistance), because the quasi-particles block the Cooper pairs from occupying some of the electron states. If the detector is placed in a resonant circuit, a photon flux that breaks the Cooper pairs can change the resonant frequency. Resonant frequencies of ~10 GHz are used, and the resonance can be made sufficiently sharp (high Q) that it is less than 1 MHz wide. Therefore, a large number of these circuits can be connected to a single pair of lines and read out by a high frequency HEMT (high electron mobility transistor) amplifier (Section 8.4.2). The outputs of the detectors are then multiplexed in the frequency domain, and the signals from each one can be extracted by demodulation of the HEMT output. Roughly a thousand detectors can be read out through a single HEMT amplifier, and without providing complex cryogenic circuitry such as must be used with the two bolometer types described above.

8.3 Heterodyne receivers for the submm- and mm-wave

We now describe the third and last major approach to photon detection, heterodyne receivers.

8.3.1 Heterodyne principles of operation

Heterodyne receivers mix the electric fields of the target source photons with those of a local source (the local oscillator (LO)) operating at a specific frequency. For simplification, we assume the target source also emits at a specific but slightly different frequency from that of the LO. If two such signals are combined within an electronic mixer, they beat against each other due to alternating constructive and destructive interference. The resulting signal contains frequencies not only from the original two signals, but also at the difference or intermediate frequency (IF). The IF signal is isolated and amplified to provide power for convenient processing of the signal. Because the IF signal retains phase information about the target source photons, heterodyne receivers are termed coherent detectors in contrast to the incoherent detectors discussed above and in Chapter 3.

8.3.2 The Antenna Theorem

Achieving interference of the source and LO fields imposes requirements on the detection process that are summarized in the Antenna Theorem. Etendue must be preserved through the telescope, that is, the $A\Omega$ product cannot decrease. Therefore, the signal photons cannot be concentrated onto the mixer in a parallel beam; even for a point source, they will strike it over a range of angles. Because of their range of tilts, the incoming wavefronts have a range of phases over the extent of the mixer. The requirement that interference occurs between the LO (at a single phase and frequency) and the signal photons sets a requirement on the useful range of acceptance angle for the heterodyne receiver: $2\theta \leq \lambda/d$, where d is the diameter of the mixer. Since the optical system must conserve $A\Omega$, this condition can be expressed as

$$2\Phi \approx \frac{\lambda}{D} \tag{8.6}$$

where D is the diameter of the telescope aperture and Φ is the angular radius of the field of view on the sky (i.e., 2Φ is the angular diameter of the FOV); this result is identical to our derivation of the diffraction limit (equation 1.13). Thus, a coherent receiver must operate at the diffraction limit of the telescope. By squaring equation (8.6) we find

$$A\Omega \approx \lambda^2 \tag{8.7}$$

A second restriction is that the interference that produces a heterodyne signal only occurs for components of the source photon electric field vector that are parallel to the electric field vector of the LO power, that is, only a single polarization of the source emission produces any signal. These two requirements together are termed the Antenna Theorem.

8.3.3 Performance description

The performance of a heterodyne receiver is quoted in terms of the noise temperature, T_N, defined as the temperature of a blackbody placed over the receiver input that would be detected at signal-to-noise ratio of 1.[17]

As with all other types of detectors, coherent receivers are subject to the noise associated with the background emission, termed the thermal limit. However, they have an additional source of noise, because retaining phase information is equivalent to measuring the time of arrival of a photon. By the

---

[17] The figures of merit used previously do not work: (1) counting statistics are not applicable because individual photons are not detected; and (2) the definition of NEP breaks down for a detector that imposes a spectral bandwidth.

Heisenberg Uncertainty Principle, there is an unavoidable minimum noise in the measurement of both the energy and time of arrival of the photon:

$$\Delta E \, \Delta t \geq h/2\pi \tag{8.8}$$

This uncertainty results in the quantum limit to the receiver performance. It is manifested by effects such as electron statistical ("shot") noise in the mixer, which scales with the LO power and therefore, unlike other noise sources, cannot be overcome with more LO output. Given that Johnson noise is also a form of electron statistical noise, we can assign a noise temperature from equation (8.3). We assume a resistor connected in a circuit with exponential response with time constant τ and hence $df = 1/4\tau$ to get $\langle I_J^2 \rangle \, R\tau \approx \Delta E = kT_N$. The time uncertainty corresponding to 1 radian in a phase measurement of a photon with frequency v is $\Delta t = 1/2\pi v$. Combining these results, we find the quantum limit of a receiver:

$$T_N = \frac{hv}{k} \tag{8.9}$$

The quantum limit is a useful metric for mixer performance in the mm- and submm-wave, but at lower frequencies it is negligible compared with other noise sources.

It is often convenient to express the power per unit frequency from a source as an antenna temperature, T_S, in analogy with the noise temperature. This concept is particularly useful at millimeter and longer wavelengths, where the observations are virtually always at frequencies that are in the Rayleigh–Jeans regime ($hv \ll kT$). In this case, the antenna temperature is linearly related to the input flux density:

$$T_S = \frac{A_e S_v}{2k} \tag{8.10}$$

where A_e is the effective area of the antenna or telescope, S_v is the flux density from the source, and the factor of ½ is a consequence of the sensitivity to a single polarization. To maintain the simple formalism in terms of noise and antenna temperatures, it is conventional to use a Rayleigh–Jeans equivalent temperature such that equation (8.10) holds by definition whether the Rayleigh–Jeans approximation is valid or not.

The achievable signal-to-noise ratio for a coherent receiver is given in terms of antenna and system noise temperatures by the Dicke radiometer equation:

$$\left(\frac{S}{N}\right)_c = K\frac{T_S}{T_N}(\Delta f_{IF}\Delta t)^{1/2} \tag{8.11}$$

where Δt is the integration time of the observation, Δf_{IF} is the IF bandwidth, and K is a constant of order 1. Of course, life is not quite as simple as this equation implies; the signal-to-noise ratio can be degraded relative to the prediction by instability in the receiver or the atmosphere, or by confusion noise (Section 1.5.4). Further discussion of performance characterization for practical receivers can be found in Section 8.5.1.

8.3.4 Comparison of incoherent and coherent detection

Equations (8.9), (8.10), and (8.11) let us compare the performance of coherent (heterodyne) and incoherent (e.g., bolometer) detection, as long as we also keep in mind the Antenna Theorem and related restrictions. From equation (8.10) and the definition of NEP, the signal-to-noise ratio with an incoherent detector system operating at the diffraction limit is

$$\left(\frac{S}{N}\right)_i = \frac{2kT_S\Delta v(\Delta t)^{1/2}}{\text{NEP}} \tag{8.12}$$

where T_S is the flux density converted to an antenna temperature, Δv is the width of the spectral band, and Δt is the integration time. Therefore, using equation (8.11), we obtain the ratios of signal to noise achievable with the two types of system under the same measurement conditions:

$$\frac{(S/N)_c}{(S/N)_i} = \frac{\text{NEP}(\Delta f_{IF})^{1/2}}{2kT_N\Delta v} \tag{8.13}$$

Suppose a bolometer is operating background limited and we compare its signal-to-noise ratio on a continuum source with a heterodyne receiver operating at the quantum limit. We set the bolometer field of view at the diffraction limit, $A\Omega = \lambda^2$ and assume that the background is in the Rayleigh–Jeans regime (i.e., thermal background at 270 K observed near 1 mm). The background-limited NEP is given in equation (8.2). The photon incidence rate, φ, can be shown to be

$$\varphi = \frac{2\eta k\, T_B\, \Delta v}{hv} \tag{8.14}$$

where T_B is the equivalent blackbody temperature of the background. If we assume the bolometer is operated at 25% spectral bandwidth, $\Delta v = 0.25v$, and that the IF bandwidth for the heterodyne receiver is $3 \times 10^9\,\text{Hz}$ (a typical value) then

$$\frac{(S/N)_c}{(S/N)_i} \approx \frac{2.4 \times 10^6}{v}(4\,\Delta f_{IF})^{1/2} \approx \frac{2.6 \times 10^{11}}{v} \tag{8.15}$$

Thus, the bolometer becomes more sensitive near 2.6×10^{11} Hz and at higher frequencies, or at wavelengths shorter than about 1 mm. Actually, this comparison is slightly unfair to it (since, for example, it does not have to work at the diffraction limit), so it is the detector of choice for continuum detection at wavelengths out to 2–3 mm. Hence, large-scale bolometer cameras have been developed for mm-wave and submillmeter telescopes. Conversely, at wavelengths longer than a few millimeters, equation (8.15) shows why coherent detectors are the universal choice. Of course, coherent detectors are preferred for high-resolution spectroscopy and for interferometry (Chapter 9).

8.3.5 Basic receiver elements

Figure 8.5 shows the components of a millimeter or submillimeter heterodyne receiver. The signal photons (from the telescope) are combined at the beam splitter (sometimes called a diplexer) with the LO signal, and then conveyed to the mixer, which downconverts the combined signal to the intermediate frequency. This signal is then amplified and sent (probably via a spectrometer stage) to a detector stage where it is converted to a slowly varying direct current (DC) signal. We now describe the individual components of this receiver.

Local oscillators start with a lower-frequency tunable oscillator and put its output through a highly nonlinear circuit element (e.g., a diode). The resulting waveform has substantial power in frequency overtones, which can be isolated and used as the input to an amplifier, then taken to another nonlinear device, from which the frequency overtones can again be isolated and amplified. The result is that the original oscillator frequency is multiplied up to the operating frequency of the receiver. As we shall see, LO power is a critical asset for a receiver, and the power that can be delivered through such a multiplier chain is limited.

Mixers are diodes or other nonlinear electrical circuit components. Figure 8.6 shows the current conducted by some hypothetical mixers as a function of

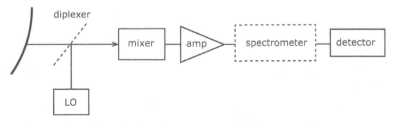

Figure 8.5. Basic submillimeter heterodyne receiver.

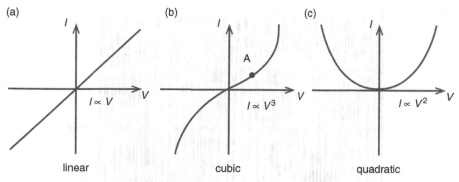

Figure 8.6. *I–V* curves of possible mixers.

the voltage placed across them. If the mixer has a linear *I–V* curve, then the conversion efficiency is zero (Figure 8.6(a)) because the negative and positive excursions of the voltage produce symmetric negative and positive current excursions that cancel. Similarly, any mixer having a symmetric characteristic curve that is an odd function of voltage around the origin will have zero conversion efficiency if operated at zero bias, although conversion can occur for operation away from zero (e.g., point A in Figure 8.6(b)). If $I \propto V^2$ (Figure 8.6(c)), the downconverted output current is proportional to the square of the voltage signal amplitude. In addition, the signal power is proportional to its electric field strength squared. $I \propto V^2 \propto E^2 \propto P$, where E is the strength of the electric field; hence, it is attractive to use *square law* devices as fundamental mixers because their output is linear with input power.

We now assume a square law device, that is,

$$I = \alpha V^2 \tag{8.16}$$

where *I* is the output current for an input voltage of *V* and α is a constant, for example the gain. Let the mixer be illuminated by signals from both a target source and a LO, of the form

$$E_S \sin(\omega_S t + \phi_S)$$
$$E_{LO} \sin(\omega_{LO} t + \phi_{LO}) \tag{8.17}$$

where ω represents angular frequencies, ϕ represents phases, and *E* represents voltage amplitude. These two signals are added in the mixer. The resulting output current is

$$I = \alpha [E_S \sin(\omega_S t + \phi_S) + E_{LO} \sin(\omega_{LO} t + \phi_{LO})]^2$$
$$= \alpha E_S^2 \sin^2(\omega_S t + \phi_S) + \alpha E_{LO}^2 \sin^2(\omega_{LO} t + \phi_{LO}) \tag{8.18}$$
$$+ 2\alpha E_S E_{LO} \sin(\omega_S t + \phi_S) \sin(\omega_{LO} t + \phi_{LO})$$

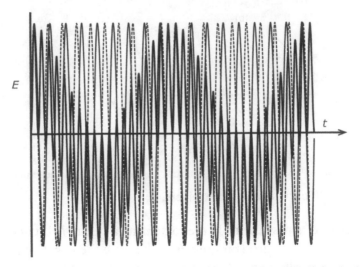

E

t

Figure 8.7. Mixing of two equal signals (S_1 light solid and S_2 light dashed) of slightly different frequencies produces a signal (heavy line) with a component at the difference frequency.

Equation (8.18) becomes[18]

$$I = \tfrac{1}{2}\,\alpha\left(E_S^2 + E_{LO}^2\right) - \tfrac{1}{2}\,\alpha E_S^2 \sin(2\omega_S t + \phi_S + \pi/2)$$
$$- \tfrac{1}{2}\,\alpha E_{LO}^2 \sin(2\omega_{LO} t + \phi_{LO} + \pi/2) + \alpha E_S E_{LO} \sin((\omega_S - \omega_{LO})t$$
$$+ (\phi_S - \phi_{LO} + \pi/2)) - \alpha E_S E_{LO} \sin((\omega_S + \omega_{LO})t$$
$$+ (\phi_S + \phi_{LO} + \pi/2)) \tag{8.19}$$

The first term is a constant that can be removed with an electronic filter. The second and third terms are the second harmonics of the signal and LO, respectively. Both are at high frequency and readily removed. The fourth term, oscillating at the IF frequency, $(\omega_S - \omega_{LO})$, is the IF current, with behavior illustrated in Figure 8.7. The fifth term is at the sum frequency and is also easily removed.

Thus, from equation (8.19), the IF current has a mean-square-amplitude proportional to the product of the signal and LO power

$$\langle I_{IF}^2 \rangle \propto I_L I_S \tag{8.20}$$

where I_L is the current in the detector from the LO signal and I_S is that from the source. Because the signal strength depends on the LO power, many forms of noise can be overcome by increasing the output of the local oscillator, putting a premium on delivering high LO power to the mixer.

[18] With use of the trigonometric identities, $\sin^2(a) = [1 - \cos(2a)]/2$, $\sin(a) = \cos(\pi/2 - a)$, and $\sin(a)\sin(b) = \tfrac{1}{2}\,[\cos(a-b) - \cos(a+b)]$.

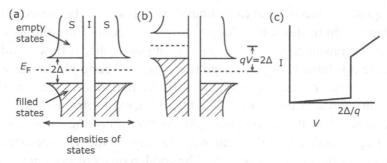

Figure 8.8. Operation of a SIS junction.

The conversion gain is defined as the IF output power that can be delivered by the mixer to the next stage of electronics divided by the input signal power. *Classical* mixers do the downconversion with conversion gain < 1. The ability to provide an increase in power while downconverting the input signal frequency (conversion gain > 1) is characteristic of *quantum mixers*, but to achieve better stability, quantum mixers are usually operated with gains less than 1 also.

The IF signal encodes the spectrum of the target source over a range of input frequencies equivalent to the IF bandwidth. This "extra" information allows efficient spectral multiplexing (many spectral elements observed simultaneously with a single receiver) and very flexible use of arrays of telescopes and receivers for interferometry. However, in the simplest case there is no way of telling in the mixed signal whether $\omega_S > \omega_{LO}$ or $\omega_{LO} > \omega_S$. Since the signal at ω_{IF} can arise from a combination of true inputs at $\omega_{LO} + \omega_{IF}$ and $\omega_{LO} - \omega_{IF}$, it is referred to as a double sideband signal. When observing continuum sources, the ambiguity in the frequency of the input signal is a minor inconvenience. When observing spectral lines, the *image* frequency signal at the off-line sideband can result in ambiguities in interpreting the data. Therefore, more complex approaches have been developed to separate the sidebands (discussed below).

SIS mixers are the highest-performance mixers in the submm- and mm-wave regimes. SIS stands for "superconductor–insulator–superconductor." The sandwich of these materials has very nonlinear electrical behavior and is the heart of a mixer. Its operation is illustrated in Figure 8.8.

The two superconductors (S) in Figure 8.8 are shown in a pseudo-bandgap diagram where the bandgap is the binding energy of the Cooper pairs. Unlike semiconductors, the number of available states is huge at the top of the "valence" and bottom of the "conduction" band because the pairs are not subject to the Pauli Exclusion Principle. The superconductor layers are separated by a thin insulating layer (I) that blocks the flow of currents, except through quantum-mechanical tunneling. Panel (a) shows the device without a bias voltage; even in this state, Cooper pairs can tunnel from one side to the other (the Josephson effect). However, only a small current flows until a bias is applied that is large enough to align the "valence" band on the left with the "conduction" one on the right (panel (b)). At that point, the tunneling of Cooper pairs through the insulator suddenly increases dramatically because of the large number of available states at the bottom of the conduction band and the large number of filled states at the top of the valence one on the opposite side of the device (panel (c)). The sudden onset of this current results in a sharp inflection in the *I–V* curve. As the bias is increased further the device behaves resistively ($I \propto V$). When the SIS junction is biased to put the operating point just at this inflection, the operation can be visualized as that of a switch; whenever the mixed signal (Figure 8.7) exceeds a threshold value, a significant current is conducted. Thus, the current is dominated by a signal just at the IF. Because the inflection in the bias curve is so sharp at $2\Delta/q$, relatively little local oscillator power is needed to get a strong IF signal.

Up to about 4×10^{11} Hz, SIS mixers can operate close to the quantum limit. However, the mixer performance degrades dramatically if used with photons capable of breaking the Cooper pairs. Operation up to 1.2×10^{12} Hz is possible with the relatively high gap energy (and hence large pair binding energy) of NbTiN, and SIS mixers remain within a factor of 2 to 3 of the quantum limit up to 10^{12} Hz. At higher frequencies alternative types of mixer can be used, with the best performance achieved with very small hot electron bolometers coupled to the signal and local oscillator energy through small antennae. Current performance levels are a factor of 4 or more above the quantum limit at these very high frequencies.

Sideband separation. We emphasized that a simple mixer produces indistinguishable outputs from two ranges of input frequencies (two sidebands). More complex arrangements can separate the sidebands. An example is to use two mixers driven either by LO signals that have a 90° phase difference (see Figure 8.9), or to use LO signals in phase but to delay the input to one of the mixers by 90°. If the upper and lower sideband signals are exactly in phase at the input for one of the mixers, they will be exactly at opposite phase for the other. Circuitry can combine these two signals so that either the upper or lower sideband is canceled; generally, there are two outputs, one with the lower and the other with the upper sideband. Although in principle this added complexity might increase the noise temperature of the receiver, in many cases state-of-the-art receivers are limited by atmospheric noise, so the overall performance is not degraded by separating the sidebands. High-performance radio telescopes may also use a pair of receivers to capture both polarizations.

(a)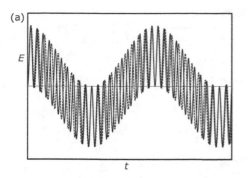

(b)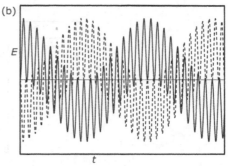

Figure 8.9. Principle of operation of a sideband separating receiver. The dashed line is the mixer signals for the lower sideband and the solid line is for the upper sideband. The LO is shifted by 90° between the left and right panels.

Amplifiers and back ends. The properties of the remaining components in Figure 8.5 have a lot in common with similar components in radio receivers, so we will discuss them in Sections 8.4.2. and 8.4.3.

8.4 Radio astronomy

8.4.1 The antenna

The basic principles of operation for super heterodyne radio receivers are similar to those for submillimeter receivers, but the longer wavelengths permit a more direct handling of the electric fields. One example is that it is generally no longer necessary to use optical methods (lenses, complex imaging optical trains) to handle the photons after they have been collected. Instead, their electric fields excite currents in antennae and the signals can be amplified and conveyed large distances through waveguides and related devices.

A half-wave dipole (Figure 8.10) is the simplest example of an antenna. It is made of two conducting strips each $\leq \frac{1}{4}$ wavelength long, with a small gap between them. The electric field of a photon excites a current in the antenna wires. The outputs of these wires are brought through a shielded cable (indicated schematically in Figure 8.10) to a resistor: in principle, the voltage due to the photon field can be sensed as a voltage across the resistor.

We can visualize the performance of such an antenna in terms of the pattern of the radiation it would *emit* if excited at an appropriate frequency. This reversal of the point of view is based on the Reciprocity Theorem. If we excite the antenna with a varying signal, the electric charges in it are accelerated and according to Maxwell's equations they will emit electromagnetic radiation. Emission at wavelength λ from one end of a half-wave dipole will

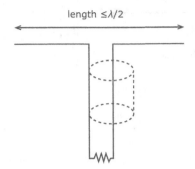

Figure 8.10. A half-wave dipole antenna.

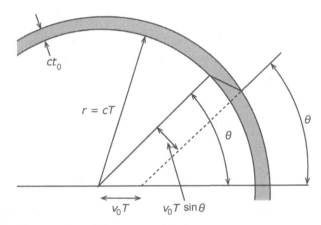

Figure 8.11 Generation of Larmor radiation.

cancel emission from the other end, yielding a pattern peaked in the direction perpendicular to the antenna. The radiation pattern is defined by the zone where the phases match sufficiently well to combine constructively.

We now demonstrate this behavior quantitatively. The motions of electrons in a wire are subrelativistic, so the emission pattern can be calculated using the approach introduced by Larmor. A particle moving at constant velocity will always have a purely radial electric field, but when the velocity changes, by continuity arguments the field must develop a tangential component. The situation is illustrated in Figure 8.11 (Purcell 1985). For simplicity, we assume that the charge was decelerated uniformly from velocity v_0 over time t_0 (acceleration $a = v_0/t_0$), a time of T ago. The field from the time of this event has expanded to $r = cT$, where to maintain continuity with the radial behavior before and after the deceleration there must be a kink in the field line spanning the distance ct_0. This kink represents the transverse component of the field; from the geometry illustrated in Figure 8.11,

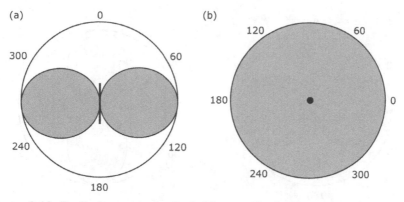

Figure 8.12. Radiation pattern of a half-wave dipole (the pattern is symmetric viewed parallel to the antenna axis as shown in (b)).

$$\frac{E_{\text{tran}}}{E_r} = \frac{a\,T\sin\theta}{c} \tag{8.21}$$

where a is the acceleration of the particle, $T \gg t_0$ is the time of the observation (after the acceleration has occurred), E_r is the usual radial field at distance r, and θ is the angle between the acceleration and the direction toward the observer. Substituting for E_r and setting $T = r/c$ yields

$$E_{\text{tran}} = \frac{qa\sin\theta}{4\pi\varepsilon_0 c^2 r} \tag{8.22}$$

The energy flux density of the radiation is given by the Poynting vector

$$\vec{S} = \frac{E_{\text{tran}}^2}{\mu_0 c}\hat{r} = \frac{q^2 a^2 \sin^2\theta}{16\pi^2 \varepsilon_0 r^2}\hat{r} \tag{8.23}$$

where μ_0 is the magnetic constant (or permeability of free space), ε_0 is the permittivity of free space, and $\hat{r}$ is a unit vector. The Poynting vector also gives the power radiated by the particle per unit solid angle. The field of the dipole antenna is therefore the integral of equation (8.23) over the length of the antenna, which gives a power output similar in angular behavior to the simple Larmor radiation, since the half-wave dipole confines the charge motions to less than the wavelength:

$$P \propto \sin^2\theta \tag{8.24}$$

The resulting polar pattern of the radiation is shown in Figure 8.12, both viewed perpendicular to the antenna axis (a) and parallel to it (b).

The directivity (or maximum gain) of the antenna is measured in dBi, the dB above the signal that would be received from a perfectly isotropic

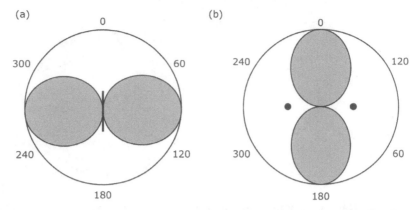

Figure 8.13. Radiation pattern of a pair of half-wave dipoles. The dipoles are shown by the line in (a) (they are projected onto each other) and by the dots in (b).

radiator. dB (short for decibel) is a logarithmic unit of signal above some reference level:

$$G_{\mathrm{dB}} = 10 \log(G) = 10 \log\left(\frac{P_{\mathrm{out}}}{P_{\mathrm{in}}}\right) \qquad (8.25)$$

where $G = P_{\mathrm{out}}/P_{\mathrm{in}}$ is the gain, G_{dB} is the gain in dB, P_{in} is the input power level, and P_{out} is the output. A parallel definition applies to losses of power; for example, a loss of half of the power in a signal corresponds to 3 dB. Losses are indicated with $L = 1/G$. Since dB are logarithmic, the gains and losses in a series of devices can be added to get their net effect.

An isotropic radiator has $G = 1$ (because the power out equals the power in). Similarly, by energy conservation, the gain averaged over all directions for any antenna is

$$\langle G \rangle = \frac{\int G \, d\Omega}{\int d\Omega} = 1 \qquad (8.26)$$

However, different antennas have different maximum gains, as a result of their beaming radiation (or isolating reception) over different solid angles. Roughly speaking, if the maximum gain of an antenna is G_{max}, then the solid angle subtended by its beam is

$$\Omega_{\mathrm{beam}} = \frac{4\pi}{G_{\mathrm{max}}} \qquad (8.27)$$

The gain for the simple lossless dipole is $G_{\mathrm{max}} \sim 1.76\,\mathrm{dBi}$; that is, its beam subtends about 8 steradians. If we use more than one dipole, the polar diagram becomes more peaked as shown for two dipoles arranged parallel to each other and $\frac{1}{2}\lambda$ apart in Figure 8.13. Viewed perpendicular to the dipoles, the pattern is identical to the single dipole. However, viewed parallel to them, and assuming that they are powered in phase, the pattern is as shown in Figure 8.13(b). Any wave launched from the right antenna toward the left one arrives there 180° out of phase compared with the wave from the left antenna, so there is no propagation in the plane defined by the dipoles. Perpendicular to this plane, the waves from both dipoles are in phase and the emission is peaked. Adding dipoles in either direction (parallel or perpendicular to the original one) will further narrow the zone of constructive interference, that is, make the system more directional in its response and increase its gain above isotropic. By driving one of the dipoles in Figure 8.13 at a constant phase difference from the other, the locus of these lobes will be shifted in direction – that is, the pointing of a dipole array can be controlled by controlling the phases from its elements.

By the Antenna Theorem (equation (8.7)), the effective area of an isolated dipole is

$$A_e = \frac{\lambda^2}{\Omega_A} \tag{8.28}$$

$\Omega_A \sim 4$ for a dipole in a dense array spaced at half-wavelengths. Therefore, at long wavelengths very large collecting areas can be realized with arrays of dipole antennas. At wavelengths of ~ 10 m, such arrays are used effectively as large-area radio telescopes (e.g., the Gauribidanur Telescope in India, LOFAR at Effelsberg, and the proposal in the US for the Long Wavelength Array (LWA)). Appropriate phase shifts on the output of each dipole can be used to point the beam and to track objects. Thus, dipole arrays have large fields on the sky within which the beam can be selected. In fact, by adding additional sets of electronics operating at different phase shifts, a number of beams can be generated simultaneously. At the low frequencies where this approach is most effective, the sky foreground is dominated by emission by the Galaxy, and extremely sensitive receivers are not needed to reach the background limit; hence, low-cost commercial electronics can be used for phased dipole arrays. However, for a large collecting area at wavelengths less than a meter, equation (8.28) shows that the number of dipoles and thus the amount of electronics to process their outputs becomes very large. In addition, it is characteristic of the symmetry of dipoles that their response has two lobes one above and one below the plane of the multiple dipoles; the lobe directed toward the ground can pick up interfering signals.

Therefore, in the cm-wave (and often at longer wavelengths), a paraboloidal reflector is placed behind the antenna (now termed a feed), which causes the beam pattern to have one main lobe along the optical axis of the

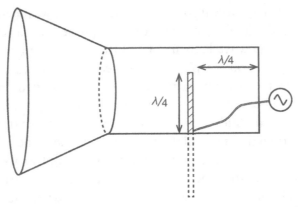

Figure 8.14. A feedhorn with a ground-plane vertical as the antenna. The ground plane vertical creates a full dipole antenna by reflection of one half of such an antenna.

paraboloid. Radio telescope designs of this type are described in Section 2.4.3. They may operate (1) at prime focus, or may have a secondary mirror (called a subreflector), allowing either (2) a classic Cassegrain or (3) an offset Cassegrain to avoid beam blockage by the subreflector, or (4) be designed to couple efficiently to the receiver directly through a waveguide.

With these telescopes, suppressing the backward lobe intrinsic to dipoles and eliminating the multiple antenna interactions in dipole arrays substantially increases the resistance to sources of unwanted interference. At high frequencies (cm-wavelengths), a feed horn (Figure 8.14) is a better approach than an antenna feed because it can give better control of the angular dependence of the signal and thus helps suppress unwanted off-axis response (sidelobes). Corrugations along the wall of a feed horn increase the surface impedance and help convey the wave without losses to a dipole antenna at the exit aperture of the horn. Figure 8.14 illustrates a horn that feeds a waveguide, within which the antenna is placed to optimize its absorption of energy.

Continuing with the thought exercise of using the antenna as a radiator, the response pattern of the feed over the telescope primary mirror aperture is referred to as the illumination pattern. In the design of a feed antenna or feed horn, a variety of illumination patterns can be created. They determine the weighting of the beam over the telescope primary mirror that is used in creating an image. A similar approach is discussed in the preceding chapter (equation (7.24) and surrounding discussion on apodization). In Figure 8.15, we illustrate a number of options. Uniform illumination, (a), yields a PSF following the familiar Airy function. Illuminations with reduced weight at the edge of the primary (e.g., Gaussian, (b)) reduce the sidelobes (the bright rings

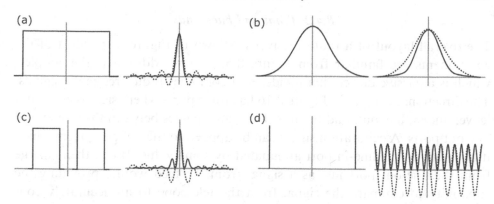

Figure 8.15. Illumination patterns (to left) compared (to right) with the resulting distribution of electric field in the telescope beam (dashed) and point spread functions (solid).

in the Airy pattern), but they also increase the width of the central maximum, that is reduce the resolution. Patterns with increased illumination at the primary edges (e.g., (c)) improve the resolution in the central image at the expense of increased sidelobes. The extreme case of illumination just from apertures at the primary mirror edge (panel (d)) yields fringes. For many radio telescopes, the primary illumination is reduced at the edge of the mirror; a Gaussian illumination pattern is generally a reasonable approximation. This practice both reduces sidelobes and minimizes the response of the telescope to sources behind the primary, whose signals otherwise can be picked up at the focal plane without being reflected by the primary mirror. This unwanted signal is called spillover radiation.

Because the receiver must work at the diffraction limit, any failure to deliver images at this limit by the telescope results in lost efficiency. The resulting requirements can be determined from the Maréchal formula, equation (2.2), called the Ruze formula by radio engineers:

$$S \approx e^{-(2\pi\sigma/\lambda)^2} \tag{8.29}$$

where in this case S is the efficiency and σ is the rms wavefront error. Thus, a surface accuracy of $\lambda/28$ rms (wavefront error of $\lambda/14$) gives an efficiency of 0.8, while a surface of $\lambda/20$ yields an efficiency of about 0.7. A radio telescope with its feed is subject to additional losses due to: (1) blockage of the aperture, for example by a Cassegrain subreflector; (2) spillover, for example energy reflected by the subreflector outside the acceptance solid angle for the receiver; (3) the feed illumination; and (4) other miscellaneous causes. The net efficiency is typically ~0.4, if the antenna surface accuracy is sufficient to make it diffraction-limited.

8.4.2 Front and back ends

The overall layout of a radio receiver is shown in Figure 8.16. Most of the components are familiar from Figure 8.5. A new addition is the coupler, which is a passive device that divides and combines radio frequency signals. The directional coupler in Figure 8.16 has three ports where signals enter and leave: line in, line out, and the tap. The signal passes between the line in and line out ports. A calibration signal can be applied to the tap port, from which it is passed to the line-in port attenuated by some value. It can then emerge from the line-out port just as a signal from the feed would, providing the capability to compare the signal from the telescope to an accurately controlled electronic calibration signal.

We now follow the signal path from the feed to the mixer. This chain is called the front end; its characteristics dominate the behavior of the system. At frequencies below about 100 GHz ($\lambda \approx 3\,\mathrm{mm}$), improved performance can be obtained by interposing high-performance amplifiers between the output of the antenna and the mixer. The effective input noise of the amplifier is lower than the mixer noise, so the receiver noise temperature is reduced with this design. A typical front end then consists of the feed, connecting cables, perhaps filters or couplers, and then one or two low-noise amplifiers, followed by the mixer.

The amplifiers used for the low-power IF output of a mm- or submm-wave mixer, or to amplify the antenna output in the cm-wave, must have very low noise. Excellent performance is obtained with high electron mobility transistors (HEMTs) built on GaAs (and with other devices of related design). The basic transistor (a metal-semiconductor field effect transistor, MESFET) is shown in Figure 8.17. The electron flow between source and drain is regulated by the reverse bias on the gate; with an adequately large reverse bias the depletion region grows to the semi-insulating layer and pinches off the current. Because this structure is very simple, MESFETs can be made extremely small, which reduces the electron transit time between the source and drain and increases the response speed. In the HEMT, the MESFET performance is further improved

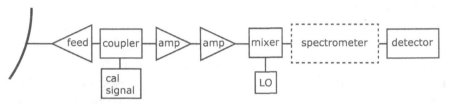

Figure 8.16. Block diagram of a radio receiver.

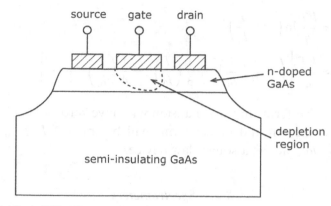

Figure 8.17. A HEMT.

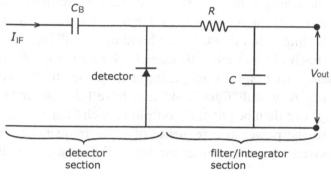

Figure 8.18. A detector stage.

by using a junction between two different zones of the semiconductor with different bandgaps so that the electrons can flow in undoped material. The high mobility in undoped GaAs makes for very fast response, to ~100 GHz.

For a radio receiver, the amplifier stages are followed by a mixer, in the form of a nonlinear circuit element such as a MESFET or diode. The mixer is described as having three ports: an input, a port for the local oscillator signal, and an output port for the IF signal. For all heterodyne receivers (submm-through m-wave), the IF signal is brought to a detector stage, shown schematically in Figure 8.18 (a variety of alternative circuit concepts can perform the same function). This stage rectifies the signal and sends it through a low-pass filter, converting it into a slowly varying indication of the signal strength suitable for interpretation by human beings. We would like the circuit to act as a square law detector because $\langle I_{IF}^2 \rangle$ is proportional to I_S (see equation (8.20)), which in turn is proportional to the power in the incoming signal. To demonstrate how the detector stage achieves this goal, we solve the diode equation (equation 3.12) for voltage and expand in I/I_0:

$$V = \frac{kT}{q} \ln\left(1 + \frac{I}{I_0}\right)$$

$$\approx \frac{kT}{q} \left[\frac{I}{I_0} - \frac{1}{2}\left(\frac{I}{I_0}\right)^2 + \frac{1}{3}\left(\frac{I}{I_0}\right)^3 - \frac{1}{4}\left(\frac{I}{I_0}\right)^4 + \cdots \right] \qquad (8.30)$$

The first and third terms in the expansion will have zero or small conversion efficiency, and the 4th and higher terms will be small if $I \ll I_0$. Thus, the detector stage *does* act as a square law device.

8.4.3 Spectrometers

Usually it is desirable to carry out a variety of operations with the IF signal itself before smoothing it in the detector stage. For example, imagine that the downconverted IF signal is sent to a bank of narrow bandpass fixed-width electronic filters that divide the IF band into small frequency intervals (Figure 8.19). Each of these intervals maps back to a unique difference from the LO frequency, that is, to a unique input frequency to the receiver from the target source. A typical "filter bank" may have 128, 256, or 512 channels. A detector stage can then be put at the output of each filter, so the outputs are proportional to the power at a sequence of input frequencies; that is, they provide a spectrum. In this manner the total IF bandwidth is divided into

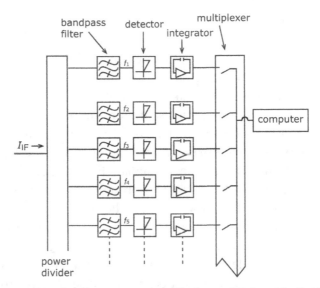

Figure 8.19. A filter bank spectrometer. The input IF signal is divided among the bandpass filters and the output of each is processed by a detector/integrator stage. The outputs of these stages are switched sequentially to the computer where the spectrum can be displayed.

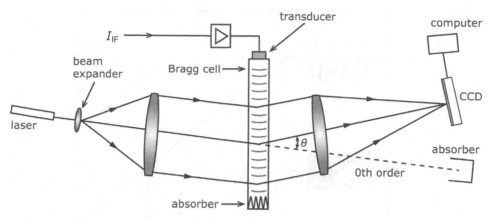

Figure 8.20. An acousto-optical spectrometer.

a spectrum, even though only a single observation with a single receiver has been made; the process is called spectral multiplexing.

Although conceptually simple, a high-performance filter bank can be an engineering challenge. The filters need to have closely matched properties and be robust against drift of those properties due to effects like temperature changes. A filter bank is also inflexible in use; the resolution must be set during design and construction. Finally, these devices are complex electronically and expensive to build if many channels are required.

An acousto-optical spectrometer (AOS) provides a many-channel spectrometer without the electronic complexity of a filter bank, since it divides the IF signal into frequency components without a dedicated unit for each component. In this device (see Figure 8.20), a piezoelectric transducer is attached to a Bragg cell, a transparent volume containing either a crystal like lithium niobate or water. When the IF signal is fed into this transducer, it vibrates to produce ultrasonic waves that propagate through the Bragg cell and produce periodic density variations. As a result, the index of refraction in the cell also varies periodically, making it act like a volume phase diffraction grating (Section 6.3.2.1). When light from a near-infrared laser diode passes through the cell, it is deflected accordingly – the zero-order path is absorbed and the first-order path is focused onto a CCD by camera optics. The light intensity is proportional to the IF power injected into the Bragg cell, while the deflection angle and hence position on the CCD is determined by the ultrasound wavelength. The output signal is basically the Fourier transform of the IF signal. AOSs are capable of resolving the IF signal into more than 2000 spectral channels.

A third method to divide the IF into a spectrum is called a chirp transform spectrometer. In this case, the LO is swept in frequency, at a rate df/dt.

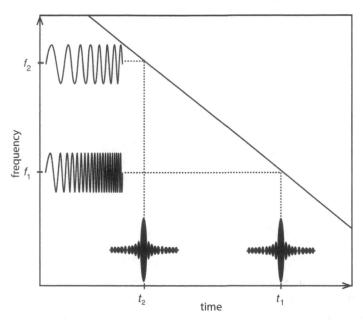

Figure 8.21. Operation of a chirp transform spectrometer.

This process is termed the expander. As a result, an input signal at a fixed frequency is modulated at linearly swept frequencies in the IF. A delay line with a delay time depending on frequency, called the compressor, is put in the IF path with a dispersion, $d\tau/dt$, that just counteracts the frequency sweep rate. The result is that the fixed input frequency is converted to a sinc function output at a specific time; a different input frequency will produce an output at a different time, as illustrated in Figure 8.21. Therefore, the spectrum emerges in time series. Since the full input bandwidth is being processed at all times, there is no compromise in the sensitivity of the receiver.

The fourth method to obtain a spectrum from the IF signal is to compute its autocorrelation, the integral of the results of multiplying the signal by itself with a sequence of equally spaced delays:

$$R(\tau) = \lim_{T \to \infty} \frac{1}{2T} \int_{-T}^{T} U(t)U(t+\tau)dt \tag{8.31}$$

The Fourier transform then gives the spectrum, according to the Wiener–Khinchin Theorem (Wilson *et al.* 2009). Because it is not possible actually to carry out the limit to infinite time in equation (8.31), autocorrelators produce spectra with "ringing" due to any sharp spectral features. Similar behavior was discussed with regard to Fourier transform spectrometers (Section 6.5). As in that case, these artifacts can be reduced by filtering the signal (e.g., with a "Hanning filter"), but with a loss in spectral resolution.

Table 8.1. *Information retained as a function of digital bits*

Bits	Information
1	64%
2	81%
3	88%
infinite	100%

To measure the autocorrelation, the IF signal is first digitized. High-speed performance is required; the Nyquist Theorem says that the digitization rate must be at least twice the IF bandwidth. Therefore, autocorrelators often digitize to only a small number of bits. As shown in Table 8.1, the loss of information is surprisingly modest if the gains are set optimally. Many systems only digitize to two bits. The digitized signal is taken to an electronics circuit that imposes the necessary delays in the signal using shift registers and then combines the results to provide the autocorrelation as an output. Autocorrelators are flexible in operating parameters and very stable, since they work digitally.

8.4.4 Receiver arrays

Because of the complexity of the supporting electronics (primarily the back end) for each receiver, most telescopes have operated with just one receiver observing at a time. Thus, maps in the radio are typically made "on the fly" (OTF), meaning that the single beam is scanned systematically over the field to sample the image pixels one at a time. Substantial gains in mapping speed can be achieved with an array of receivers. Making such instruments practical depends on creating cost-effective forms of multiple back end electronics. Arrays with of order 100 receivers are in operation at a number of telescopes (e.g., APERATIF at Westerbork operating in the GHz range; Oosterloo *et al.* (2010) and Super-Cam at the University of Arizona Submm Telescope, operating in the mm-wave; Groppi *et al.* (2008)).

8.5 Characterizing heterodyne receivers

8.5.1 Noise temperature

Although the final astronomical results from heterodyne receivers are often quoted in familiar radiometric units (e.g., Jy), a temperature-based system of characterization is also used, as introduced in Section 8.3.3 where we defined noise temperature and antenna temperature. This approach is also employed

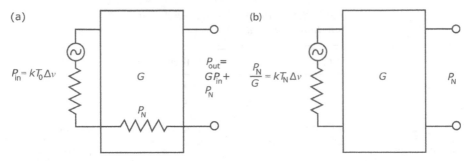

Figure 8.22. (a) Two-port with internal noise, and (b) idealized version (with input signal removed).

for the components of the receiver. We have also already used a common terminology in radio electronics, in which circuits are considered as black boxes without detailed consideration of their components, and they are characterized in terms of the inputs and outputs, called ports.

For example, a two-port device is shown in Figure 8.22(a). It has a gain for signals, G,

$$G = \frac{P_{\text{out}}}{P_{\text{in}}} \tag{8.32}$$

where P_{in} is the input signal power and P_{out} is the output. In addition, the device will add any internal noise to the output. It is convenient to abstract it to be ideal, that is, lossless and noise free.

Thus, to reproduce the sources of noise internal to the device we assume an imaginary input that gives the noise, P_{N}, as the output. That is, as shown in Figure 8.22(b), we put on the input a resistor that produces a Johnson noise (equation (8.3)) of P_{N}/G. The resistance value is set to the input impedance of the device. The power that can be delivered to the matched load (equal resistance) is then

$$P_{\text{N,in}} = kT\,\Delta v \tag{8.33}$$

where Δv is the frequency bandwidth. We have to adjust the temperature to obtain the required noise. This temperature is called the noise temperature of the device, T_{N}:

$$T_{\text{N}} = \frac{P_{\text{N}}}{kG\,\Delta v} \tag{8.34}$$

A complex system can be represented as a linear chain of devices; see Figure 8.23. We show the input (i.e., from the feed) not as power, but as the equivalent temperature, T_0, and each two-port device is labeled with its

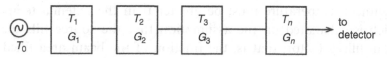

Figure 8.23. A series of two-ports; the T_i are the noise temperatures and the G_i are the gains of each one.

noise temperature and gain (this notation includes the possibility that some of the gains are less than 1 – e.g., a lossy cable or an attenuator). The corresponding system temperature is

$$T = T_0 + T_1 + \frac{T_2}{G_1} + \frac{T_3}{G_1 G_2} + \frac{T_4}{G_1 G_2 G_3} + \cdots + \frac{T_0}{G_1 G_2 G_3 \dots G_{n-1}} \quad (8.35)$$

The noise temperatures of the following devices add to the external temperature (the antenna temperature), but each stage after the first one is divided by the total gains of the preceding stages. To achieve minimum system temperature, low noise in the first amplifier stage is critical.

As an example, we analyze the system in Figure 8.16, assuming a telescope of 25-m aperture and taking values appropriate for operation at 6 cm ($\sim$5 GHz). Typical values for the noise inputs that constitute T_0 are the Cosmic Microwave Background (CMB), $T_{CMB} = 3$ K; sky emission, the product of temperature and opacity, $T_{sky} = 3$ K; and a contribution from scattering of ground radiation off the antenna and of spillover radiation from beyond the geometric aperture of the telescope, $T_{spill} = 7$ K. Therefore, $T_0 = 13$ K. In analyzing the receiver, we assume that the cable and coupler together impose a loss of 0.3 dB. This corresponds to $L = 1.0715$ and, for a physical temperature of 290 K, a noise temperature of 20.7 K (see exercise 8.5). We assume the amplifier stages have noise temperatures of 20 K and gains of 25 dB. The noise contribution of the first amplifier is then $1/G_1$ times T_2, or 1.0715×20 K $= 21.4$ K. We assume the following cables are short and their losses are negligible. The second amplifier than adds a noise of 20 K times $1/G_1$ and $1/G_2$, $G_2 = 316$, so 0.1 K. Clearly, any following electronics will not contribute significantly to the noise. The quantum limit is only 0.2 K. Therefore, the noise temperature of the entire unit is the sum of the values above, or 55.2 K.

8.5.2 Source characteristics

The brightness temperature of an astronomical source, T_b, is the temperature of a blackbody that subtends the same solid angle and matches its flux density. For sources that are unresolved by the beam of a radio telescope, the brightness temperature must be greater than the antenna temperature, T_S, which is calculated on the assumption that the source fills the beam. The beam filling factor, that is, the fraction of the beam area filled by the source, is the ratio

$$filling\ factor = \frac{T_S}{T_b} \tag{8.36}$$

This quantity is also called the beam dilution.

To illustrate, assume a source that yields a flux density of 0.01 Jy at 6 cm is observed with the telescope discussed above, which has an efficiency of 0.4. From equation (8.10), the antenna temperature from this source is 0.0007 K. From equation (8.11) (the Dicke Radiometer equation), and if the system has an IF bandwidth of 2 GHz, then it will require about 310 seconds to detect it at a signal-to-noise ratio of 10. This estimate is very optimistic, since it ignores time spent observing sky for a zero reference and other overheads.

8.6 Observatories

8.6.1 Submillimeter

Observations from the ground in the submillimeter are extremely sensitive to the amount of water vapor overlying the observatory site. Figure 1.6 demonstrates that the windows at 350 and 450 μm close and the one at 850 μm is impaired at total water vapor levels of 2 mm and above. Even observatory sites considered for other purposes to be high and dry only allow submm observations under the most favorable circumstances. For example, over Paranal, the site of the VLT and at an altitude of 2635 m, for half of the nights the level is above 2 mm and it reaches 0.5 mm only on very rare winter nights. Therefore, the selection of sites with extremely low water vapor is critical to success for a submm observatory, requiring that the observatory be either placed at high latitude (the Antarctic has been the site of a number of successful observations) or extremely high altitude (e.g., 5000 meters elevation for the Atacama Large Millimeter Array (ALMA)).

Wind-blown fluctuations in the water vapor content of the air over a submm telescope produce a fluctuating level of thermal emission into the beam, resulting in noise. This noise rises rapidly with decreasing frequency. For modern broadband continuum cameras (e.g., bolometric detectors), it dominates the detector noise below about 0.5 Hz (e.g., Sayers *et al.* 2008). As in the infrared (Section 2.4.2), the noise can be reduced by nutating the secondary mirror at a frequency of 1 Hz or higher, so the

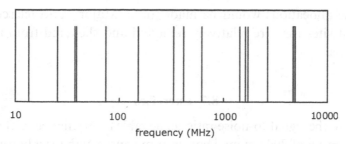

Figure 8.24. Radio frequencies allocated to astronomy between 10 MHz and
10 GHz. The astronomy "preserves" are the vertical lines (seldom of a
resolved width); all the open spaces between them are used for other purposes.

measurements are made between the source and a blank region next to
the source in rapid sequence. The effect largely arises close to the telescope,
owing to the small-scale height for water vapor. For small nutation angles,
the two positions sample nearly the identical path through the low-lying
atmosphere, thus suppressing the noise while modulating the source signal
completely. Fitting the common-mode fluctuations and spatial behavior
over the pixels of an imaging array can also be effective in reducing the
impact of the sky signal.

8.6.2 *Radio*

Over much of the cm- and m-wave spectrum, the atmosphere is quite transparent. At very long wavelengths, radio signals are reflected by the ionosphere
of the earth; this effect increases in proportion to λ^2, making the atmosphere
opaque as viewed from the outside at a frequency of about 10 MHz and below
(wavelength of about 30 m and longer). Absorption by atmospheric oxygen
makes groundbased radio observations impossible between 52 and 68 GHz and
in a narrow range near 118 GHz. Water vapor absorption becomes increasingly
strong at frequencies above 10^{11} Hz and is dominant above 3×10^{11} Hz, as
discussed in the preceding section.

However, the dominant obstacle to untrammeled radio astronomy is man,
in the form of radio transmission devices. Spectral multiplexing becomes a
problem when there is a strong source of interference, since a strong signal
at a single frequency within the range of frequencies downconverted to the
IF can overwhelm the entire output of a receiver. Often, notch filters that
suppress a single frequency must be employed to circumvent such issues. Radio
frequency allocations are subjects of intense negotiation because substantial
amounts of money can be involved in radio communications. Figure 8.24
shows the frequency allocations for radio astronomy. Clearly, operating only

within these allocations would be inadequate. Major radio telescopes must be placed at sites that are relatively removed and sheltered from man-made transmissions.

8.7 Exercises

8.1 Compare the signal-to-noise ratio achievable (at negligible background) at a wavelength of 300 μm on a continuum source with (a) a bolometer with an NEP of $4 \times 10^{-16}\,\mathrm{W\,Hz^{-1/2}}$ operated through a spectral band of 30% the center frequency; and (b) a heterodyne receiver operated double sideband but with single sideband noise temperature of 1500 K, and an IF bandwidth of $3 \times 10^{9}\,\mathrm{Hz}$. At what spectral bandwidth would the two systems give equal signal-to-noise ratio?

8.2 Show that the gain of a simple dipole antenna is 1.76 DBi. Hint: By conservation of energy, the integral of the gain over all solid angles is

$$\int G\, d\Omega = 4\pi$$

8.3 The quiet sun has a brightness temperature T_b (10 GHz) ~ 11 000 K and an angular diameter of 32 arcmin. What is the flux density we receive from the sun? A VLA (Very Large Array) telescope is 25 m in diameter. What is its beam size at 10 GHz? What flux density will it measure at the center of the sun?

8.4 Prove that the antenna temperature equals the source brightness temperature multiplied by the fraction of the beam solid angle filled by the source.

8.5 Show that the noise temperature of a transmission line is

$$T_N = (L - 1)T$$

where L is the loss in the line and T is its physical temperature. Hint: Use the relation (from the equation for Johnson noise) that the maximum power that can be extracted from a resistor at temperature T is

$$P_{\max} = kT\, df$$

8.6 Consider a 10-m aperture radio telescope operating at 230 GHz with an IF bandwidth of 4 GHz. Consulting Figure 8.5, the noise temperatures are: sky, 80 K; spillover and scattered radiation, 7 K; diplexer and relay optics, 20 K; mixer, 40 K; amplifier, 8 K with 30 dB gain. The mixer is operated with a conversion loss of 5 dB and the aperture efficiency is 0.4. How long will be required to achieve 5 : 1 signal-to-noise ratio on a source of 0.01 Jy?

Further reading

Burke, B. F. and Graham-Smith, F. (2009). *Introduction to Radio Astronomy*, 3rd edn. Cambridge: Cambridge University Press.

Condon, J. J. and Ransom, S. M. (2010). Essential radio astronomy: www.cv.nrao.edu/course/astr534/ERA.shtml

Kraus, J. D. (1986). *Radio Astronomy*, 2nd edn. Powell, OH: Cygnus-Quasar Books.

Mazin, B. A. (2009). Microwave kinetic inductance detectors: The first decade. *AIP Conf. Proc.*, **1185**, 135–142.

Rieke, G. H. (2003). *Detection of Light from the Ultraviolet to the Submillimeter*, 2nd edn. Cambridge: Cambridge University Press.

Wilson, T. L., Rohlfs, K., and Hüttemeister, S. (2009). *Tools of Radio Astronomy*, 5th edn. Berlin, New York: Springer.

9

Interferometry and aperture synthesis

9.1 Introduction

The Byrd Green Bank Telescope is the largest fully steerable filled-aperture radio telescope, with a size of 100×110 meters. The runners-up are the Effelsberg telescope, with a diameter of 100 meters, and the Jodrell Bank Lovell Telescope, 76 meters in diameter. The collecting areas of these telescopes are awesome, but their angular resolutions are poor: at $\lambda = 21$ cm, λ/D for a 100-m telescope is about 7 arcmin. These enormous telescope structures are difficult to keep in accurate alignment and are subject to huge forces from wind. Proposals for larger telescopes (e.g., the Jodrell Bank Mark IV and V telescopes at 305 and 122 m respectively) pose major engineering challenges for modest improvements in resolution and have also proven to press the limits of what other humans are willing to purchase for astronomers. The Arecibo disk circumvents some of these issues by being fixed in the ground and pointing by moving its feed; this concept allows a diameter of 259 m. However, the dish must be spherical, limiting the optical accuracy of the telescope, and it can reach only a fraction of the sky (roughly $\pm$ 20° from the zenith). The angular resolution is still modest, a bit worse than 2 arcmin at 21 cm (1.4 GHz). This resolution is only equivalent to that of the *unaided* human eye!

Even if we could build larger telescopes, they would be limited by source confusion: in very deep observations, the background of distant galaxies is so complex to these relatively large beams that it becomes impossible to distinguish an object from its neighbors (see Section 1.5.4).

Clearly some other approach is needed to make radio images at resolutions complementary to those in the visible. The way to do so is indicated in panel (d) of Figure 8.15. If we mask off all of our telescope mirror except for two small apertures at the edge, the "central peak" of our image becomes

266

sharper than the Airy function but with huge sidelobes, that is, fringes. In general, if apertures of diameter d are separated by a distance D', the fringes projected up onto the sky will be spaced at λ/D' with envelopes of λ/d (corresponding to the diffraction limit of the individual apertures). If we put the apertures at the edges of the mirror, so $D' \sim D$, the spacing will be $\sim\lambda/D$. Since this angular distance corresponds to adjacent maxima, the fringe full width at half maximum (or "beam" width) is $\sim\lambda/2D$. Higher resolution can be obtained with an interferometer, where we separate our small apertures by a large distance, B, without filling in between them with a mirror. The λ/d images are then called the primary beam of the interferometer, impressed on which are fringes providing resolution (FWHM) of $\sim\lambda/2B$. Although we have justified interferometry with the example of radio telescopes, it is used generally to push to angular resolutions beyond those obtainable with a conventional telescope.

In principle, we could try to determine if a source is resolved with an interferometer by measuring the width of the fringe "peaks," but it is easier and more quantitative to base the measurement on the filling-in of the dark fringes for resolved sources, the fringe contrast. We describe the contrast in terms of the visibility, $V(r)$:

$$V(r) = \frac{I_{\max} - I_{\min}}{I_{\max} + I_{\min}} \tag{9.1}$$

where $I_{\max}$ and $I_{\min}$ are the maximum and minimum fringe intensities. The structure of a resolved source can be probed by varying the spacing of the apertures, r, and measuring the visibility as a function of this spacing.

We now illustrate the behavior of the visibility. It can be shown (Van Cittert–Zernicke Theorem; Thompson *et al.* 2001) that the output signal of the interferometer is the Fourier transform of the observed brightness distribution of the source. More simply, the behavior of an extended source can be viewed as a superposition of fringes from its components, each with an individual amplitude and phase. Thus, if we observe two equal-brightness stars that are far enough apart to resolve the pair, the fringes will have smaller amplitude than expected for a single star of the equivalent total brightness because of the phase differences of the fringes (which encode the positions of the stars).

Visibilities give a somewhat abstract view of the source structure; true images are more user-friendly. As a thought experiment, one could measure through pairs of apertures one at a time at all possible distances and clock angles over the primary mirror of a telescope and use the information (rather painfully) to reconstruct the image that would have been obtained with the telescope without apertures. At least the various patterns of fringes should have all the information in that image. This procedure is the basis of aperture synthesis. By using many interferometer spacings and orientation angles, we can collect information (and make images) approaching what we would have obtained from a single huge telescope.

9.2 Radio two-element interferometry

Radio heterodyne receivers provide great flexibility for interferometry because their outputs retain phase information about the incoming signal. Therefore, the functions underlying the interference can be conducted off-line. As a result, very large baselines allow efficient imaging at arcsec or subarcsec resolution, and large collecting areas can be accumulated by combining the outputs of many telescopes of modest size and cost.

9.2.1 Basic operation

The basic radio two-element interferometer is shown in Figure 9.1. For simplicity, we assume that the signal is at a single frequency, v, and instant of time so the path difference to the two telescopes can be described by a single phase difference. We also assume that the two telescopes are identical, so their signals are of equal amplitude.

As shown in Figure 9.1, the signals from telescopes 1 and 2, V_1 and V_2, consist of voltages from the source, V_{S1} and V_{S2}, and voltages from the noise of the receivers, V_{R1} and V_{R2}: for example, $V_1 = V_{S1} + V_{R1}$. They are combined in a cross-correlator, where they are multiplied:

$$\langle V_1 \cdot V_2 \rangle = \langle (V_{R1}V_{S2} + V_{R1}V_{R2} + V_{S1}V_{R2} + V_{S1}V_{S2}) \rangle \approx \langle V_{S1}V_{S2} \rangle \quad (9.2)$$

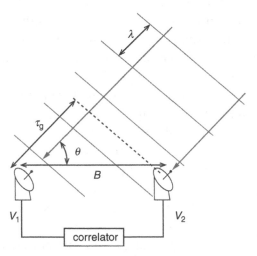

Figure 9.1. A radio two-element interferometer.

where the angular brackets designate a time average. The simplification is possible because the terms other than the final one represent multiplication of uncorrelated quantities and hence average approximately to zero, leaving only small noise residuals. As a result, many potential sources of noise, such as fluctuations in the receiver gains or outside interference, drop out of the signal. Since the noise terms drop out of the relation, the output of the correlator becomes[19]

$$
\begin{aligned}
\langle V_1 \cdot V_2 \rangle &= a_{s1}\cos(\omega(t - \tau_g))a_{s2}\cos(\omega t) \\
&= a_{s1}a_{s2}(\cos^2(\omega t)\cos(\omega\tau_g) + \cos(\omega t)\sin(\omega t)\sin(\omega\tau_g)) \qquad (9.3) \\
&= \tfrac{1}{2}a_{s1}a_{s2}(\cos(\omega\tau_g) + \cos(2\omega t)\cos(\omega\tau_g) + \sin(2\omega t)\sin(\omega\tau_g))
\end{aligned}
$$

where $\omega = 2\pi v$ and a_{s1} and a_{s2} are the amplitudes and τ_g is the time delay between the signals into the two telescopes.

The terms with arguments of ωt are at very high frequency and average to zero; we can also simplify because the amplitudes are equal, giving

$$
\langle V_1 \cdot V_2 \rangle \approx \langle V_s^2 \rangle \approx \left(\frac{V^2}{2}\cos\omega\tau_g\right) \equiv R_C \qquad (9.4)
$$

R_C is defined as the cosine response.

Returning to Figure 9.1, different directions in the plane of the figure yield different values for τ_g and so by equation (9.4), the response of the interferometer varies sinusoidally across the sky in the direction along the line joining the two telescopes, yielding interferometer fringes. At the same time, the individual signals from a source will fall as it is put increasingly off-axis to the telescopes, defining the primary beam of the interferometer within which the fringes have reasonable amplitudes. The fringe phase is

$$
\phi = \omega\tau_g = \frac{\omega}{c}B\cos\theta \qquad (9.5)
$$

where θ is the direction toward the source measured relative to the line joining the telescopes (Figure 9.1) and B is the distance between the telescopes. The change of phase with direction toward the source is

$$
\frac{d\phi}{d\theta} = \frac{\omega}{c}B\sin\theta = 2\pi\left(\frac{B\sin\theta}{\lambda}\right) \qquad (9.6)
$$

A change in the phase by 2π, corresponding to the fringe period, occurs for a change in direction toward the source, $\Delta\theta$, by $\lambda/B \sin\theta$. This behavior is as

---

[19] The first equality uses the trigonometric identity $\cos(a - b) = \cos(a)\cos(b) + \sin(a)\sin(b)$, while the second uses $\cos^2(a) = (1/2)(1 + \cos(2a))$ and $\sin(2a) = 2\cos(a)\sin(a)$.

expected, since $B \sin\theta/\lambda$ is the distance between the telescopes as viewed from the source in units of the wavelength.

Thus, the period is inversely proportional to the baseline, B, and the fringe phase is a sensitive measure of the source position, particularly if the baseline, B, is large. We also have that the beamwidth, θ_S, is half the fringe period on the sky, that is:

$$\theta_S \approx \frac{\lambda}{2B \sin\theta} \tag{9.7}$$

It is important that these quantities depend only on the phase difference between the telescopes, and therefore on the baseline and measures of time – issues in imaging with the individual filled apertures such as tracking accuracy and wind buffeting are not critical (except to the relatively relaxed tolerances required for pointing of the individual telescopes).

9.2.2 Operational considerations

In practice, as the interferometer tracks a source a constantly varying delay must be imposed on the signal from the leading telescope (to the right in Figure 9.1) to compensate for the change in path length. This correction can only be precise in one direction, called the delay center, as can be understood by visualizing the interferometer fringes over a range of wavelengths. At the center of the field, they will all be at the same phase and the visibility will be large. However, they will have different fringe periods and as one goes away from this central field point the fringes will become displaced in phase relative to one another and the visibility will decline, thus imposing a limit on the useful interferometer field. A modest range of frequencies can be accommodated with an acceptable loss of efficiency, but for large bandwidths it is necessary to divide the signal into frequency sub-bands and impose appropriate delays on each one. If the inaccuracy of this delay is $\Delta\tau_g$, the resulting inaccuracy of the path is $c\Delta\tau_g$; for reasonable efficiency in the interference, this path variation must be small compared with the range of wavelengths in the instrument bandwidth:

$$\frac{c\Delta\tau_g}{\Delta\lambda} = \Delta\nu \, \Delta\tau_g \ll 1 \tag{9.8}$$

From Figure 9.1,

$$c \, \tau_g = B\cos\theta \tag{9.9}$$

or

$$|c\Delta\tau_g| = B \sin\theta \, \Delta\theta \tag{9.10}$$

where $\Delta\theta$ is the angular distance from the delay center. Combining equations (9.7), (9.8), and (9.10), we find the requirement that

$$\Delta\theta \, \Delta v \ll \theta_S v \tag{9.11}$$

to avoid "bandwidth smearing" that broadens the beam in the radial direction. A related requirement is that the correlator must sample the signals fast enough that the change in beam direction is less than the beamwidth, θ_S, or the image will be subject to "time smearing." In the following, we will ignore these complexities (or assume that the engineering staff has provided sufficient electronics to banish them) and discuss basic interferometer performance.

The cosine form in equation (9.4) is sensitive on an extended source only to the symmetric part of the source distribution. The anti-symmetric part can be recovered by dividing the output of the telescopes into two signals and applying a 90° phase shift to the signal from one of the telescopes in one of these pairs. This pair is also cross-correlated to yield an output as in equation (9.4) but with a sine response:

$$R_S \equiv \left(\frac{V^2}{2} \sin\omega\tau_g \right) \tag{9.12}$$

Correlators with these dual outputs are called complex correlators because their outputs are generally manipulated as complex numbers combining the amplitude and phase of the signal. The complex visibility is

$$V = R_C - jR_S = Ae^{j\phi} \tag{9.13}$$

with the visibility amplitude

$$A = \sqrt{R_C^2 + R_S^2} \tag{9.14}$$

and phase

$$\phi = \tan^{-1}\left(\frac{R_S}{R_C} \right) \tag{9.15}$$

The complex visibility is the response of an interferometer with a complex correlator and encodes the full structure of an extended source. For the case of a linear interferometer, the Van Cittert–Zernicke Theorem shows that the intensity distribution of the source on the sky is the Fourier transform of the visibility (to be discussed further in Section 9.3.2); a generalization of this theorem allows extension to three-dimensional interferometers (e.g., the Very Large Array (VLA) and Atacama Large Millimeter/Submillimeter Array (ALMA).

9.3 Aperture synthesis

9.3.1 Basic operation

Aperture synthesis increases the power of interferometry to produce true images. The limited information about the structure of a source provided by a two-element interferometer is expanded by moving the telescopes to change the baseline spacing and repeating the observations. This process can be accelerated by putting a number of telescopes along the baseline; if there are N telescopes, then their outputs can be combined to yield $N(N-1)/2$ unique baselines (of course, with $N(N-1)/2$ correlators to carry out the processing). Each baseline adds a new Fourier component (unique fringe spacing and/or direction) to the image.

The sensitivity also grows with the increasing number of telescopes. V_1 and V_2 in equation (9.4) are proportional to the electric fields generated from the signals from each telescope, that is, the gains of the two telescopes. Since the power is proportional to the autocorrelation of the field, the fields are proportional to the square roots of the telescope areas; therefore, V^2 is proportional to the flux density of the source times $(A_1 A_2)^{1/2}$, where A_1 and A_2 are the collecting areas of the two telescopes. Thus, the effective area of the interferometer is

$$A_{\text{effective}}^{\text{interferometer}} = \sqrt{A_1 A_2} \tag{9.16}$$

For example, for an interferometer with two identical elements, the effective area is that of one of the elements. The noise from each of the two elements is, however, independent, so the noise in the correlator output is reduced by the square root of 2 compared with that from a single telescope. That is, the signal-to-noise ratio achieved with a two-element interferometer is equal to that of a single telescope with area equal to the square root of 2 times the area of one of the elements (not the combined area of the two elements). Since an array of N identical telescopes provides $N(N-1)/2$ independent two-element interferometers, its net sensitivity is given by a modification of the radiometer equation (8.11):

$$\left(\frac{S}{N}\right)_{\text{c}} = K \frac{T_{\text{S}}}{T_{\text{N}}^{\text{S}}} [N(N-1)(\Delta f_{\text{IF}} \Delta t)]^{1/2} \tag{9.17}$$

(with terms of order unity subsumed into K).

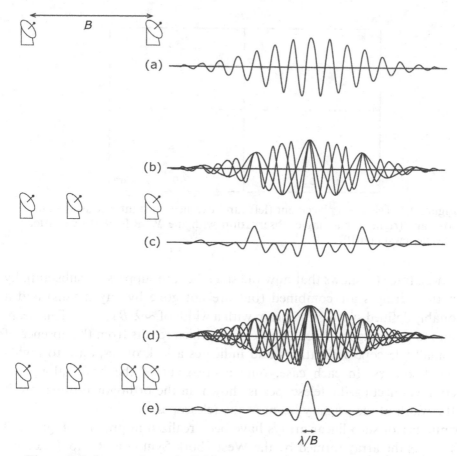

Figure 9.2. Improvement in field pattern quality (the images are the auto-correlation) with increasing number of interferometer baselines. Based on Condon and Ransom (2010).

Since even the most complex combination of multiple radio telescopes can be treated as a large number of two-element interferometers, our analysis of the case in Figure 9.1 sets the foundation for the discussion of more complex telescope arrays used for aperture synthesis. Figure 9.2 illustrates the process.

The number and placement of the telescopes is indicated by the icons. Trace (a) is the field pattern obtained from a two-element interferometer with elements separated by B. It is the product of the fringes and the primary beams of the constituent telescopes. Trace (b) shows the fringe patterns resulting if one more telescope is added to provide three baselines. As shown in trace (c), these fringes can be combined to provide a central response peak, called the synthesized beam, but with a number of large sidelobes. Trace (d) is the fringe patterns from an interferometer with four telescopes (and six

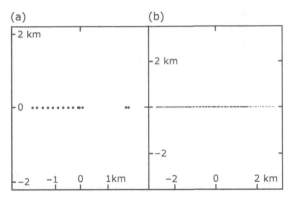

Figure 9.3. Telescope placement (left) and overhead instantaneous *uv* plane coverage (right) for a single observation with the WSRT. Redrawn after Mioduszewski (2010).

baselines); trace (e) shows that now the sidelobes are suppressed substantially when these fringes are combined (but are not gone by any means) and a reasonably defined synthesized beam with a width of $\sim \lambda/B$ results. This beam appears to be sitting in a broad depression, which results from the absence of very small telescope separations and indicates a lack of response to highly extended sources. In each case, only a cross-cut of the beam along the direction connecting the telescopes is shown; in the orthogonal direction, it has the width of the primary beam of the telescopes.

A number of such linear arrays have been realized in practice. Figure 9.3 (left) shows the array formed by the Westerbork Synthesis Radio Telescope (WSRT), consisting of 14 telescopes along a 2.7 km baseline oriented east–west; 10 telescopes are fixed, and four are movable on railroad tracks to allow repeated measurements of a source to generate new baselines. Its receivers cover the range from 115 MHz to 8.6 GHz (these telescopes are being converted to another application: rapid, wide-angle surveys). The Australian Telescope Compact Array (ATCA) was originally another east–west interferometer, consisting of five 22-m telescopes that could be moved along a 3-km railroad track and a sixth telescope 3 km to the west of the western end of the track. Its configuration has been modified to improve the imaging performance near the equator. It is the prime southern-hemisphere cm-wave telescope array and operates over the 0.6 to 90 GHz range.

The right side of Figure 9.3 shows the multiple baselines among the 14 WSRT elements as they might be viewed from a radio source. We describe this view in terms of a coordinate system in *u*, *v*, and *w*; *w* is the direction toward the source and *u* and *v* are Cartesian coordinates on a plane perpendicular to *w* and with *u* to the east and *v* to the north (Figure 9.4). Dimensions

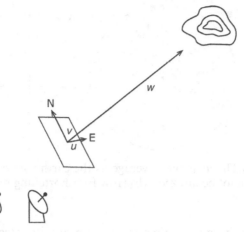

Figure 9.4. Definition of the *uv* plane.

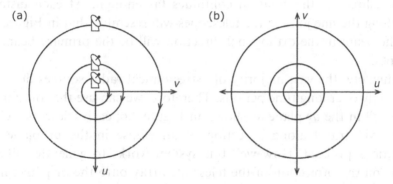

Figure 9.5. The rotation of a linear interferometer can fill out the *uv* plane coverage.

in these coordinates are measured in wavelengths. Thus, the *uv* plane coverage of the WSRT at a single time is as given in Figure 9.3 (right).

As described up to now, telescopes like the WSRT would produce beams well confined in one coordinate but with widths equal to the primary beamwidth of the constituent telescopes in the other. Fortunately, much better performance is possible. To form images, we would like to have our telescopes completely cover the area that would be occupied by a filled telescope mirror – in the case of Westerbork, a 2.7-km diameter disk pointing toward the source.

An approximation to this complete *uv* plane coverage can be achieved by rotating the array, as shown in Figure 9.5 for the ideal case where the source is at the zenith. In Figure 9.5, a three-element linear array similar to that in Figures 9.2(b) and (c) is shown to the left as it might be viewed from the

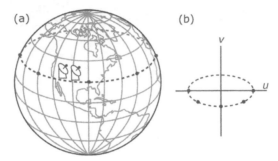

Figure 9.6. (a) The *uv* plane coverage as the earth rotates (the track behind the horizon cannot be utilized). (b) How foreshortening reduces the coverage in the *v* direction.

source. There are three possible baselines among its telescopes; as the array is rotated, these three baselines provide coverage in arcs in the *uv* plane that describe ellipses if the rotation continues far enough. At each instant the beam along the line joining the telescopes will resemble that in Figure 9.2(c), while the beam in the orthogonal direction will be the primary beam of the telescopes.

Fortunately, the vast majority of astronomical radio sources do not vary significantly over long time periods. Therefore, we can use the rotation of the earth to fill in the *uv* plane as shown in Figure 9.6; each telescope–telescope baseline will travel along a section of an ellipse in the *uv* plane as the observations proceed. How well this system works to generate full images depends on the projection of the telescope array onto the *uv* plane; in some directions, the ellipses are significantly foreshortened. The foreshortening for Westerbork as a function of the declination of the source is illustrated in Figure 9.7. The very different patterns of *uv* plane coverage as a function of declination with linear arrays is avoided with two-dimensional arrays of telescopes such as the "Y" configuration of the 27 telescopes of the Very Large Array (VLA), as shown in Figure 9.8. Although the earth may block the view of the source over part of its rotation, sufficient *uv* plane coverage can be obtained to make good images at all accessible declinations.

For the VLA, the 25-m diameter telescopes can be moved to provide configurations ranging from D-array (minimum baseline 35 m, maximum 1 km) to A-array (minimum baseline 0.68 km, maximum 36 km). The VLA has receivers to provide data from 75 MHz to 50 GHz. As another example, the Giant Meterwave Radio Telescope (GMRT) has 30 telescopes each 45 meters in diameter, with 14 telescopes in a compact central core and the remaining in a Y configuration similar to that of the VLA (but fixed, so the flexibility to fill in the *uv* plane is limited). It has receivers operating from

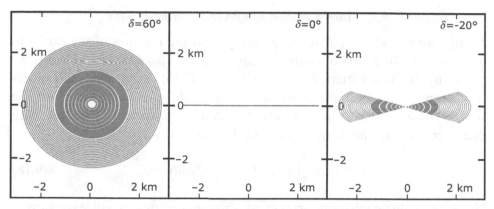

Figure 9.7. The *uv* plane coverage with Westerbork for full synthesis. At the celestial equator, the rotation of the earth does not change the array orientation in celestial coordinates. At the celestial pole, the linear array describes circles. Other declinations are intermediate in behavior (as sinδ, where δ is the declination). Redrawn after Mioduszewski (2010).

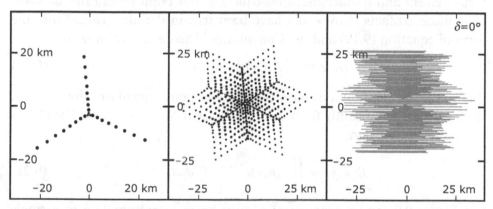

Figure 9.8. Arrangement of the telescopes in the VLA (left), the resulting instantaneous *uv* coverage (center) and full coverage with tracking (right) at the equator and with the source above 10° elevation, to be compared with the center panel in Figure 9.7. Redrawn after Mioduszewski (2010).

50 to 1420 MHz. A variety of other two-dimensional configurations have specific advantages; for example, spiral placement of the telescopes can yield many short baselines and good performance on extended emission, whereas placing them in a circle yields many long baselines. ALMA will have 54 12-m telescopes and 12 7-m telescopes (covering 86 to ~900 GHz) that can be put into a variety of configurations to tailor the performance for different applications.

9.3.2 *Interpretation of aperture synthesis data*

Ideally, we would use our telescope array (and the rotation of the earth) to measure visibilities for our target source densely spaced over the entire *uv* plane. By the Van Cittert–Zernicke Theorem and its extension, the Fourier transform of this visibility function $V(u,v)$ would give us an image of the source just as if it had been observed with a filled aperture telescope with diameter equal to the diameter of our telescope array:

$$I(x, y) = \iint V(u, v)e^{2\pi j(ux+vy)} \, du \, dv \tag{9.18}$$

However, realistic telescope arrays and observing strategies will leave gaps in the *uv* plane coverage, and as a result we have a "dirty" image:

$$I_\mathrm{D} = \iint V(u, v)S(u, v)e^{2\pi j(ux+vy)} \, du \, dv \tag{9.19}$$

where $S(u,v)$ is the sampling function over the *uv* plane, 1 where we have a measurement and 0 otherwise. Consequently, our point spread function will have more artifacts than would have been true in the ideal case. From the form of equation (9.19) and the Convolution Theorem, our image is

$$I_\mathrm{D}(x, y) = I(x, y) \otimes B(x, y) \tag{9.20}$$

that is, the true distribution convolved with our point spread function, $B(x, y)$. It may depart significantly from the ideal image if $S(u, v)$ is sparse. The PSF is termed the dirty beam:

$$B(x, y) = \iint S(u, v)e^{2\pi j(ux+vy)} \, du \, dv \tag{9.21}$$

See Figure 9.9 for an example of a dirty beam, corresponding to a single orientation of the VLA (e.g., the center panel of Figure 9.8). Even if the target area is the simple case of a field of unresolved sources, the superposition of the dirty beams for all of them will leave our image looking like a mess, appropriately termed a dirty map.

The simplest approach to converting such messy data into a high-quality image is the Högbom (1974) CLEAN algorithm. To implement it, one:

1. assumes that the image can be approximated by a field of point sources;
2. locates the position of the brightest point in the dirty map;
3. subtracts a scaled version of the dirty beam from this position – the subtraction should account for only a modest fraction of the brightness at this point;

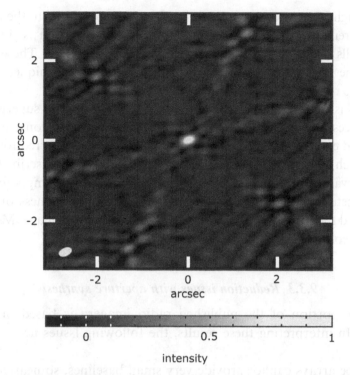

Figure 9.9. The instantaneous beam for the VLA (e.g., from the center panel in Figure 9.8) at 8.4 GHz. The synthesized beam ($0.27'' \times 0.18''$) is the white ellipse to the lower left. The sidelobes reach 20% of the peak response. After Fomalont (2006).

4. records the position and subtracted intensity in a "CLEAN component" file;
5. finds the brightest position in the dirty map left from the subtraction;
6. repeats steps 3–5 until no subtraction is possible without making part of the dirty map negative.

Practical CLEAN algorithms and the supporting analysis steps have a number of adjustable parameters that can improve the performance of this simple approach. One of the most useful is analogous to apodization. In this case, the process is called tapering, and it consists of convolving the visibility function with a function, often a Gaussian, that reduces the terms at the largest baselines, for example:

$$T(u, v) = e^{-(u^2+v^2)/s^2} \tag{9.22}$$

As with other forms of apodization, the image artifacts are suppressed at the expense of a larger synthesized beam diameter. Another approach, called density weighting, generates a weighting factor proportional to the number of u, v measurements in a given area of the uv plane (which is divided into gridded cells for the purpose of determining the weights). These enhancements come with the price that the products are not unique, and some judgment is required to select the best version.

CLEAN is centered on the assumption that the image is a superposition of point sources and hence may give an overly structured reconstruction of a smooth and extended source. Alternative methods are preferred for extended sources, such as the Maximum Entropy Method (MEM) (described for radio data by Bryan and Skilling 1980, for example). MEM attempts to optimize two parameters simultaneously, one representing the goodness of the fit to the data and the other representing the smoothness of the image. More on the MEM method can be found in Section 7.7.

9.3.3 Reduction issues with aperture synthesis

A large proportion of the published radio imagery is based on aperture synthesis. In interpreting these results, the following issues need to be kept in mind.

Telescope arrays cannot provide very small baselines, so nearly any aperture synthesis image will be missing low spatial frequencies. The result is that extended source structures may be completely missing; they cannot be recovered by smoothing the image or any of the other usual approaches to bring out extended, low-surface-brightness features because the instrument does not respond to them. The lost flux can be shockingly large. For example, the best WSRT and VLA images of the nearby galaxy M33 capture only about 15% of the flux seen with maps using filled-aperture telescopes (Viallefond *et al.* 1986; Tabatabaei *et al.* 2007).

The phase of the signals can be corrupted by the atmosphere of the earth, as illustrated in Figure 9.10. In the frequency realm up to a GHz or so, the charged particles in the ionosphere can produce phase errors. Refractive index variations due to atmospheric water vapor are significant at 1 GHz and produce increasingly large phase effects as the frequency increases above about 10 GHz, with coherence times ranging from minutes at the lower end to seconds in the submillimeter. The impact of these effects can be reduced by taking data over short time intervals and by observing nearby calibrator sources very frequently (on timescales of tens of seconds or even more often). Another way to reduce their effects is to use the information from at least three telescopes to determine the closure phase. Referring to Figure 9.10, the measured phase over each baseline consists of that due to the source, for

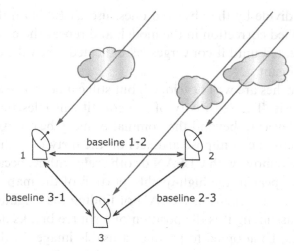

Figure 9.10. Generation of phase errors over three telescopes.

example ϕ_{12}, plus the difference of the phases induced by the atmosphere in the two individual beams, for example ε_1 and ε_2:

$$\Phi_{12} = \phi_{12} + \varepsilon_1 - \varepsilon_2 \tag{9.23}$$

$$\Phi_{23} = \phi_{23} + \varepsilon_2 - \varepsilon_3 \tag{9.24}$$

$$\Phi_{31} = \phi_{31} + \varepsilon_3 - \varepsilon_1 \tag{9.25}$$

The sum of equations (9.23)–(9.25) is

$$\Phi_{12} + \Phi_{23} + \Phi_{31} = \phi_{12} + \phi_{23} + \phi_{31} \tag{9.26}$$

In this closure phase the atmospheric-induced phase errors have all canceled. With four telescopes, it is also possible to deduce a quantity analogous to the closure phase for the signal amplitudes. By modeling the closure phases and amplitude information in a telescope array, it is possible to deduce the source structure independent of the phase errors. These procedures are most effective on relatively bright sources, so that there is enough signal in a coherence time that the atmospheric effects are stationary.

Self-calibration is a powerful approach that is invoked wherever possible in radio interferometer data reduction. It requires that there is a point source (or well-understood sharp source structure) in the primary beam that is bright enough to be detected in all the interferometer baselines within a coherence time. If so, then one can use a beginning model of the source, compute its visibilities and divide by the observed ones, use CLEAN on the residuals to deduce the implied correction in the model, and repeat the comparison. This iteration is repeated until it converges, as indicated when the corrections are no longer significant.

These approaches are well developed, but still cannot solve all the issues in aperture synthesis. The sensitivity of the constituent telescopes in an array extends at low levels to beyond the nominal primary beam. As a result, very strong radio sources can impose artifacts on the dirty map that are particularly difficult to remove with CLEAN or other algorithms because the source itself does not appear in the high-quality portion of the map – only its dirty artifacts do. Thus, the focus in CLEAN of finding the brightest point in the dirty map and assuming it is the position of a source breaks down and other approaches must be adopted to produce a usable image. In this regard, it is helpful to obtain short exposures on nearby bright sources to aid in removing the associated artifacts. However, the results will be imperfect because of differences in atmosphere-induced phase and amplitude errors. Effects such as bandwidth and time smearing can also produce artifacts that are hard to remove. As a result, high-quality reduction of very deep radio images is challenging.

9.3.4 Very long baseline interferometry

The receiver outputs from each element of a telescope array contain all the information required by the correlator to implement the interferometer imaging. Therefore, signals can be recorded and analyzed later. This feature allows implementation of interferometers among telescopes that are very widely spaced, of which the most extreme is very long baseline interferometry (VLBI) with telescopes spread across a continent or even further. These networks allow milli-arcsec resolution (at cm wavelengths), although some care is required in constructing images because of the unavoidable holes in the coverage of the *uv* plane. Good images are, however, possible if the objects observed are relatively compact.

VLBI can measure absolute source positions to about one milli-arcsec and relative ones significantly more accurately. A network of 212 bright, compact radio sources with VLBI positions is the foundation of the International Celestial Reference System (ICRS) in astrometry that we introduced in Section 4.6.4. These sources have been selected to be bright and point-like at optical wavelengths also, so they can be tied in accurately with astrometry there. VLBI also has applications outside of astronomy. For example, through repeated measurements of sources that we are confident are fixed in coordinates (e.g., compact quasars), it is possible to determine the telescope separations to precisions of about a millimeter and measure continental drift. VLBI also provides very accurate measurements of precession and nutation (Section 4.6.6) and of the rotation rate of the earth. Accurate time keeping requires reconciliation of the length of the day with universal coordinated time (UTC) and the occasional slight adjustments ("leap seconds" – Section 4.6.2) are determined from VLBI.

Equation (9.16) was discussed in terms of two telescopes of equal diameter, but it holds in general. As a result, there are interesting options for VLBI using a modest-sized telescope in orbit or even on the moon to provide a huge baseline. If a very large groundbased telescope is used, the effective area can be reasonably large and hence good sensitivity can be obtained.

9.3.5 The future

The VLA has been the flagship aperture synthesis array for cm-wave radio astronomy since its commissioning in 1980. The transmission of signals from the VLA receivers to the correlators has had limited IF bandwidth, a problem that has now been solved by installation of optical fiber links between the telescopes that enable an IF bandwidth of 8 GHz (per polarization) – an 80-fold improvement. Similar upgrades have been made in MERLIN and the European VLBI network. These latter links enable bandwidths of $\sim$2 GHz, improving the achievable signal-to-noise ratio by factors of 5–10 (equation (9.17)). The broader frequency/wavelength coverage can also substantially improve the coverage of the *uv* plane; since *u* and *v* are measured in wavelength units, a given physical baseline corresponds to a range of positions in *uv* if the range of wavelengths is significant. We have already mentioned ALMA in the context of telescope configurations for interferometers. The dramatic reduction in the water vapor over ALMA both enhances the atmospheric transparency in the submm and reduces the phase errors. ALMA incorporates state-of-the art sideband-separating receivers and broadband IF links (as with the VLA, 8 GHz IF bandwidth) for very high sensitivity over a range of 86 to $\sim$900 GHz.

Our discussion of the limitations to the sizes of filled-aperture radio telescopes, plus equation (9.17) and the discussion surrounding it, suggest a new approach to obtaining very large collecting areas. One could build a huge number of modest-sized telescopes (thus providing large primary beams, making imaging and mapping efficient) into an aperture synthesis interferometer. Since the number of correlators grows as the number of baselines, for example as $N(N-1)/2$ for N telescopes, one is trading mechanical challenges for electronic ones. Judging by the growth of optical telescopes over the last century, the doubling time for telescope diameters is at least 30 years, whereas the well-known Moore's Law places the doubling time for electronic capability at about two years, so the trade is a good one. Nonetheless, it must be done carefully because the electronic complexity grows rapidly with the number of telescopes for a given total collecting area (Cornwell 2004). A realization of this concept is the Allen Telescope Array. Operations started with 42 6-m telescopes in 2007, but currently expansion plans and even operations are suspended due to lack of funding. A far more ambitious plan is for the Square Kilometer Array (SKA), so named because the plan is to provide a square kilometer of collecting area. The SKA is envisioned to have a combination of different antenna types with the most ambitious being some thousands of 12-m telescopes operating from 500 MHz to 10 GHz and with baselines up to 3000 km. Predecessor arrays in Australia (the Australian Square Kilometer Array Pathfinder (ASKAP) and South Africa (the Karoo Array Telescope (MeerKAT)) will be powerful radio facilities in their own right.

9.4 Optical and infrared interferometry

9.4.1 General properties

Heterodyne interferometry, as described for the radio regime in the preceding section, can also be employed in the infrared and optical. However, the limitations on IF bandwidth severely restrict the spectral band and the sensitivity achievable is subject to the quantum limit, so only very bright sources can be observed successfully. Instead, most effort for these spectral regions has been in the development of *homodyne* instruments where the light from the interferometer elements is brought to a common station where it interferes. The necessity to interfere the light from the interferometer beams directly rather than converting each beam to an electronic signal that can be used to reproduce the interference places strong demands on the interferometer designs.

Figure 9.11 shows two different ways of bringing the interferometer beams together. To provide an imaging field, the interferometer must obey the Abbe sine condition requiring that the image plane have the same geometry as the object plane. That is, the relation of the interferometer apertures as viewed from the source must be preserved optically when their outputs are

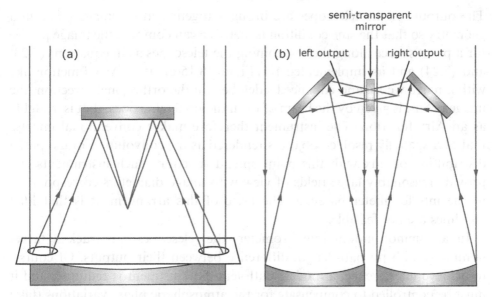

Figure 9.11. (a) Image-plane beam (also called Fizeau) combination, and (b) pupil-plane (also called Michelson) interferometers.

superimposed to cause them to interfere. This requirement was given in equation (1.22); it is satisfied if

$$\frac{\sin \alpha_1}{\sin \alpha_2} = \frac{h_1}{h_2} \tag{9.27}$$

where h_1 and h_2 are the distances of the incoming beams from the axis of the system and α_1 and α_2 are the angles at which the corresponding beams are brought to the focal plane. Such interferometers are called image-plane (or Fizeau) devices. Figure 9.11(a) shows the simple case, used as an example at the beginning of this chapter, of an interferometer constructed by masking all but two small apertures at the edge of the primary mirror of a telescope. If the telescope obeys the sine condition, then the interferometer can be made to do so also. Interference can also be obtained at other convenient points along the optical path of an instrument, but in general (e.g., Figure 9.11(b)) if the sine condition is not met, the field of view over which useful levels of interference can be obtained is very small. Such instruments are called pupil plane (or Michelson) interferometers.

9.4.2 Common-mount interferometry

A powerful implementation of image-plane interferometry is to place two telescopes on a common mount, such as in the Large Binocular Telescope.

The outputs of the telescopes are brought together at an image plane in a geometry so that the sine condition is met. We can compute the image profile for a point source along the axis joining the telescopes as in equations (7.23) and (7.24) and as implemented for Figure 8.15(c). It is Airy-function-like with a narrow core and enhanced sidelobes. In the orthogonal direction, the image profile is given by the beam of the individual telescopes, that is, roughly as an Airy function. The instrument therefore makes conventional images; that of a spatially resolved object is rendered as the convolution of its spatial distribution of flux with this point spread function. Such instruments can provide reasonably large fields of view with image diameters corresponding to the interferometer baseline. The price of this arrangement is that long baselines are not feasible.

In a common-mount interferometer, the telescopes can track a target nominally with no path-length difference between their outputs. Of course, to achieve interference, a modest path-length adjustment is required, and it must be controlled to compensate for the atmospheric phase variations (fast) and flexure in the telescope (slow). Because common-mount interferometers are imaging instruments, a full AO correction is desirable to provide reasonably constant spatial phase over their fields of view. The fringes are still somewhat unstable and they need to be tracked by measuring the fringe position in the imaging plane and stabilizing it by feeding back a correction into the path-length compensator. Earth rotation is used to fill in the uv plane and probe the structure of a source in two dimensions on the sky, just as discussed for radio interferometers.

9.4.3 Nonredundant and partially redundant masks

This chapter began with a thought experiment where we considered the results on images of masking the mirror of a telescope. This procedure can be used to customize the spatial frequencies delivered to an image. For example, the Airy function is dominated by low spatial frequencies because of the dominance of short baselines over a circular mirror aperture (see Figure 2.13). Noise associated with the strong low-spatial-frequency signals can overwhelm the high-spatial-frequency information. However, with a suitable set of masks to control the baselines across the mirror we can customize the spatial frequency content of the image and optimize it for high-resolution information (at the loss of much of the light, of course). In particular, masking makes it much easier to reach the ultimate resolution limit of a telescope, $\lambda/2D$, as well as improving the contrast near bright, unresolved sources.

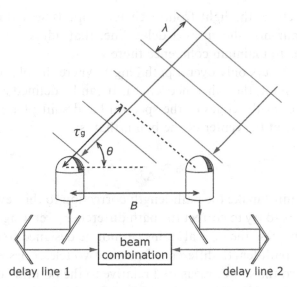

Figure 9.12. An infrared (or optical) two-element interferometer connecting two telescopes on separate mounts.

This approach is implemented in masks placed either over the primary mirror or at a pupil (for convenience), and with a limited set of baselines defined by apertures. If the baselines are selected so there are none in common, the mask is nonredundant (i.e., a NRM); partially redundant masks have some baselines duplicated. Data obtained with such masks are treated as the outputs of multiple interferometers. The NRM has the advantages that the different visibilities do not cancel each other and that separating the baselines by Fourier transformation is relatively simple because each yields a different spatial frequency (or direction). The principle of phase closure (Section 9.3.3) is used to track and correct phase variations in the wavefronts.

9.4.4 Interferometry with separate telescopes

Interferometry with separately mounted telescopes has the advantage that large baselines and hence high angular resolution can be obtained. The Very Large Telescope Interferometer (VLTI) and Keck Interferometer (KI) are of this general design, as are a number of smaller-scale instruments. A form of interferometer similar to that in Figure 9.1 but combining the light directly is shown in Figure 9.12. The light from each of the two telescopes is brought by a train of mirrors to a common point, or beam combiner, where it is merged in a way that it interferes. To make this possible, it is necessary to control the path length; as the telescopes track a source a large and variable adjustment is required. Therefore, the light from each telescope is sent through a device with moving mirrors, shown as a delay line, that adjusts its arrival at the beam combiner to maintain coherence there.

Interference occurs only over a path length where the phase of the light is reasonably constant, the coherence length. It can be defined as the path over which the waves at the edges of the spectral band shift phase by one radian relative to those at the center of the band, that is:

$$\ell_{coh} \approx \frac{\lambda^2}{\Delta\lambda} \qquad (9.28)$$

The delay line must make the path-length correction to this level of accuracy. It is too complex to try to correct the path difference over a significant field of view, and in any case the optical trains cannot be designed to satisfy the sine condition. The path-length difference into the two telescopes is $B\theta$ where θ is the angle toward the source measured relative to the pointing direction of the interferometer elements. If we equate this difference to the coherence length, we find that the usable field is very small,

$$\theta_{max} \leq \left(\frac{\lambda}{B}\right)\left(\frac{\lambda}{\Delta\lambda}\right) \qquad (9.29)$$

In addition, if the correcting optics are in air, the dispersion of air (equation (7.1)) results in different path-length corrections for different wavelengths, or, more practically, places a restriction on the spectral band that can be observed for a single correction. This requirement can be calculated by setting the path-length difference equal to the coherence length. Assuming the air path is equal to the baseline (see Figure 9.12), for a 100-m baseline the maximum permitted spectral band at 2.15 µm is 0.34 µm (i.e., almost the full K_S photometric band), but is only 0.012 µm at 0.56 µm (V). Thus, the restriction is not serious in the infrared but to operate with a broad band (and the resulting sensitivity) in the visible requires either correcting for the atmospheric dispersion with a suitable optical element in the beam, or placing the interferometer beams in vacuum for most of their length. In addition to the reduction in spatial phase errors into the infrared, the coherence time increases as $\lambda^{6/5}$, so the requirements for interferometry are much more easily met in the infrared (and the farther into the infrared the better) than in the visible.

Because the imaging field is severely restricted by the requirements for path-length correction, interferometers of separate telescopes use pupil–plane combination, that is, they superimpose afocal beams. The superimposed beams can be brought to a single detector or small array; the fringes are

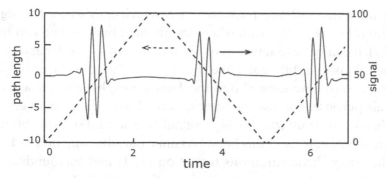

Figure 9.13. Output of a pupil-scanning interferometer (solid, right scale) and the corresponding scanning of the path length (dashed, left scale). The width of the sets of fringes is a result of the coherence length, ℓ_{coh}, over which significant interference can occur.

shifted to modulate the signal by varying the path length. Getting good fringe visibilities also depends on controlling the spatial- and time-varying phase shifts imposed by the atmosphere, which we discussed with regard to adaptive optics (Section 7.1). Temporal phase variations occur on the coherence timescale as in equations (7.9)–(7.12). Spatial phase errors can be reduced by tip-tilt corrections, with further improvement with higher-order AO corrections. Another approach is to filter the signal frequency light, by passing the light through either a single-mode optical fiber or a pinhole aperture. This approach results in variable intensity as the cost of reducing the fluctuations in the visibility.

A full measurement requires determination of three parameters: intensity, visibility, and phase. To obtain this information and reduce the effects of atmospheric-induced phase variations, pupil–plane interferometers modulate the fringes by scanning the path length rapidly so that slight variations in phase do not cause loss of signal, but change its timing within the scan (see Figure 9.13). High-quality interferometry requires that the phase be tracked dynamically and corrected. There are two general approaches. One is to track the general range where coherence holds and adjust the path to maintain the operation there. For example, in group delay tracking, one disperses the interference fringes; when the path length is not correct, the fringe spacing will vary with wavelength and after dispersion the fringes will be tilted relative to the zero-path-difference configuration. This approach allows relatively long integrations – several seconds – and hence can be applied to relatively faint targets. More accurate stabilization of the fringes requires fringe tracking, in which the fringes are measured directly and therefore at high frequency. In general, a SNR ≥ 10 is needed to measure the phase to 10% of the fringe width.

When the object is being measured at a wavelength where the signal-to-noise ratio is low (e.g., at 10 μm where the thermal background can limit the sensitivity), it can be advantageous to divide the light spectrally and carry out these operations at a different wavelength where the source can be detected more easily. Above, we showed that the 2 μm atmospheric window is attractive for this purpose, because the atmospheric dispersion is small and nearly the full band can be utilized for high signal-to-noise ratio. How bright does our target star have to be in this band? Assume our sampling time is 1–2 ms to follow the atmospheric variations (equation (7.11) and surrounding discussion), and that we want a signal-to-noise ratio of 10:1. If we are using 8-m telescopes, it is a straightforward calculation that a star of $m_K = 10$–11 will be required.

As a consequence of the operational restrictions (very small field, need to have a bright source to track and stabilize phase), multi-telescope optical/infrared photometry is most easily conducted where the source itself has a bright, unresolved component. Fortunately, there are many high-priority astronomical problems that present just this situation: (1) resolving the structure of protoplanetary disks; (2) searching for massive planets; (3) studying outflows from young and old stars; (4) measuring stellar diameters; or (5) probing the regions around active galactic nuclei. More complex designs allow operation with a reference source within the isoplanatic patch.

So far we have only discussed interferometers that produce constructive interference of the target source. For some applications, it is more desirable to produce destructive interference – that is, to stabilize the phase on fringes for a constructively interfered image, but to obtain the science data where the path length has been adjusted to make this bright source disappear so far as possible. These nulling interferometers are a powerful way to look for faint extended structures around bright sources and are an alternative in this regard to the approaches discussed in Section 7.8.

9.5 Exercises

9.1 Compute the visibility function for two equal-brightness stars separated by angle α. (Hint: Represent the source brightness distribution as the sum of two delta functions; use a result from exercise 2.7.) Compare the visibility if one star is twice as bright as the other.

9.2 Show how equation (9.17) follows from equations (8.11) and (9.16).

9.3 Given an array of antenna locations (in meters) in the table below, operating at 1.4 GHz, construct plots of the instantaneous u, v coverage for a source at the meridian: (1) at declination $\delta = 90°$, and (2) at declination $\delta = 0°$. Place your array at the latitude of the VLA, 37° N. In your plots, plot u on the horizontal axis, and v on the vertical axis, with $(0, 0)$ at the center of the plot, and use equal u and v scales in units of kilowavelengths.

Antenna	East	North
1	0 m	0 m
2	700 m	–300 m
3	0	1000 m
4	–700 m	–300 m
5	1500 m	–700 m
6	0	2200 m
7	–1500 m	–700 m

9.4 Assume that a secondary "blob" can be distinguished in VLBI when it is one fringe FWHM displaced from the fringe peak of the primary object. One proposal for a ground-space VLBI capability would achieve a maximum baseline of 100 000 km and operate up to 86 GHz. For a quasar at $z = 2.5$, what is the minimum physical displacement at which this VLBI concept could identify a blob? (You will have to use cosmology to convert angles to physical dimensions – see, for example, http://www.astro.ucla.edu/~wright/CosmoCalc.html.)

9.5 Assuming the beams are conveyed through air, compute the maximum baseline that could be used for an interferometer operating in the N band at 10.6 μm with a bandwidth of 5 μm without compromise of the coherence by atmospheric dispersion.

9.6 Compute the shape of the Large Binocular Telescope Interferometer beam along the line between the mirror centers. Assume a wavelength of 10 μm, that the two mirrors are 8 meters in aperture, and that they are separated by 14 meters, center to center. Treat the problem in one dimension only.

Further reading

Burke, B. F. and Graham-Smith, F. (2009). *Introduction to Radio Astronomy*, 3rd edn. Cambridge: Cambridge University Press.

Condon, J. J. and Ransom, S. M. (2010). *Essential radio astronomy*: http://www.cv.nrao.edu/course/astr534/ERA.shtml

Glindemann, A. (2011). *Principles of Stellar Interferometry*. Berlin, New York: Springer.

Kellermann, K. I. and Moran, J. M. (2001). The development of high-resolution imaging in radio astronomy. *ARAA*, **39**, 457–509.

Lebeyrie, A., Lipson, S. G., and Nisenson, P. (2006). *An Introduction to Optical Stellar Interferometry*. Cambridge: Cambridge University Press.

Monnier, J. (2003). Optical interferometry in astronomy. *Rep. Prog. Phys.*, **66**, 789–857.

Quirrenbach, A. (2001). Optical interferometry. *ARAA*, **39**, 353–401.

Taylor, G. B., Carilli, C. L., and Perley A. (eds.) (1999). *Synthesis Imaging in Radio Astronomy*. San Francisco, CA: Astronomical Society of the Pacific; see also http://www.aoc.nrao.edu/events/synthesis/2010/lectures10.html

Thompson A. R., Moran J. M., and Swenson G. W. (eds.) (2001). *Interferometry and Synthesis in Radio Astronomy*, 2nd edn. New York: Wiley.

Wilson, T. L., Rohlfs, K., and Hüttemeister, S. (2009). *Tools of Radio Astronomy*, 5th edn. Berlin, New York: Springer.

10

X- and gamma-ray astronomy

10.1 Introduction

For many years, X-ray astronomy depended on gaseous detectors: basically, capacitors or series of capacitors with a voltage across them and filled with gas. Depending upon the value of the voltage, these devices: (1) just collect the charge freed when an energetic particle interacts with the gas (ionization chamber); or (2) provide gain [in order of increasing voltage, exciting the gas to produce ultraviolet light (scintillation proportional counter); creating modest-sized ionization avalanches to provide gain but with signals still in proportion to the original number of freed electrons (proportional counter); providing gain to saturation (Geiger counter); yielding a visible spark along the path of the avalanche (spark chamber)]. The absorbing gas is typically argon or xenon, for which high absorption efficiency in the 0.1–10 keV range requires a path of order 1 cm. Very thin windows are required to admit the X-rays to the sensitive volume – for example, 1 μm of polypropylene to provide > 80% transmission for energies above 0.9 keV. Proportional counters with multiple anode wires provide spatial resolution of a few hundreds of microns.

Because the atmosphere is opaque to them, X-rays and gamma rays require telescopes and detectors to operate from balloons, or more commonly from space (with the exception of the highest-energy gamma rays). Initially, the detectors were used without collecting optics; the large detector areas then resulted in high spurious detection rates due to cosmic rays. Anti-coincidence counters were required to identify charged particles coming from random directions, allowing probable X-ray events to be isolated. The necessity to operate at the top of or above the atmosphere plus these requirements on the detector systems were very limiting in terms of the angular resolution and sensitive areas that could be achieved.

Focusing optics have revolutionized these approaches (see Section 2.4.4). However, achieving the full resolution gain from an X-ray telescope is not possible with the few-hundred-micron positional resolution of a proportional counter. Getting the full benefit from X-ray optics requires the finer resolution of microchannel plates and CCDs. The combination of focusing telescope and one of these detector types not only provides far better angular resolution than was achievable previously, but also allows shrinking the detector size to minimize the rate of spurious cosmic ray events. For example, Chandra has a collecting area of about $600\,cm^2$ at $1.5\,keV$, whereas the signal from a 16.9×16.9 arcmin imaging field of view can be collected by four CCDs with a total area of $24\,cm^2$, 25 times less.

Missions using X-ray focusing optics up to an energy of nearly $100\,keV$ are nearing completion. The very high-energy detectors are based on different technologies, but they too have migrated away from gaseous devices to solid-state ones. Although optics for even higher energies are in development, for now the detectors continue to define the collection areas above $100\,keV$. The resulting telescopes are large, very massive and with more limited angular resolution than can be achieved at lower energies. Because of such technical issues, this spectral region has not been explored in the detail that has been achieved in most other ranges, leaving a high discovery potential that is being exploited by the Fermi mission.

10.2 X-rays from 0.5 keV to 10–15 keV

10.2.1 Detectors

At low X-ray energies (up to $\sim$10–15 keV), CCDs (Section 3.6) meet the goals for X-ray detectors well. Microchannel plates (Section 3.7) are also used for X-ray astronomy, but less widely than CCDs. Compared with CCDs, they have lower quantum efficiency ($\sim$30%) and energy resolution, but much faster time response. Many specifications of CCDs are similar whether they are used in the optical or the X-ray: similar pixel sizes ($\sim$10–20 μm) and array sizes, low read noise (2–4 electrons desired), and backside illumination (particularly for low-energy X-rays). However, X-ray detectors are generally operated in a single-photon-detection mode: each X-ray that strikes the CCD generates a number of electrons that are collected under the gates and soon after are transferred to the output amplifier. The number of electrons is proportional to the energy of the X-ray, so by reading out the CCD sufficiently rapidly that each X-ray is recorded separately, one also gets a low-resolution spectrum. Although operation in an integrating mode, as in the optical, would be feasible, it loses information (and scrambles the spatial and spectroscopic properties of the source). At typical detection rates with existing X-ray observatories, it is necessary to read out every few seconds (e.g., 3.2 s for the AXAF CCD Imaging Spectrometer (ACIS)). These rates are obtained while preserving high charge transfer efficiency and low read noise by using frame transfer devices.

The most important differences between operating a CCD in the X-ray rather than in the optical derive from the absorption properties of X-rays in semiconductors. There are three basic interaction mechanisms. The *photo-electric effect* occurs when the energy of the X-ray is consumed by freeing an electron tightly bound to the atom. So far, in our discussions of the detection of photons with energies close to 1 eV we have been concerned only with the most easily freed electrons, corresponding in silicon to the bandgap at 1.11 eV. However, silicon has only four valence electrons with ionization energies this low; there are ten additional electrons and the most tightly bound require 1.839 keV for ionization. The second interaction type is Thompson scattering, where an X-ray is scattered by an atom but with no change in energy. The third is Compton scattering, where the X-ray gives up some but not all of its energy to an electron. The latter two types of inter-action have mass absorption coefficients given by the Bragg–Pierce Law:

$$\mu_{\mathrm{m}} = K Z^4 \lambda^3 \tag{10.1}$$

where K is a constant and Z is the atomic number of the absorbing atom.

Because photoelectric absorption is strong when the X-ray energy matches the binding energy of an inner shell electron, but below this energy the X-ray only undergoes Compton and Thompson scattering, the absorption coefficient falls dramatically at the inner shell binding energy. It then increases as λ^3 toward lower energy (longer wavelength), as shown in Figure 10.1.

The overall absorption is exponential:

$$\frac{I}{I_0} = e^{-\mu_{\mathrm{m}}\rho x} = e^{-\mu_{\mathrm{l}}x} \tag{10.2}$$

where I_0 is the incident intensity, I is the intensity at depth x, ρ is the density of the material and μ_{l} is the linear absorption coefficient. The absorption or attenuation length, $1/\mu_{\mathrm{l}}$, is the distance a beam of X-rays traverses before being attenuated to $1/e$ of its initial intensity. The attenuation length for silicon as a function of X-ray energy is shown in Figure 10.2. Relatively thin CCDs – 10 to 20 μm thick – can have high absorption up to 3–5 keV. Normal front-illuminated CCDs are suitable for operation from 1 to ~6 keV, and back-illuminated devices are preferred below 1 keV because there is no absorption of the X-rays by the gate structure. Figure 10.3(a) and (b) show typical quantum efficiencies for these types of device.

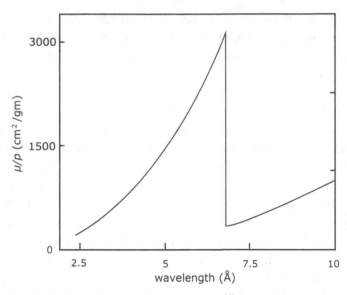

Figure 10.1. Behavior of mass absorption coefficient with wavelength.

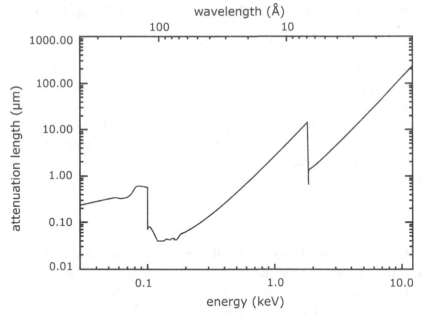

Figure 10.2. X-ray absorption length in silicon.

However, depletion thicknesses of ~100 μm are desirable to cover the full range of operation of traditional Wolter telescopes such as Chandra and XMM and even greater depths are necessary to extend the energy response toward 20 keV. Deep depletion CCDs can be manufactured by using

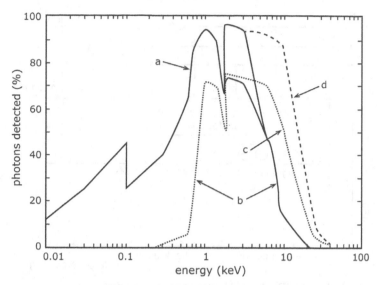

Figure 10.3. Quantum efficiencies of X-ray CCDS: (a) is back illuminated, (b) is front illuminated, (c) is front illuminated deep depletion, and (d) is back illuminated pnCCD.

high-purity silicon for the absorbing layer; the field from the CCD electrodes then penetrates farther into the material, allowing charge collection from depths of 40 to 50 μm. Further improvements are possible by adding an electrode to the backside of the device and applying a bias to increase the depletion depth, as in the fully depleted devices discussed in Section 3.6.2. An even deeper depletion region can be achieved with the pnCCD (Figure 10.4) (see, e.g., Strüder and Meidinger 2008). These devices can have useful quantum efficiency (> 20%) to 20 keV. The example in Figure 10.4 is formed on a wafer of high-purity silicon, 450 μm thick. The 7 μm layer is grown on this wafer, and includes the n-doped output contact that maintains contact to the bulk of the silicon. The CCD gates are deposited on strongly p-doped structures and the backside has a monolithic p-doped surface, thus forming back-to-back pn junctions across the wafer. By a suitable choice of voltages on the front and back p-type electrodes and the n contact layer, it is possible to backbias the two junctions both to deplete the entire wafer and to bring the minimum electron potential close to the CCD gates and the n+ layer. Conventional manipulation of the gate electrodes creates potential wells to collect charge and then transfers the collected charge to an output amplifier.

Another approach under development is based on Depleted P-channel Field Effect Transistors (DEPFETs). The basic architecture resembles that of an infrared direct hybrid array readout, with an amplifier integral with each pixel and the capability for full random access. The FET (Figure 3.12) is

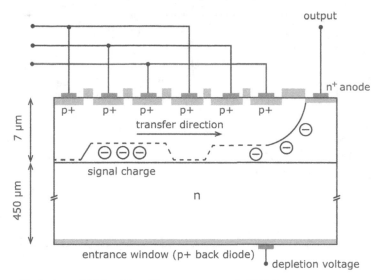

Figure 10.4. A pnCCD. After Hartmann *et al.* (2006).

grown on an n-doped wafer with a p+ doped contact on the backside and, similarly to the pnCCD, the biases are manipulated to bring the potential minimum for electrons close to the FET channel. As X-rays free electrons in the depleted bulk wafer and they collect in this minimum, they create a field that modulates the channel current just as the field from electrons on the gate does. A separate electrode can be used to clear the collected electrons. The DEPFET has the advantage of providing a long absorption path, as with the pnCCD. However, by avoiding the need to transfer the charges, it can be read out faster and is not subject to degradation in performance due to reduced charge transfer efficiency.

Returning to CCDs, virtually all of the absorbed X-rays lead to energetic free electrons. These electrons in turn collide with and free additional electrons, to yield the cloud of free electrons that are driven to the CCD gate and read out. The number of electrons in this cloud is

$$N_e = \frac{E_X}{w} \qquad (10.3)$$

where E_X is the X-ray energy and w is the effective ionization energy, $w \sim$ 3.65 eV. The value of w is significantly higher than the 1.11 eV silicon bandgap because most of the electrons are freed from states with energies well above the minimum energy one. The standard deviation of N_e is

$$\sigma_e = \sqrt{F N_e} \qquad (10.4)$$

The Fano factor, F, reflects the fact that the generation of the free electrons is not a Poisson-distributed process. In general, F is less than 1 when the electrons are produced in a correlated manner. In this case, the average size of the electron cloud is determined by the X-ray energy with relatively minor statistical fluctuations around this value, so $F \sim 0.12$. If the CCD read noise is σ_{read} and there are no other sources of extraneous noise, the spectral resolution is

$$R = \frac{E_X}{\Delta E_X} = \frac{N_e}{2.36 \, w \sqrt{\sigma_e^2 + \sigma_{\text{read}}^2}} \qquad (10.5)$$

where ΔE_X is the full width at half maximum of the distribution of output signals for an unresolved spectral line.

In practice, the response to an unresolved line is more complex than implied above because the absorption of the X-ray may occur in layers of the CCD other than the depleted region from which charge is collected – for example in the gate polysilicon or oxide layers. Consequently, the response described by equation (10.5) is accompanied by a low-energy tail. There may also be a spectral feature 1.839 keV below the line energy (assuming it is larger than this value) due to escape of an X-ray emitted by an excited silicon atom. A second spurious feature called the fluorescence peak occurs at 1.739 keV when the silicon emits a fluorescent X-ray that is detected as a separate event. Fortunately, these issues account for only a modest part of the response of the CCD.

There are some additional cautions for interpreting X-ray CCD signals. If the X-ray flux onto the detector is too high (or the readout rate too low), more than one X-ray will frequently be absorbed by a pixel within a readout time. As a result, the signal will correspond to that for the sum of the energies of the two X-rays; that is, the spectrum of the source deduced from the CCD outputs will be distorted toward apparently enhanced high energies. In addition, the point spread function will be affected because the X-ray flux is highest at its center. This problem is called pileup and can be difficult to remove from the data; hence, observations should be designed carefully to avoid it if possible. A more benign problem occurs when an X-ray is detected while the CCD charge is being transferred; the signal then appears as a streak across the reconstructed image. Because of its unique shape, such a signal can be readily identified and removed from the data. Yet another issue arises if the charge transfer efficiency (CTE) of the CCD is too low, resulting in increased noise. Unfortunately, the CTE can be degraded by cosmic ray protons (and energetic electrons and photons). These particles can break bonds in the silicon crystal permanently, leading to crystal lattice damage that can trap free electrons as they are transferred over the damaged spot. As discussed in Section 3.6.1, the trapped electrons are released randomly with a long time constant and thus the statistics of this process increase the noise.

It would be wonderful if every detected source came with a low-resolution spectrum, but such is not the case. The intrinsic X-ray backgrounds are very low, although excess events can come from soft solar protons (very prominent during flares) and cosmic-ray-induced effects in the instrumentation. However, under good conditions and at the resolution of a high-quality X-ray telescope, the rate of background detections is so low that only 3–4 X-rays from the direction of a source suffice to detect it (see exercise 10.1). Virtually no spectral details can be derived from so little information. Instead, spectral properties are characterized by the relative counts, or flux, in the soft X-ray band (SX), ∼0.5–2 keV and the hard band (HX), ∼2–10 keV. These bands are extremely broad compared with their center energies, so bandpass corrections similar to those discussed in Section 5.4.3 are essential to interpret the fluxes in terms of the underlying source spectrum. Generally, the results are interpreted by fitting either a power law or thermal continuum spectrum, in the former case with the amplitude and power law index as variables and in the latter the amplitude and temperature. When comparing such results from different X-ray telescopes, there may be substantial systematic differences (see, e.g., Tsujimoto *et al.* 2011).

A complication in such fits is that the soft band can be strongly affected by interstellar absorption. It is conventional to estimate the absorption from measurements of the HI column along the line of sight. However, the interstellar medium is highly structured (see, e.g., Elmegreen and Scalo 2004) down to size-scales much smaller than the typical resolution of the HI data (see, e.g., Miville-Deschênes *et al.* 2003; Kim *et al.* 2008). Consequently, priority should be given to use of HI data with high angular resolution, and a careful statistical analysis is desirable to confirm any conclusions that depend on the assumed absorption level.

10.2.2 Dispersive spectroscopy

Diffraction gratings can provide substantially higher spectral resolution than in equation (10.5) – of order $R = E/\Delta E = \lambda/\Delta\lambda \sim 100$ to 1000, where E is the photon energy (Paerels and Kahn 2003). X-ray gratings follow the same basic principles as in the optical; the grating imposes a phase delay on slices of the incoming beam that are brought together at the diffraction plane, producing spectra due to interference effects. The behavior is described by the grating

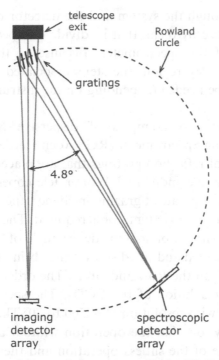

Figure 10.5. Layout of the grating spectrometer for XMM (den Herder *et al.* 2001). Some of the X-rays proceed past the gratings to the imaging detector array, whereas those that strike the gratings are deflected to the spectroscopy detector array where the dispersed spectrum is imaged.

equation (6.7). The limitations of X-ray optics do not allow for the conventional collimator–grating–camera designs used at longer wavelengths. Instead, the gratings are placed in Rowland configurations; an example is shown in Figure 10.5. By placing the spectrometer gratings at appropriate angles, it is possible to reproduce the geometry for a conventional Rowland circle spectrograph (Section 6.3.3.1), where the light would be assumed to enter through a slit at the position of the imaging detector array and be reflected off a concave grating to the spectroscopic detector array diffraction gratings.

The data analysis with these instruments may feel a little different than in the optical because the photons are detected one at a time rather than integrated on the detector. Nonetheless, the reduction steps follow closely those listed in Section 6.3.5: (1) reject any spurious events, that is, reduce the image optimally; (2) identify the position of zero order and of the dispersed images (according to the orders), that is, trace the spectrum; (3) measure the dispersion angle for each "good" event, that is, calibrate the wavelength/ energy scale; (4) bin the events by energy into a spectrum; and (5) calculate the effective area through the system and the detector response function for each bin and normalize by them, that is, divide the spectrum by the relative response function of the spectrometer. Because of the very low level of background in the X-ray region, the steps associated with subtracting the sky counts may not be needed (depending on the instrument and how bright the source is).

We now consider two examples of dispersive X-ray spectrometers. The Reflection Grating Spectrometer (RGS) (den Herder *et al.* 2001 – Figure 10.5) utilizes a suite of reflection gratings that are placed permanently in the beam emerging from the individual Wolter telescopes of XMM-Newton. Because the telescopes operate at grazing incidence, the range of input angles to the grating is small and no collimator is required. The gratings also operate in grazing incidence with a nominal incidence angle of $1.58°$.[20]

The gratings are shaped and tilted so that they bring the X-rays to a CCD detector array placed on the Rowland circle. The orders are separated using the inherent energy resolution of the CCDs. The spectra cover the energy range 0.3 to 2.1 keV, with resolution $R \sim 500$ to 70. The backgrounds in the X-ray are sufficiently low that the operation without a slit is not a serious shortcoming. Because of the slitless operation and the design to accommodate the relatively large image size of XMM-Newton, this instrument can yield useful spectra of sources extended up to about 30 arcsec diameter.

The Chandra High (Medium and Low) Energy Transmission Grating Spectrometers (H(M&L)ETGS – Canizares *et al.* 2005) are based on gratings mounted on hinges so they can be rotated into the optical path, where they are also configured to reproduce the Rowland geometry. Figure 10.6 shows the medium-energy grating; the high-energy grating is similar in concept with different dimensions, while the low-energy grating has free-standing gold bars to avoid absorption in the plastic carrier. Together, the three gratings cover from 0.07 to 10 keV, although the effective areas are low at either end of this full range. The tiny gold bars are thin enough to be partially transparent to the X-rays. Because the index of refraction of the gold is less than 1, photons that pass through these bars have their phase advanced relative to those that pass through the spaces between the bars. The dimensions provide destructive interference at zeroth order, which is therefore suppressed in favor of more energy in the higher orders (where the spectra lie). This design goal can only be obtained for a limited energy range, so the full Chandra range is obtained with the three different grating optimizations. The spectral resolution is

---

[20] Be warned that α and β, the angles of the incoming and outgoing beams for these gratings, are measured relative to the plane of the grating rather than to the grating normal, resulting for example in a different form for the grating equation than in our equation (6.7).

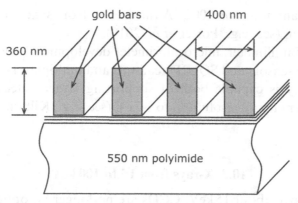

Figure 10.6. The design of the transmission grating for the Chandra METGS.

$R = E/\Delta E = \lambda/\Delta\lambda \sim 1000$ at the low-energy end and ~ 100 at the high-energy end of the range for each grating. As with the RGS on XMM-Newton, the instrument is slitless; in this case, its spectral resolution is critically dependent on the source angular size.

10.2.3 Calorimetric spectroscopy

The bolometer technology discussed in Chapter 8 has been adapted to X-ray spectroscopy with microcalorimeters, which provide much higher efficiency than can be obtained with gratings. Early devices used HgTe absorbers attached mechanically to an etched frame and carrying a thermometer of silicon doped by ion implantation. Junction field effect transistors (operating at a much higher temperature) provided the first stage of amplification (see, e.g., Stahle *et al.* 1999; Kelley *et al.* 2007). To reduce heat capacities so the absorption of a single X-ray would produce a well-measured signal, the devices were cooled to ~ 60 mK using a multi-stage refrigerator system, with the final cooling stage through an adiabatic demagnetization refrigerator (ADR[21]). Arrays with about 30 pixels and energy resolution of 5 to 8 eV were built for use on Astro-E and Suzaku; sadly, owing respectively to a launch failure and early loss of the cryogen, neither instrument returned useful astronomical data. However, similar detector arrays were used successfully on sounding rockets to measure the spectrum of the soft X-ray background

---

[21] An ADR uses a crystal of paramagnetic salt – a material in which the molecules have magnetic moments. The salt is placed in a strong magnetic field, aligning the magnetic moments of its molecules; when the field is removed, the moments remain aligned and the salt can cool by absorbing energy that randomizes the molecular orientations. When the moments are sufficiently disordered, no more cooling is possible until the salt is again subjected to the strong field.

(see, e.g., McCammon *et al.* 2002). A third effort from space is planned on the Astro-H mission (see, e.g., Porter *et al.* 2010).

As in the submillimeter, microcalorimeter development is now centered on transition edge sensors and SQUID readouts and multiplexers (Section 8.2.3). This technology is capable both of supporting larger detector arrays and providing better energy resolution, to ~3 eV (see, e.g., Kilbourne *et al.* 2008).

10.3 X-rays from 15 to 100 keV

For X-rays above about 15 keV, CCDs are no longer an optimum choice as detectors: their quantum efficiency is low and the lattice damage by the energetic X-rays reduces the charge transfer efficiency, causing their already mediocre performance to degrade with time. In their place, missions like INTEGRAL and NuStar use X-ray detectors based on absorption in CdTe or CdZnTe. The inner shell binding energy for cadmium is 26.71 keV; for tellurium it is 31.81 keV, and for zinc it is 9.66 keV, giving these materials a significant advantage over silicon for photoelectric absorption. Furthermore, their much higher atomic numbers (48 for Cd, 52 for Te) increase the interaction by Thompson and Compton scattering as shown in equation (10.1). The bandgaps of these materials are ~1.50 eV for CdTe and 1.57 eV for CdZnTe, so they can operate at room temperature without a significant rate of dark events.

Fabricating these detectors is challenging because two component semi-conductors (such as CdTe) tend to have a high rate of crystal lattice imper-fections that can trap the free charge carriers and degrade the detector performance. The introduction of Zn to replace some of the Cd reduces the concentration of large-scale imperfections, but traps remain as a problem. In addition, the free electron lifetimes in these detectors are short and the electron mobilities are small, making the diffusion lengths short (see equa-tions (3.10) and (3.11)) – the detectors must be thin for good charge collection efficiency.

These detectors are read out by application-specific integrated circuits (ASICs) that include a significant amount of circuitry for each detector. An example is the hard X-ray detectors provided for the NuStar mission (Rana *et al.* 2009). They are in a 32 × 32 array of CdZnTe with total dimensions of 2 cm square by 0.2 cm thick, with a pixel pitch of 605 μm. The individual pixels are bonded with conductive epoxy to the inputs of an ASIC, with 32 × 32 data-handling circuits on a single wafer with the same pitch as the detector pixels. Each of these circuits has a preamplifier, shaping amplifier,

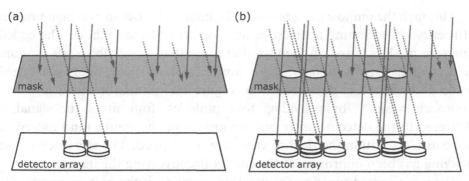

Figure 10.7. Principle of operation of a coded mask. The left image is an X-ray pinhole camera, while the right shows the behavior with four pinholes that create overlapping images.

discriminator, sample-and-hold circuit, and analog-to-digital converter, whose output is delivered to the higher-level electronics.

CdZnTe detectors have prominent low-energy tails. In principle, at high count rates they are subject to pileup, but their use of a dedicated electronics data chain for each pixel makes them more resistant to this issue than CCDs. With care to correct for various forms of non-ideal behavior, they can achieve energy resolution of 1 to 2% (at energies approaching 100 keV).

10.4 100 keV to 5 MeV

Wolter-type telescopes provide huge advantages for imaging up to 100 keV. However, between 100 keV and about 5 MeV, ingenious systems of baffles, masks, and grids must be employed to measure the positions of sources.[22] The most powerful of these methods is coded aperture masks, used for example on the INTEGRAL and SWIFT missions. Figure 10.7 shows how one operates.

Panel (a) shows X-rays from sources in two directions incident on a mask. As they strike the mask, there is a mixing of spatial and angular information that prevents deriving any information about the directions to them. However, the mask has a single hole that limits the spread in the spatial domain so the X-rays from the two sources fall onto two distinct regions on the array of detectors, separated by incident angle. The device in panel (a) is an X-ray pinhole camera.

---

[22] Laue diffraction (Bragg diffraction on transmission of X-rays through a crystal (Figure 2.18)) can be used to manufacture weak lenses (i.e., of long focal length) in the 0.1–1 MeV range. To do so, a large number of crystals are arranged in concentric rings and accurately oriented so they diffract radiation to a focus. By optimizing the geometry of the rings and selecting crystals with an angular spread in their diffracting planes, such lenses can operate over a reasonably broad energy range. Soft gamma-ray telescopes operating in this manner are under development.

Although the pinhole camera can make images, it does so at a high price in efficiency; its collecting aperture is just the area of the pinhole. The coded aperture mask is based on the idea that more than one pinhole lets in more X-rays, at the expense of providing a number of overlapping images that need to be deconvolved in data reduction. Figure 10.7 panel (b) shows how this approach works – by providing four pinholes, four times the signal is delivered to the detector array but the images of the source (and indeed of the entire sky within the field of view) are overlapped. The efficiency of the imaging has been improved, at the cost of deconvolving the image.

The point spread function for the deconvolution is the pattern cast by the transparent regions of the mask on the detector array and can be quite complex, as in the case of the Imager on Board the Integral Satellite (IBIS) instrument illustrated in Figure 10.8. As another example, the coded mask used in the SWIFT Burst Alert Telescope (BAT) has 52 000 lead absorbing elements in a random pattern. The image of a field of point sources consists of many mask patterns superimposed, one for each source. The mask must be designed so that this image has a unique deconvolution, that is, no two sources in different directions should produce identical exposed patterns on the detector array. To test whether this requirement is met, one represents the mask with a matrix of ones and zeros, the former for the open apertures and the latter for the opaque regions. This mask represents the point spread function – every unresolved source produces this pattern centered at some position on the detector array. The autocorrelation of the mask systematically tests the imaging for sources at every conceivable input angle, since it computes the result of every conceivable superposition of the mask onto itself:

$$\Lambda_{i,j} = \sum_m \sum_n a_{m,n}\, a_{m+i,n+j} \tag{10.6}$$

where $a_{k,l}$ is the transmission of the mask (1 or 0) at position k,l. The autocorrelation function of a good mask pattern should give a delta function, indicating that deconvolution can locate a source precisely and uniquely. Masks with random placement of the pinholes can be used to meet this condition (Dicke 1968; Ables 1968), but even they can have autocorrelation sidelobes. Another approach is based on cyclic difference sets (Gunson and Polychronopulos 1976, Fenimore and Cannon 1978), as is shown in Figure 10.8. Further optimization addresses maximizing the signal-to-noise ratio achieved in a given application and controlling the imaging properties for off-axis sources whose shadows are not fully captured by the detector array. Discussions of these issues can be found in Caroli *et al.* (1987) and In't Zand (1996 on).

(a)   (b)

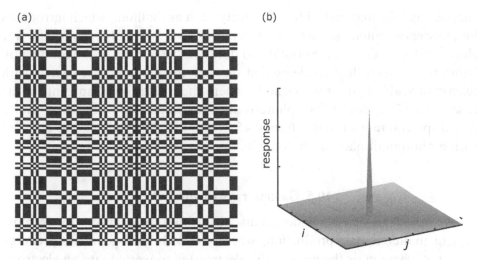

Figure 10.8. Coded aperture mask from the IBIS instrument on INTEGRAL (left) and its autocorrelation function (right). After Goldwurm *et al.* (2003).

The process of deconvolving the images from a coded mask has many parallels to the description of deconvolution in Section 7.7. In principle, the deconvolution could be done in Fourier space, but as in Section 7.7 this approach is unsatisfactory because the noise is amplified unacceptably in regions where the Fourier transform of the PSF is small. The iterative removal of sources (IROS) method (Hammersley 1986) is similar to CLEAN. One cross-correlates the data and mask, finds the strongest point source, subtracts its signal from the image, and repeats the process until there are no more sources to remove (the signal goes negative). The maximum entropy method has also been used successfully (Willingale *et al.* 1984).

The detectors used in this energy range include CdTe and CdZnTe, similar to those employed at lower energies. Germanium is also used, for example in the Spectrometer for INTEGRAL (SPI). Its atomic number of 32 is a significant improvement over silicon and it can be grown as large, high-purity, low-defect crystals. It has a bandgap of 0.66 eV and hence must be cooled to suppress dark current.

Alternatively, scintillators can be used, of which CsI (Tl) is perhaps the most common example. CsI has a bandgap of 6.3 eV. When an X-ray is absorbed in it, many electrons are lifted from the valence band to the conduction band. If they recombined from those energies, the intrinsic absorption by the crystal would absorb the resulting photon over a very short path length. However, significant light output can be achieved by "activating" the material with an impurity such as thallium, which introduces luminescence centers in the form of impurity levels in the bandgap. The free electrons lose energy to the crystal and decay to the bottom of the conduction band, from which they are de-excited through the luminescence centers with release of visible light at wavelengths where the crystal is transparent. In the case of CsI(Tl), about 64 000 photons are emitted per MeV absorbed, over a broad spectral range roughly from 0.45 to 0.65 μm. This light can be detected with a photomultiplier or photodiode.

10.5 Gamma rays to 100 GeV

Above about 5 MeV, the dominant absorption mechanism for gamma rays in matter is pair production, where 1.022 MeV of gamma-ray energy ($= 2m_e c^2$ where m_e is the mass of the electron) is converted into an electron–positron pair. To preserve energy and momentum, this process must take place near an atomic nucleus that recoils to balance the momentum of the gamma ray with the momenta of the two particles. Because of the high mass of the nucleus, it carries away little of the remaining gamma-ray energy, which instead appears as kinetic energy of the electron–positron pair (but can be divided unequally between them). The characteristic angle of the electron–positron pair is

$$\theta_C = \frac{m_e c^2}{E} \tag{10.7}$$

where E is the particle energy. At energies above 10 MeV, the angular distribution is strongly peaked in the same direction as the gamma ray was taking. The directional nature of the process simplifies the problem of determining the direction from which the gamma ray came, since one can use particle physics tracking techniques to determine the directions taken by the electron and positron.

A useful parameter to describe this process is the radiation length, X_0 (typically in gm/cm^2), which characterizes the amount of material traversed by energetic electrons before their energy drops by the factor $1/e$ due to energy loss by bremsstrahlung. Bremsstrahlung is energy radiated by an electron that is being strongly accelerated as it passes an atomic nucleus, and is the dominant loss mechanism for electrons at high energy. The distance for pair production by gamma rays is similar to X_0; the radiation length is 7/9 of the mean free path for pair production. The cross-section for pair production is negligible for energies below 2 MeV. In fact, this process is dominant only for energies above the critical energy

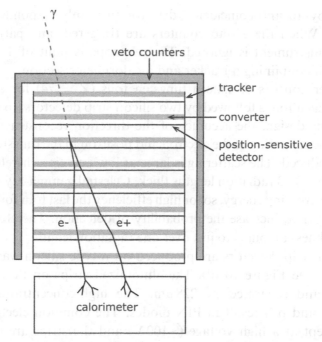

Figure 10.9. Schematic layout of the large area telescope for Fermi.

$$E_C \approx \frac{800 \, \text{MeV}}{Z + 1.2} \qquad (10.8)$$

where Z is the atomic number of the absorbing material.[23]

Below E_C ionization loss dominates over pair production; equivalently, E_C is where the ionization loss rate per radiation length is the energy of the electron. Thus, below the critical energy, particles quickly lose energy and are lost to the chain of interactions. Above E_C, the dependence of the absorption on energy is modest but the cross-section is proportional to $Z(Z + 1)$. Therefore, it is desirable to use high-Z absorbing materials in gamma-ray detectors.

We illustrate gamma-ray detection by describing the Large Area Telescope (LAT) on the Fermi Gamma-Ray Space Telescope, which is illustrated schematically in Figure 10.9 (Atwood *et al.* 2009). The gamma-ray detection process in the LAT occurs in two stages: a tracker and a calorimeter. The tracker determines the direction of the gamma ray, but absorbs only a small fraction of its energy. The calorimeter is designed to absorb completely and determine the gamma-ray energy. In addition, the tracker sections are

---

[23] Equation (10.8) is an approximate fit to the relevant energy loss equations and depends somewhat on the absorbing material.

surrounded by an anti-coincidence detector that only responds to charged cosmic rays. When these veto counters are triggered, any particle passing through the instrument is ignored. The telescope is built of 4×4 identical modules, each containing a tracker and a calorimeter section.

The tracker consists of sixteen tungsten foils ($Z = 74$) for efficient pair production. Each foil is followed by two silicon strip detectors (to be described below). In this design, the accuracy of the direction reconstruction requires minimizing the scattering of the electron and positron in the tungsten foil where they were produced. The scattering is large at low energies; therefore, the first 12 foils are only 0.03 radiation lengths thick. Celestial gamma-ray numbers fall rapidly with increasing energy, so for high efficiency the last four foils were made six times thicker to increase the probability of detection. Therefore, the total tungsten thickness amounts to just over one radiation length.

The silicon strip detectors are produced on n-type silicon wafers that are $400\,\mu m$ thick (see Figure 10.10). The aluminized strips on these wafers are $58\,\mu m$ wide and separated by $228\,\mu m$. The high-concentration n and p implants (n^+ and p^+) result in PIN diodes. The common electrode (at the bottom) is kept at a high voltage ($\sim 100\,V$) and the strips are near ground potential; the backbias field through the device then depletes the entire volume. When an energetic charged particle passes through the wafer carrying the strips, the free charge created there produces the output signal (holes are collected at the strips and electrons at the common electrode). The signals are sensed by electronics connected to the end of each strip. Two such devices are used in series with their strips oriented orthogonally; the position at which an ionizing particle passed through the pair can be reconstructed from the strips that generated signals.

The veto counters are a plastic scintillator material that emits light when the electrons freed by an energetic charged particle are recaptured by the material atoms. In this case, fluorescent additives shift the emission of the plastic from the ultraviolet to the visible, where the transparency of the material is excellent and the light can be collected efficiently. The light from the plastic scintillators is detected by photomultipliers.

The calorimeter is based on CsI(Tl) scintillator crystals. Each LAT calorimeter module has 96 CsI(Tl) scintillators, $2.7\,cm \times 2.0\,cm \times 32.6\,cm$ in size and stacked like logs in eight layers, 12 to a layer and alternating direction between layers by $90°$. PIN photodiodes at the ends of the logs measure the scintillation light; the ratio of the outputs from the two ends of a log can be used to determine where along the log the light was produced to an accuracy of about a millimeter. The total thickness of the stack of scintillator logs is 8.6 radiation lengths, over which the gamma ray is decomposed into successively

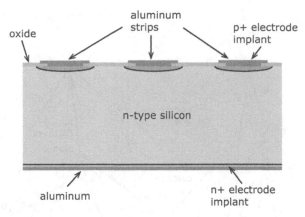

Figure 10.10. A silicon strip detector.

less energetic electrons and positrons until the individual particles reach the range where they lose energy quickly by ionization loss, that is, through multiple small interactions that strip the outer electrons from the crystal atoms. At this point, their kinetic energy is absorbed by the CsI crystal as heat. Thus, the total light output is a measure of the gamma-ray energy, while the 192 individual measurements give a picture of the growth and eventual decay of the number of particles it produces.

The interaction of the gamma ray with the LAT to produce many daughter particles is termed an electromagnetic shower. Figure 10.11 shows the early stages of such a shower. Accurate models require calculations in which the output energies of each interaction are computed and tracked. However, a heuristic picture of the process can make the simple assumption that, at the end of each radiation length of the material, (1) a pair production event has occurred for each gamma ray, dividing its energy equally between the electron and positron; (2) each positron has annihilated to produce a pair of equal-energy gamma rays, and (3) each electron has contributed half its energy to a gamma ray in a major bremsstrahlung event. All these processes multiply the number of particles without removing energy from the overall shower; ionization is the only true energy loss mechanism. Therefore, when the particle energies are degraded to the critical energy, E_C, ionization losses become dominant and the shower is quenched.

With our simplistic assumptions (and ignoring losses), the number of particles doubles every radiation length into the material. Then, if t is the distance in radiation lengths, the number of particles in the shower prior to losses becoming significant is

$$N(t) = 2^t \tag{10.9}$$

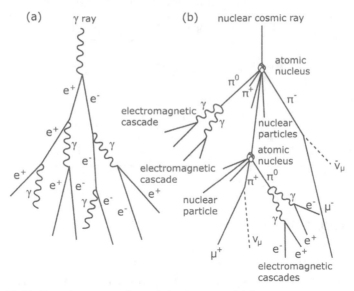

Figure 10.11. Development of particle showers: (a) for a gamma-ray primary particle and (b) for a cosmic ray (proton). The shower in (a) is purely gamma rays (γ), electrons (e⁻) and positrons (e⁺), whereas in (b) there are also muons (μ), pions (π), and neutrinos (ν).

For a gamma ray of initial energy E_0, the average energy of a shower particle is

$$E(t) = \frac{E_0}{2^t} \tag{10.10}$$

and the maximum number of particles in the shower occurs at

$$t_{\max} = \frac{\ln{(E_0/E_C)}}{\ln 2} \tag{10.11}$$

since losses will dominate beyond this depth. Beyond $t_{\max}$ this model fails badly. It would predict that the shower would end abruptly a radiation length beyond $t_{\max}$, whereas the wide range of particle energies produced in the interactions results in a gradual reduction in the shower intensity.

Equation (10.11) shows that the gamma-ray energy can be estimated from the position of the shower maximum even when the energy is sufficiently high that the calorimeter does not absorb it all. Showers are also initiated by energetic nuclear particles (typically protons), but many of the interaction lengths are substantially longer than a radiation length (e.g., for muons), so these showers are slower to develop and penetrate deeper (Figure 10.11). Thus, because of their differing shower development, the calorimeter can also identify and reject some of the charged particles that elude the veto counter.

Other gamma-ray telescopes are similar in concept to the LAT but use different approaches for some of the functions we have described. For example, the Energetic Gamma Ray Experiment Telescope (EGRET) on the Compton Gamma Ray Observatory (CGRO) tracked the electron–positron pairs by applying high voltage to the gas they had traversed and partially ionized to yield a spark that was imaged to determine the direction of motion (this type of detector is called a spark chamber). The inability to gather high-energy gamma rays with relatively light weight focusing optics drives designs for both the LAT and EGRET so that the "detector" effectively becomes synonymous with the "telescope" and the need for large absorption paths to measure the particle energies makes these detector/telescopes massive as well as large. Thus, EGRET weighed about two metric tons to achieve an effective area of a sixth of a square meter, while the LAT is about three metric tons and $1.8 \times 1.8 \times 0.72$ meters in size to achieve an effective telescope area of about 1 square meter (its larger field of view and other characteristics increase the gain over EGRET). In comparison, the AGILE mission has a gamma-ray telescope that is similar in many operating principles to the LAT, but with a number of compromises to reduce mass, which is only 55 kg (see Morselli *et al.* 2000) (it has about 1/16 the area and a much thinner calorimeter so energies above ~10 GeV are not measured accurately).

The Fermi satellite carries a second instrument, the Gamma Ray Burst Monitor. It consists of twelve NaI and two Bismuth Germanate (BGO) scintillators viewed by photomultipliers to identify possible bursts of gamma rays. Similar technology was used in the Burst And Transient Source Experiment (BATSE) on CGRO.

10.6 Gamma rays from 100 GeV to 100 TeV

Above 100 GeV, the gamma-ray fluxes are too low to allow the existing detector/telescopes to provide interesting information. For example, above 100 GeV, the flux from the brightest celestial source, the Crab Nebula, produces less than one detection per week in the LAT, with the rate falling rapidly with increasing energy. Since launching a much bigger telescope of many more metric tons into space is not feasible, a different approach is used: Cherenkov radiation from air showers. The process illustrated in Figure 10.11 occurs in the atmosphere, leading to large cascades of particles. All of the interactions occur at relativistic velocities, so the particles in the cascade travel in nearly the direction of the incident gamma ray (or nuclear particle) (equation (10.7). The speed of light in the atmosphere is $v_{light} = c/n$, where $n > 1$ is the refractive index of the air. If it is less than the velocity of a relativistic particle, v, then Cherenkov light is emitted in the direction

$$\theta = \cos^{-1}\left(\frac{1}{n\beta}\right) \tag{10.12}$$

where $\beta = v/c$. Since n is only slightly greater than 1, the light is emitted only about 1° off the direction of the particles in the cascade. The maximum of the shower (in terms of numbers of Cherenkov-emitting particles) occurs at an altitude of roughly 9 km for a TeV gamma-ray primary particle, and somewhat lower ($\sim$7 km) for a nuclear primary; the latter type of shower also develops more slowly and penetrates closer to the ground. By the time the light reaches the ground, it appears as a short (3–5 ns) pulse that has spread to fill a diameter of about 200 meters. A light collector anywhere in this area with a fast detector can record this pulse to detect the gamma ray (or nuclear particle). The collecting area is then of order 30 000 square meters, large enough to allow studying celestial gamma-ray sources to 100 TeV.

Four major facilities have been purpose-built to use this technique: the High Energy Stereoscopic System (HESS) in Namibia, the Collaboration of Australia and Nippon (Japan) for a GAmma Ray Observatory in the Outback (CANGAROO-III) at Woomera, Australia, the Major Atmospheric Gamma-ray Imaging Cherenkov Telescope (MAGIC) on La Palma, and the Very Energetic Radiation Imaging Telescope Array System (VERITAS) in Arizona, USA.

The air shower instruments are constructed along very similar principles to implement strategies to distinguish gamma-ray-initiated showers from the far more numerous (by three to four orders of magnitude) showers initiated by nuclear particles. Each has 2 to 7 collectors of segmented mirrors distributed over an area nearly as large as the pool of Cherenkov light (separations of 80 to 120 m – Figure 10.12). Since the image quality can be far worse than for a normal telescope, the mirror segments are identical, with spherical figures, aligned so their foci coincide. The total collecting area is 75–500 m². At the collector foci, there are arrays of 500 to 1000 photomultipliers. The gamma-ray showers, with maxima relatively high in the atmosphere, yield relatively compact images that are roughly elliptical in shape. The nuclear showers, with their lower maxima and slower development, have larger and more irregular images. For the same reasons, the gamma-ray showers spread their light more uniformly compared with a significant drop with increasing distance from the shower core for the nuclear ones, and the gamma-ray shower pulses are of shorter duration. Finally, nuclear showers produce very penetrating individual particles (e.g., muons) that can mimic some of the characteristics of gamma-ray showers to a single telescope, but they tend

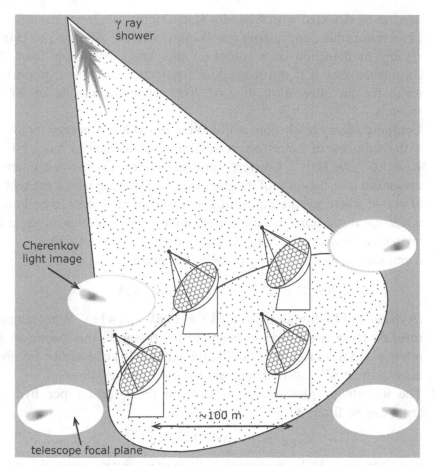

Figure 10.12. An atmospheric air shower Cherenkov light telescope array.

not to emit over a wide enough area to produce signals in the rest of the array. The resulting high levels of discrimination have revolutionized this pursuit, allowing detection to levels of less than 1% of the flux from the Crab Nebula in integration times that were inadequate to detect the Crab before the discrimination features were added. The energy thresholds for these facilities are about 50 GeV, complementing space missions such as the LAT well. In addition, they can locate strong sources to a few arcseconds within a field of order 3° diameter and provide energy resolution of about 15 to 30%.

10.7 Exercises

10.1 Given an X-ray background of 2 counts/cm² second ster keV over a band from 0.5 to 2 keV, compute the minimum number of X-rays that need to be detected with Chandra (for a point source within a field of 12 arcmin radius, and taking a collecting area of $1000\,\text{cm}^2$) to claim a significant detection of a source in this band at greater than 99% confidence that it is not a random confluence of counts. Report the result for an integration time of 3000 seconds and one of 300 000 seconds.

10.2 Compute energy resolution as a function of the X-ray energy for a CCD with read noise of 3 electrons rms. Assume a calorimeter has a NEP of $1 \times 10^{-17}\,\text{W/Hz}^{1/2}$. Convert this parameter to its X-ray energy resolution as a function of its time constant, making use of the relation $df = 1/4\tau$ where τ is the exponential time constant. Take τ to be 1 ms for a realistic resolution, and use it to compare the resolution of the calorimeter as a function of X-ray energy with that for the CCD.

10.3 Compute the spectra of an optically thin plasma:

$$S = C\,E^{-0.4}\,e^{-E/kT}\,\text{erg}\,\text{cm}^{-3}\,\text{s}^{-1}\,\text{keV}^{-1}$$

(Rybicki and Lightman 1979) for $kT = 1\,\text{keV}$ and $4\,\text{keV}$. What temperatures do these two curves correspond to? Integrate the spectra to give relative signals in soft (0.5–2 keV) and hard (2–10 keV) bands. Fit power laws to these SX and HX numbers for both temperatures.

10.4 The interstellar photoelectric extinction cross-section per hydrogen atom can be fitted by

$$\sigma_{\text{ISM}} = 2 \times 10^{-22} E^{-2.383}\,\text{cm}^2$$

(the absorption is the exponential of the cross-section times the column of atoms). The average density of hydrogen atoms in the interstellar medium is about $1\,\text{cm}^{-3}$. At what distance would the intrinsic power law spectral indices in exercise 10.3 be expected to change by 0.5 due to interstellar aborption?

10.5 Determine a simple formula for the probability of two events on a CCD pixel as a function of the pixel read time and the count rate. For the case of Chandra, with readouts every 3.2 seconds, above what count rate will the pileup (two or more events) rate exceed 10% of the count rate?

10.6 Estimate the width of the photon pulse due to Cherenkov light emitted by an air shower and received at the ground, using our simple shower description. The radiation length in air is $37\,\text{g/cm}^2$. You can obtain the refractive index of air from equation (7.1); assume a pressure at the ground of 1000 mbar and a density at the ground of $1.3\,\text{kg/m}^3$, a scale height of 8 km, and a temperature of -20 °C (just to plug into the equation; ignore the change of temperature with altitude). Also, assume that the radiating particles are all moving at the speed of light – only the light is slowed due to the refractive index. Let the initial gamma ray have an energy of 10 TeV.

10.7 Design a linear coded aperture mask. Make it have 64 elements (either transmitting or blocking). Evaluate its performance using autocorrelation (extend the mask in both directions by adding another 64 elements so it is periodic, for calculational convenience). Remember, the goal is to have the largest possible number of transmitting elements with the lowest autocorrelation for other positions.

Further reading

Aharonian, F. A. (2004). *Very High Energy Cosmic Gamma Radiation*. Singapore: World Scientific Publishing.

Caroli, E., Stephen, J. B., Di Cocco, G., Natalucci, L., and Spizzichino, A. (1987). Coded aperture imaging in X- and gamma-ray astronomy. *SSRv*, **45**, 349–403.

Giacconi, R. (2003). Nobel lecture: The dawn of X-ray astronomy. *Rev. Mod. Phys.*, **75**, 995–1010.

In't Zand, J. (1996) (with updates). Coded aperture camera imaging concept: http://www.sron.nl/~jeanz/cai/coded_intr.html

Paerels, F. B. S. and Kahn, S. M. (2003). High-resolution X-ray spectroscopy with CHANDRA and XMM-NEWTON. *ARAA*, **41**, 291–342.

Trumper, J. E. and Hasinger, G. (2008). *The Universe in X-rays*. Berlin, New York: Springer.

Weekes, T. C. (2003). *Very High Energy Gamma Ray Astronomy*. Bristol: Institute of Physics Publishing.

11

Epilogue: cosmic rays, neutrinos, gravitational waves

11.1 Introduction

We have concentrated on methods to detect and analyze the information from photons, the approach used in the vast majority of astronomical investigations. In this chapter, we will take advantage of what we have discussed about these approaches to discuss a few other ways to study the cosmos: (1) cosmic rays; (2) neutrinos; and (3) gravitational rays. Curiously, the efforts to expand astronomy into new spectral regions have often been led by physicists and engineers rather than astronomers (Jarrell 2005; Low *et al.* 2007). Thus, it is no surprise that the methods we are about to discuss are rooted in physics.

11.2 Cosmic rays

The term cosmic ray is applied to very energetic charged particles that circulate in interstellar space. We chose the term "circulate" advisedly, since they are trapped by the Galactic magnetic field and stored in the Galaxy. The gyro-radius for an energetic charged particle can be expressed as:

$$r_\mathrm{g}(pc) \approx 1 \times 10^{-21} \frac{E(\mathrm{eV})}{Z\, B(\mathrm{gauss})} \tag{11.1}$$

where Z is the charge of the particle in atomic units. The interstellar magnetic field near the earth is $\sim 6 \times 10^{-6}\,\mathrm{G}$, so r_g for a $10^{12}\,\mathrm{eV}$ particle is only $\sim 0.0002\,\mathrm{pc}$. We can estimate how long the particles circulate by analyzing the abundances of the elements in the cosmic ray flux and comparing with calculations of the yields from collisions with atomic nuclei in the interstellar medium (ISM); their typical age is about 10 million years, 100 times the time to traverse the Galaxy in a straight line. As a result, by the time cosmic rays reach Earth, all evidence about their places of origin has been obliterated. Direct studies of the cosmic ray flux must content themselves with determining the atomic abundances and the energy spectrum; both parameters should give some indication of the places and conditions where the particles are accelerated to relativistic energies, and the conditions in the ISM they have traversed. They can also be studied indirectly through observations of "nonthermal" radiation, that is, synchrotron radiation emitted by relativistic electrons when they are accelerated in magnetic fields (the electron component of the cosmic ray flux is greatly attenuated by energy loss in the ISM).

High-energy physicists were astounded in the early 1990s by the detection of two air showers initiated by cosmic rays of 2–3 $\times$ 10^{20} eV, one by the Fly's Eye air shower detector, and a confirming one by the Akeno Giant Air Shower Array. From equation (11.1), r_g for such particles substantially exceeds the dimensions of the Milky Way, even if they are massive nuclei with large Z. Thus, the particles are unlikely to have originated in the Milky Way; on the other hand, they could not have traveled far in intergalactic space because of energy losses when they collide with photons in the cosmic microwave background:

$$p + \gamma \rightarrow \Delta(1230) \rightarrow p + \pi \tag{11.2}$$

where $\Delta(1230)$ is a delta baryon, consisting of three quarks with co-aligned spin axes (compared with protons and neutrons where one quark has its spin axis opposite to those of the other two). The energy loss length for this pion production mechanism drops two orders of magnitude to only about 10 Mpc between 10^{20} and 10^{21} eV (Arkhipov 2006).

Studying such events is challenged by their very low rate, about 1 km^{-2} yr^{-1} for showers of 10^{19} eV and greater. As with the TeV gamma rays discussed in the preceding chapter, the detection approach depends on the atmosphere and the particle showers resulting when a cosmic ray strikes it. To understand the approach, we will make some rough estimates combining results from gamma-ray and hadron showers. From equations (10.8) and (10.11), assuming an atmosphere with an exponential scale height of 8 km and a density at sea level of 1.3 $\times$ 10^{-3} g cm^{-3}, the maximum of a shower initiated by a 10^{17} eV gamma-ray incident at normal incidence to the surface of the earth is only ∼500 m above sea level. From equation (10.9), there are roughly 10^9 particles at this maximum. The lateral distribution for hadronic showers can be fitted by (Klepser *et al.* 2008):

$$S(r) \approx S_{\text{ref}} \left(\frac{r}{R_{\text{ref}}} \right)^{-3 - 0.303 \log(r/R_{\text{ref}})} \tag{11.3}$$

where $S(r)$ is the particle surface density and S_{ref} and R_{ref} affect the normalization of the curve. Therefore, although the particle density is very strongly peaked (our nominal 10^{17} eV shower yields a density $> 10\,000/\mathrm{m}^2$ within 20 meters of its center), there is a significant density at 500 m ($\sim 1/\mathrm{m}^2$) and even at 1000 m ($\sim 0.1/\mathrm{m}^2$). Therefore, arrays of energetic particle detectors with large detector-to-detector spacing are suitable for studying these events.

The huge Pierre Auger Array has been constructed at a site in Argentina ~ 1350 m above sea level to observe very energetic air showers over a large enough area to provide a tolerable detection rate (Pierre Auger Collaboration 2004). It has a collecting area of ~ 3000 km^2 and is at an altitude close to the maxima of air showers from 10^{17}–10^{19} eV and within 45° of the zenith. We will describe this array as an example of the methods used in general to observe high-energy cosmic rays.

1600 surface detectors (Pierre Auger Collaboration 2004) are distributed over the array area, at spacings of about 1.5 km and designed to be autonomous, operating using solar power. Each detector consists of a tank of water with an area of 10 m^2 and a depth of 1.2 m. From the estimates above, the spacing and areas of these tanks result in any shower with energy $> 10^{17}$ eV yielding energetic particles in a number of the water tanks surrounding its core. The tanks are light-tight and have reflective inner surfaces. The relativistic particles that enter the tanks produce a pulse of Cherenkov radiation that is sensed by three photomultipliers viewing the interior of each tank. Electronic logic applied to the photomultipler signals identifies if the event is probably associated with a massive air shower. If so, then the result is sent by radio transmission to a central station, time-tagged to a resolution of 8 ns. Analysis of the timing outputs from the surrounding detectors determines whether there is a consistent solution for a single shower from one direction. If so, the intensity of the signals and the number of detectors triggered can be used to determine the size and energy associated with the shower, while the relative times of arrival determine its direction.

There is a second type of detector in the Pierre Auger Array (Pierre Auger Collaboration 2009). We have discussed studying such showers through detection of the Cherenkov radiation emitted by their relativistic particles; in addition, they leave in their trails large numbers of ionized and excited atmospheric atoms, particularly nitrogen. When the nitrogen returns to its ground state, it emits fluorescently, primarily in the near ultraviolet between 300 and 400 nm. About 5 photons are emitted per Mev of energy deposited by ionization loss in the atmosphere; the total emission from a 10^{17} eV shower is about 0.04 W. The amount of ionization and hence the light emitted through fluorescence is proportional to the shower energy, and hence is a measure of the energy of the incident cosmic ray. On clear, moonless nights the fluorescent trails of air showers are imaged with four detector units, each consisting of six Schmidt cameras. Each camera has a collecting area of $10\,m^2$ and a field of view of 30°; they are placed at the edges of the surface detector array and arranged to overlook the full distribution of surface detectors by offsetting their fields by ~30° to give a net coverage of $30° \times 180°$. There are 440 photomultipliers at the focus of each camera, providing images with resolution of about 1.4 degrees. In addition to providing another determination of the shower energy, the air fluorescence measurements document its development in the atmosphere. Since protons penetrate more deeply than heavy nuclei before interacting and produce a lower ratio of muons to electrons, this information can determine the composition of the extremely high-energy cosmic rays.

The Telescope Array Project in Utah is similar in concept but smaller than the Pierre Auger array. It has three atmospheric fluorescence units and 507 ground detectors. The latter, distributed at 1.2 km spacing over $762\,km^2$, are based on plastic scintillators rather than Cherenkov radiation in water. The IceTop Array as part of IceCube (next section) at the South Pole has detectors similar to the water tanks for Pierre Auger, but using blocks of solid ice rather than water for the absorber. There are a number of other air shower arrays operating with a mixture of techniques (e.g., GRAPES-3, Yakutsk, KASCADE-Grande).

The primary instrument for measuring the composition of the cosmic ray flux is the Cosmic Ray Isotope Spectrometer (CRIS) on the Advanced Composition Explorer (ACE) satellite (Stone *et al.* 1998). CRIS makes use of the dependence of the energy loss rate for a particle on the square of its atomic number (in the classic Bethe equation). Therefore, measurements of how far a particle travels before losing all its energy – its range – along with the rate of energy loss along this track can be used to determine its energy and atomic number. CRIS carries out these measurements with four stacks of 10-cm-diameter silicon detectors. It determines the direction of the particles entering the silicon detectors using a system of scintillating optical fibers.

11.3 Neutrinos

The unique feature of neutrinos for astronomy is their very low interaction cross-section of only $\sim10^{-52}\,cm^2$ on hydrogen. Consequently, neutrino fluxes can emerge from the cores of stars virtually unattentuated, providing a unique perspective on processes such as the fusion reactions in the sun, or the collapse of a massive star core to a neutron star during a supernova explosion. Of course, the bad news is the difficulty in getting enough neutrino reactions in a detector to produce a usable signal. Nonetheless, there are two notable results in this field. The first is the apparent underproduction of neutrinos from the fusion reactions in the core of the sun compared with theoretical predictions for the relative role of different fusion possibilities (Davis *et al.* 1989). It now appears this discrepancy lay in the neutrino physics, not in the models of the solar interior (Ahmad *et al.* 2001). The second is the detection of the burst of neutrinos emitted as a massive stellar core collapsed while the rest of the star exploded as supernova 1987A (Hirata *et al.* 1987, Bionta *et al.* 1987).

We discuss IceCube (Halzen and Klein 2010) as an example of a neutrino detector. The ice in the Antarctic is highly transparent with an absorption length of 100 m or more; in IceCube, it is used as a source of Cherenkov light from relativistic particles, analogously to the water tanks in the Pierre Auger surface detectors. IceCube is at the Amundsen–Scott South Pole Station; 86 holes have been drilled deep into the ice at a spacing of 125 m, and strings of 60 sealed photomultiplier modules have been lowered into each of these holes to a depth of 1450 to 2450 m, that is, with a photomultiplier module roughly every 17 m. The photomultiplier modules are then immersed and frozen in place to establish an efficient optical interface with the ice. The entire array is therefore a cube that has a cross-sectional area of $\sim 1 \, \text{km}^2$ and a thickness of 1 km. The IceTop air shower array is on the ice surface, also with a $1 \, \text{km}^2$ area; it consists of two ice tanks at the location of each IceCube string, each tank viewed by two photomultipler modules.

The vast majority of neutrinos will pass through the IceCube detector without interaction. When one interacts with an atomic nucleus in the ice, muons as well as electromagnetic and hadronic particle showers are created. The timing information from the photomultipliers determines whether a Cherenkov-emitting particle is moving upward or downward, and the imaging indicates its direction relative to the surface of the earth. If these particles are found by the photomultiplier images to be moving upward, away from the center of the earth, they are ascribed to neutrinos; the entire Earth is used as a shield to stop other cosmic ray particles from creating false positives. IceCube detects several hundred neutrinos per day with energy >100 GeV; a section with a denser array of photomultipliers can extend the energy threshold down by an order of magnitude. Since the muons and showers that constitute the detected signal are directed along the path of the neutrino creating them, the incident direction can be measured to better than a degree. IceTop detects air showers and can provide information that complements the detection of deeply penetrating muons that reach the buried IceCube. For example, the comparison of IceTop and IceCube measurements of a shower can provide information on the composition of its initiating cosmic ray.

Other neutrino experiments work on similar principles to IceCube, but with different implementations. The Super-Kamiokande detector is a tank 39 m in diameter and 42 m tall, filled with very pure water and lined with about 13 000 photomultipliers. To reduce the background from other types of cosmic ray, the tank is placed at a depth of 1000 m in a mine. The Astronomy with a Neutrino Telescope and Abyss Environmental Research (ANTARES) instrument has vertical strings of photomultipliers placed deep in the Mediterranean Sea, which acts as the absorber and Cherenkov radiator. The Baikal Underwater Telescope is similar, located deep in Lake Baikal in Russia. The Sudbury Neutrino Observatory is based on a large vessel of heavy water; if the deuterium in this water absorbs a neutrino, a neutron is freed that emits a burst of gamma rays when it is recaptured. The Gran Sasso Large Volume Detector uses large tanks of liquid scintillator to detect charged particles that are the products of neutrino interactions.

11.4 Gravitational waves

According to general relativity, gravitational waves are emitted when a mass is accelerated. They appear as a quadruple (or higher order) disturbance of space-time; dipole radiation is not possible because there is a single sign of gravitational mass. In some ways, the process is analogous to Larmor radiation (Section 8.4.1), although the latter is a non-relativistic effect. The most readily detectable example is expected to be the merger of the two components of a binary neutron star. The orbital energy of such a binary is radiated away as gravitational waves, an effect used by Hulse and Taylor to confirm the predictions of general relativity. As this process progresses, the stellar orbits shrink, the accelerations rise, and the energy loss due to gravitational radiation rises. Eventually, the orbital period rises to about 1 kHz, the stars merge, and the process abruptly stops. The strong accelerations during this sequence appear as a varying strain in space-time that propagates outward at the speed of light. Converted to audio, the strain might sound like a "chirp" and hence it is assigned this name. There are many other possible sources, such as massive binaries in general (before any mergers). Some events, such as core-collapse supernovae or merging black holes, may give a single burst of gravitational waves. They will be more difficult to identify uniquely unless simultaneous detections are made in more than one gravitational wave facility, and preferably in coincidence with an observation of the event in some other way (e.g., optical detection of a supernova).

The Laser Interferometer Gravitational-wave Observatory (LIGO) and its upgrade (advanced LIGO) seek to identify the physical distortions caused by this strain in space-time. LIGO is a huge interferometer (legs 4 km long) similar in principle to the one in Figure 6.21 introduced as a Fourier transform spectrometer. LIGO operates within a vacuum; the light from a powerful laser is divided at a beamsplitter and sent down the legs. Test masses at either end of the legs carry mirrors, and the beam is reflected between them and then recombined at the beamsplitter. The space-time distortions initiated by a gravitational event will shift the positions of the test masses subtly, leading to changes in the optical path lengths. Thus, the observation of gravitational waves takes the form of measuring a change in the phase of the returning light, through very high signal-to-noise measurements of their interference. The sensitivity of LIGO to such changes is enhanced by reflecting the light back and forth between the test masses about 100 times before recombining the beams.

The rate of coalescing neutron stars is low enough that they must be observed to a distance of about 20 Mpc for a reasonable detection rate. At this distance, the strain is about 10^{-21}, which produces a displacement of order 10^{-18} m over a distance of 4 km (Bradt 2004). Success clearly requires great care in the detector design and operation. The very high accuracy in phase measurement can in principle be achieved by operating at very high signal-to-noise ratio, that is, with a very strong laser and good detectors. Nonetheless, shot noise in the detection process sets the limit for gravitational wave detection at frequencies above 100 Hz (e.g., a neutron star pair shortly before coalescence). Below this frequency range, the limiting noise is set by a combination of thermal effects, seismic effects, and human-caused disturbances. To guard against false positives, there are two widely separated LIGO systems (in Louisiana and Washington) and data are coordinated with other systems (e.g., VIRGO in France). A general upgrade in LIGO, advanced LIGO, is expected to achieve detections of neutron star mergers at a rate of about one per month. The ultimate escape from the low-frequency seismic noise is to go to space, and the Laser Interferometer Space Antenna (LISA) was conceived for that purpose. The greatly improved low-frequency performance would open gravitational wave research to a broad range of astronomical questions. Unfortunately, NASA has withdrawn from the project due to lack of funds and at present LISA is being replanned in Europe.

Further reading

Cella, G. and Giazotto, A. (2011). Interferometric gravity wave detectors. *Rev. Sci. Inst.*, **82**, 101101.

Halzen, F. and Klein, S. R. (2010). IceCube: An instrument for neutrino astronomy. *Rev. Sci. Inst.*, **81**, 081101.

Pierre Auger Collaboration (2004). Properties and performance of the prototype instrument for the Pierre Auger Observatory. *Nucl. Inst. & Methods (A)*, **523**, 50–95.

Pierre Auger Collaboration (2009). The fluorescence detector of the Pierre Auger Observatory. *Nucl. Inst. & Methods (A).*, **620**, 227–251.

Stone, E. C., Cohen, C. M. S., Cook, W. R. *et al.* (1998). The Cosmic-Ray Isotope Spectrometer for the Advanced Composition Explorer. *SSRv.*, **86**, 285–356.

Waldman, S. J. (2011). The Advanced LIGO Gravitational Wave Detector. eprint arXiv1103.2728.

Appendix A Useful constants

Table A.1.

Symbol	Name	Value
k	Boltzmann's constant	$1.381 \times 10^{-23}\,\mathrm{J\,K^{-1}}$ $8.617 \times 10^{-5}\,\mathrm{eV\,K^{-1}}$
q	electronic charge	$1.602 \times 10^{-19}\,\mathrm{C}$
h	Planck's constant	$6.626 \times 10^{-34}\,\mathrm{J\,s}$
c	speed of light	$2.998 \times 10^{8}\,\mathrm{m\,s^{-1}}$
ε_0	permittivity of free space	$8.854 \times 10^{-12}\,\mathrm{F\,m^{-1}}$
m_e	rest mass of electron	$9.109 \times 10^{-31}\,\mathrm{kg}$
σ	Stefan-Boltzmann constant	$5.669 \times 10^{-8}\,\mathrm{W\,m^{-2}\,K^{-4}}$
$\mathring{\mathrm{A}}$	Angstrom	$10^{-10}\,\mathrm{m}$
eV	electron volt	$1.602 \times 10^{-19}\,\mathrm{J}$
AU	astronomical unit	$1.496 \times 10^{11}\,\mathrm{m}$
pc	parsec	$3.086 \times 10^{16}\,\mathrm{m}$

Appendix B Common Fourier transforms and relations

Table B.1. *Fourier transforms*

Function	Transform
$f(x)$	$F(u)$
$f(-x)$	$F(-u)$
$f^*(x)$	$F^*(-u)$
$f(x)$ is purely real	$F(u) = F^*(-u)$ even/symmetry
$f(x)$ is purely imaginary	$F(u) = -F^*(-u)$ odd/antisymmetry
even/symmetry $f(x) = f^*(-x)$	$F(u)$ is purely real
odd/antisymmetry $f(x) = -f^*(-x)$	$F(u)$ is purely imaginary
$af(x)$	$aF(u)$
$f(ax)$	$\frac{1}{\|a\|}F\left(\frac{u}{a}\right)$
$\frac{1}{\|a\|}f\left(\frac{x}{a}\right)$	$F(au)$
$f(x) + g(x)$	$F(u) + G(u)$
$f(x - x_0)$	$e^{-j2\pi ux_0}F(u)$
$e^{j2\pi u_0 x}f(x)$	$F(u - u_0)$
$f(x) \otimes g(x)$	$F(u)G(u)$
$f(x)g(x)$	$F(u) \otimes G(u)$
1	$\delta(u)$
$\delta(x)$	1
$e^{-\pi x^2}$	$e^{-\pi u^2}$
$e^{-a\|x\|}$, $a > 0$	$2a/(a^2 + (2\pi u)^2)$
e^{-x}, $x > 0$	$(1 - j2\pi u)/(1 + (2\pi u)^2)$
$\text{rect}(x)$	$\text{sinc}(u)$
$\text{sinc}(x)$	$\text{rect}(u)$
$\text{triang}(x)$	$\text{sinc}^2(u)$
$\text{sinc}^2(x)$	$\text{triang}(u)$
$\text{sgn}(x)$	$-j/(\pi u)$
$\text{step}(x)$	$1/j2\pi u + \delta(u)$
$\sin(2\pi u_0 x + \phi)$	$(j/2)[e^{-j\phi}\delta(u + u_0) - e^{j\phi}\delta(u - u_0)]$
$\cos(2\pi u_0 x + \phi)$	$\frac{1}{2}[e^{-j\phi}\delta(u + u_0) + e^{j\phi}\delta(u - u_0)]$
$\|x\|^{-1/2}$	$\|u\|^{-1/2}$

Table B.2. *Function definitions*

Function	Definition
$\delta(x)$	$0 \quad x \neq 0$
	$\displaystyle\int_{-\infty}^{\infty} \delta(x)dx = 1$
$\text{sinc}(u)$	$\frac{\sin(\pi u)}{\pi u}$
$\text{rect}(x)$	$1 \quad \|x\| \leq 1/2$
	$0 \quad \|x\| > 1/2$
$\text{triang}(x)$	$1 - \|x\| \quad \|x\| \leq 1$
	$0 \qquad\quad \|x\| > 1$
$\text{sgn}(x)$	$1 \quad x \geq 0$
	$-1 \quad x < 0$
$\text{step}(x)$	$1 \quad x \geq 0$
	$0 \quad x < 0$

References

Preface

Harwit, M. (1984). *Cosmic Discovery*. Cambridge, MA: The MIT Press.

Chapter 1

APEX atmospheric transmission calculator: http://www.apex-telescope.org/sites/chajnantor/atmosphere/transpwv/

Barbastathis, G. (2004). *Optics, MIT Open Course*: http://ocw.mit.edu/courses/mechanical-engineering/2–71-optics-fall-2004/

Condon, J.J. (1974). Confusion and flux density error distributions. *ApJ*, **188**, 279.

Leinert, Ch., Bowyer, S., Haikala, L.K. *et al.* (1998). The 1997 reference of diffuse night sky brightness. *A&AS*, **127**, 1–99.

Maihara, T., Iwamuro, F., Yamashita, T., Hall, D.N.B., Cowie, L.L., Tokunaga, A.T., and Pickles, A. (1993). Observations of the OH airglow emission. *PASP*, **105**, 940–944.

Rieke, G.H. (2003). *Detection of Light from the Ultraviolet to the Submillimeter*, 2nd edn. Cambridge: Cambridge University Press.

Rohde, R.H. (n.d.) for Wikipedia: http://en.wikipedia.org/wiki/File:Atmospheric_Transmission.png

Rousselot, P., Lidman, C., Cuby, J.-G., Moreels, G, and Monnet, G., (2000). Night-sky spectral atlas of OH emission lines in the near-infrared. *A&A*, **354**, 1134–1150.

Wall, J.V. and Jenkins, C.R. (2003). *Practical Statistics for Astronomers*. Cambridge: Cambridge University Press.

Chapter 2

Bracewell, R.N. (2000). *The Fourier Transform and its Applications*, 3rd edn. Boston, MA: McGraw-Hill.

Born, E. and Wolf, M. (1999). *Principles of Optics: Electromagnetic Theory of Propagation, Interference, and Diffraction of Light*, 7th edn. Cambridge: Cambridge University Press.

Burrows, C.J., Burg, R., and Giacconi, R. (1992). Optimal grazing incidence optics and its application to wide-field X-ray imaging. *ApJ*, **392**, 760–765.

Epps, H.W. and Fabricant, D. (1997). Field correctors for wide-field CCD imaging with Ritchey-Chretien telescopes. *AJ*, **113**, 439–445.

Fabricant, D. *et al.* (2004). The 6.5-m MMT's f/5 wide-field optics and instruments. *SPIE*, **5492**, 767–778.

Koren, Norman (n.d.) http://www.normankoren.com/Tutorials/MTF5.html

NASA (n.d.) Images. http://www.nasaimages.org/

Press, W.H., Teukolsky, S.A., Vetterling, W.T. and Flannery, B.P. (2007). *Numerical Recipes*, 3rd edn. Cambridge: Cambridge University Press.

Wynne, C.G. (1974). A new wide-field triple lens paraboloid field corrector. *MNRAS*, **167**, 189–198.

Chapter 3

Kitchin, C.R. (2008). *Astrophysical Techniques*, 5th edn. Boca Raton, FL: CRC Press.

Oluseyi, H., Karchner, A., Kolbe, W.F. *et al.* (2004). Characterization and deployment of large-format fully depleted back-illuminated p-channel CCDs for precision astronomy. *SPIE*, **5570**, 515–524.

Rieke, G.H. (2003). *Detection of Light from the Ultraviolet to the Submillimeter*, 2nd edn. Cambridge: Cambridge University Press.

Chapter 4

Calabretta, M.R. and Greisen, E.W. (2002). Representations of celestial coordinates in FITS. *A&A*, **395**, 1077–1122.

Figer, D.F., Rauscher, B.J., Regan, M.W. *et al.* (2003). Independent detector testing laboratory and the NGST detector characterization project. *SPIE*, **4850**, 981–1000.

Fruchter, A.S. and Hook, R.N. (2002). Drizzle: a method for the linear reconstruction of undersampled images. *PASP*, **114**, 144–152.

Greisen, E.W. and Calabretta, M.R. (2002). Representations of world coordinates in FITS. *A&A*, **395**, 1061–1075.

Greisen, E.W., Calabretta, M.R. Valdes, F.G., and Allen, S.L. (2006). Representations of spectral coordinates in FITS. *A&A*, **446**, 747–771.

Hodapp, K.W., Jensen, J.B., Irwin, E.M. *et al.* (2003). The Gemini Near-Infrared Imager (NIRI). *PASP*, **115**, 1388–1406.

Hook, R.N. and Fruchter, A.S. (2000). Dithering, sampling and image reconstruction. In ASP Conf. Ser., Vol. **216**, *Astronomical Data Analysis Software and Systems IX*, ed. N. Manset, C. Veillet, D. Crabtree. San Francisco, CA: PASP, pp. 521–530.

Kattunen, H., Kroger, P., Oja, H., Poutanen, M., and Donner, K.J. (2007). *Fundamental Astronomy*. 5th edn. Berlin, New York: Springer.

Lang, K.R. (2006). *Astrophysical Formulae*, 3rd edn. Berlin, New York: Springer.

Lindegren, L. (2005). The Astrometric Instrument of GAIA: Principles. In *Proceedings of the GAIA Symposium, 'The Three-Dimensional Universe with GAIA,'* ed. C. Turon, K.S. O'Flaherty, and M.A.C. Perryman. ESA SP-576.

Pier, J.R., Munn, J.A., Hindsley, R.B. *et al.* (2003). Astrometric calibration of the Sloan Digital Sky Survey. *AJ*, **125**, 1559–1579.

Perryman, M.A.C., de Boer, K.S., Gilmore, G. *et al.* (2001). GAIA: Composition, formation and evolution of the Galaxy. *A&A*, **369**, 339–363.

Chapter 5

Absil, O., Di Folco, E., Mérand, A. *et al.* (2006). Circumstellar material in the Vega inner system revealed by CHARA/FLUOR. *A&A*, **452**, 237–244.

Alard, C. (2000). Image subtraction using a space-varying kernel. *A&AS*, **14**, 363.

Alard, C. and Lupton, R.H. (1998). A method for optimal image subtraction. *ApJ*, **503**, 325–331.

Aufdenberg, J.P., Mérand, A., Coudé du Foresto, V. *et al.* (2006). First results from the CHARA Array. VII. Long-baseline interferometric measurements of Vega consistent with a pole-on, rapidly rotating star. *ApJ*, **645**, 664–675.

Baars, J.W.M., Genzel, R., Pauliny-Toth, I.I.K., and Witzel, A. (1977). The absolute spectrum of CAS A – an accurate flux density scale and a set of secondary calibrators. *A&A*, **61**, 99–106.

Bautz, M. Pivovaroff, M.J., Kissel, S.E. *et al.* (2000). Absolute calibration of ACIS X-ray CCDs using calculable undispersed synchrotron radiation. *SPIE*, **4012**, 53–67.

Bertin, E. and Arnouts, S. (1996). SExtractor: Software for source extraction. *A&AS*, **117**, 393–404.

Bessell, M.S. (2005). Standard photometric systems. *ARAA*, **43**, 293–336.

Brown, T.M., Charbonneau, D., Gilliland, R.L., Noyes, R.W., and Burrows, A. (2001). Hubble Space Telescope time-series photometry of the transiting planet of HD 209458. *ApJ*, **552**, 699–709.

Carpenter, J.M. (2001). Color transformations for the 2MASS second incremental data release. *AJ*, **121**, 2851–2871.

Elias, J.H., Frogel, J.A., Matthews, K., and Neugebauer, G. (1982). Infrared standard stars. *AJ*, **87**, 1029–1034.

Hillenbrand, L.A., Foster, J.B., Persson, S.E., and Matthews, K. (2002). The Y band at 1.035 microns: Photometric calibration and the Dwarf Stellar/Substellar Color Sequence. *PASP*, **114**, 708–720.

Kidger, M.R. and Martin-Luis, F. (2003). High-precision near-infrared photometry of a large sample of bright stars visible from the northern hemisphere. *AJ*, **125**, 3311–3333.

Lord, S.D. (1992). A new software tool for computing Earth's atmospheric transmission of near- and far-infrared radiation. NASA Technical Memorandum **103957**.

Mason, B.S., Leitch, E.M., Myers, S.T., Cartwright, J.K., and Readhead, A.C.S. (1999). An absolute flux density measurement of the supernova remnant Cassiopeia A at 32 GHz. *AJ*, **118**, 2908–2918.

Mateo, M. and Schechter, P.L. (1989). The DoPHOT two-dimensional photometry program. ESO/ST-ECF Data Analysis Workshop, 69–83.

Mellish, Bob (n.d.) for Wikipedia, http://en.wikipedia.org/wiki/Half_wave_plate

McLean, I.S. (2008). *Electronic Imaging in Astronomy*, 2nd edn. Berlin, New York: Springer.

Peng, C.Y., Ho, L.C., Impey, C.D., and Rix, H.-W. (2002). Detailed structural decomposition of galaxy images. *AJ*, **124**, 266–293.

Peng, C.Y., Ho, L.C., Impey, C.D., and Rix, H.-W. (2010). Detailed decomposition of galaxy images. II. Beyond axisymmetric models. *AJ*, **139**, 2097–2129.

Price, S.D., Paxson, C., Engelke, C., and Murdock, T.L. (2004). Spectral irradiance calibration in the infrared. XV. Absolute calibration of standard stars by experiments on the Midcourse Space Experiment. *AJ*, **128**, 889–910.

Rieke, G.H., Blaylock, M., Decin, L. *et al.* (2008). Absolute physical calibration in the infrared. *AJ*, **125**, 2245–2263.

Serkowski, K. (1974). Polarization techniques. In *Methods of Experimental Physics*, Vol. **12**, Part A, *Astrophysics*, ed. N.P. Carleton. New York: Academic Press, pp. 361–414.

Stetson, P.B. (1987). DAOPHOT – A computer program for crowded-field stellar photometry. *PASP*, **99**, 191–222.

Stetson, P.B. (1990). On the growth-curve method for calibrating stellar photometry with CCDs. *PASP*, **102**, 932–948.

Wardle, J.F.C. and Kronberg, P.P. (1974). The linear polarization of quasi-stellar radio sources at 3.71 and 11.1 centimeters. *ApJ*, **194**, 249.

Weiland, J.L., Odegard, N., Hill, R.S. *et al.* (2011). Seven-year Wilkinson Microwave Anisotropy Probe (WMAP) observations: Planets and celestial calibration sources. *ApJS*, **192**, 19–40.

Young, A.T. (1974a). Photomultipliers, their cause and cure. In *Methods of Experimental Physics*, Vol. **12**, Part A, *Astrophysics*, ed. N.P. Carleton. New York: Academic Press, pp. 1–94.

Young, A.T. (1974b). Other components in photometric systems. In *Methods of Experimental Physics*, Vol. **12**, Part A, *Astrophysics*, ed. N.P. Carleton. New York: Academic Press, pp. 95–122.

Young, A.T. (1974c). Observational technique and data reduction. In *Methods of Experimental Physics*, Vol. **12**, Part A, *Astrophysics*, ed. N.P. Carleton. New York: Academic Press, pp. 123–192.

Chapter 6

Fellgett, P. (1971). The origins and logic of multiplex Fourier and interferometric methods in spectrometry. In *Aspen International Conference on Fourier Spectroscopy*, ed. G.A. Vanasse, A.T. Stair, Jr., and D.J. Baker. Optical Physics Laboratory, Air Force Cambridge Research Laboratories, U.S. Air Force, pp. 139–141.

Ramsey, L.W. (1988). Focal ratio degradation in optical fibers of astronomical interest. In *Fiber Optics in Astronomy*, ed. S.C. Barden. San Francisco, CA: PASP, pp. 26–39.

Saptari, V. (2003). *Fourier Transform Spectroscopy Instrumentation Engineering*. Billingham, WA: SPIE.

Scaduto, R. (n.d.) rscaduto@fluorescence.com

Treffers, R.R. (1977). Signal-to-noise ratio in Fourier spectroscopy. *App. Opt.*, **16**, 3103–3106.

Volume phase gratings: http://www.astro.ljmu.ac.uk/~ikb/vph.html

Wells, M., Hastings, P.R., and Ramsay-Howat, S.K. (2000). Design and testing of a cryogenic image slicing IFU for UKIRT and NGST. *SPIE*, **4008**, 1215–1226.

Chapter 7

Aime, C. (2005). Radon approach to shaped and apodized apertures for imaging exoplanets. *A&A*, **434**, 785–794.

Bello, D., Conan, J.-M., Rousset, G. *et al.* (2003a). Numerical versus optical layer oriented: a comparison in terms of SNR. *SPIE*, **4839**, 612–622.

Bello. D., Vérinaud, C., Conan, J.-M. *et al.* (2003b). Comparison of different 3D wavefront sensing and reconstruction techniques for MCAO. *SPIE*, **4839**, 554–565.

Bracewell, R. N. (1956). Strip Integration in Radio Astronomy. *Aus. J. Phys.*, **9**, 198–217.

Codona, J.L. and Angel, R. (2004). Imaging extrasolar planets by stellar halo suppression in separately corrected color bands. *ApJL*, **604**, 117–120.

Cornelissen, S.A., Hartzell, A.L., Stewart, J.B., Bifano, T.G., and Bierden, P.A. (2010). MEMS deformable mirrors for astronomical adaptive optics. *SPIE*, **7736**, 80.

Cox A. (2000). *Allen's Astrophysical Quantities*, 4th edn. New York: AIP Press, Springer.

Dravins, D., Lindegren, L., Mezey, E., and Young, A.T. (1997). Atmospheric intensity scintillation of stars, I. Statistical distributions and temporal properties. *PASP*, **109**, 173–207.

Egner, S.E. (2006). Multiconjugate adaptive optics for LINC-NIRVANA. Ph.D. thesis, Ruperto-Carola University of Heidelberg.

Guyon, O. (2003). Phase-induced amplitude apodization of telescope pupils for extrasolar terrestrial planet imaging. *A&A*, **404**, 379–387.

Hardy, J.W. (1998). *Adaptive Optics for Astronomical Telescopes*. Oxford, New York: Oxford University Press.

Kenworthy, M., Quanz, S., Meyer, M., *et al.* (2010). New mode for VLT's NACO to image exoplanets: http://www.eso.org/public/announcements/ann1037/.

Kolmogorov, A.N. (1941). The local structure of turbulence in incompressible viscous fluid for very large Reynolds' numbers. *Dokl. Acad. Nauk., SSSR*, **30**, 301–305.

Lloyd-Hart, M. (2000). Thermal performance enhancement of adaptive optics by use of a deformable secondary mirror. *PASP*, **112**, 264–272.

Martin, F., Conan, R., Tokovinin, A. *et al.* (2000). Optical parameters relevant for High Angular Resolution at Paranal from GSM instrument and surface layer contribution. *A&AS*, **144**, 39–44.

Narayan, R. and Nityanandan, R. (1986). Maximum entropy image restoration in astronomy. *ARAA*, **24**, 127–170.

Noll, R. (1976). Zernike polynomials and atmospheric turbulence. *JOSA*, **66**, 207–211.

Puetter, R.C., Gosnell, T.R., and Yahil, A. (2005). Digital image reconstruction: deblurring and denoising. *ARAA*, **43**, 139–194.

Quanz, S.P., Meyer, M.R., Kenworthy, M. *et al.* (2010). First results from Very Large Telescope NACO apodizing phase plate: $4\,\mu m$ images of the exoplanet β Pictoris b. *ApJL*, **722**, 49–53.

Ragazzoni, R. (1996). Pupil plane wavefront sensing with an oscillating prism. *J. Mod. Opt.*, **43**, 289–293.

Roddier, F. (1988). Curvature sensing and compensation: a new concept in adaptive optics. *Appl.Opt.*, **27**, 1223–1225.

Salinari, P. and Sandler, D.G. (1998). High-order adaptive secondary mirrors: where are we? *SPIE*, **3353**, 742–753.

Shepp, L.A. and Vardi, Y. (1982). Maximum likelihood reconstruction in positron emission tomography. *IEEE Trans. Medical Imaging*, **1**(2), 113 122.

Slepian, D. (1965). Analytic solution of two apodization problems. *JOSA*, **55**, 1110–1115.

Soummer, R., Ferrari, A., Aime, C., and Jolissaint, L. (2007). Speckle noise and dynamic range in coronagraphic images. *ApJ*, **669**, 642–656.

Stapelfeldt, K.R. (2006). Exoplanets and star formation. In *The Scientific Requirements for Extremely Large Telescopes*, IAU Symposium **232**, ed. P.A. Whitelock, M. Dennefeld, and B. Leibundgut. Cambridge, New York: Cambridge University Press, pp. 149–158.

Tatarski, V.I. (1961). *Wavefront Propagation in a Turbulent Medium*. Mineola, NY: Dover.

Tubbs, R.N. (n.d.) in Wikipedia, http://en.wikipedia.org/wiki/Astronomical_seeing

Tyson, R.K. (2000). *Introduction to Adaptive Optics*. Bellingham, WA: SPIE Press.

Vanderbei, R.J. (2004). Shaped pupils: http://www.orfe.princeton.edu/~rvdb/talks/review.pdf

Vanderbie, R.J., Karsdin, N.J., Spergel, D.N., and Kuchner, M. (2003). New pupil masks for high-contrast imaging. *SPIE*, **5170**, 49–56.

Yang, W. and Kostinski, A.B. (2004). One-sided achromatic phase apodization for imaging of extrasolar planets. *ApJ*, **605**, 892–901.

Chapter 8

Benford, D.J., Allen, C.A., Chervenak, J.A., *et al.* (2000). Superconducting bolometer arrays for far-infrared and submillimeter astronomy. *ASP Conf. Series*, **217**, 134–139.

Billot N., Agnèse P, Auguères J-L, *et al.* (2006). Recent achievements on the development of the HERSCHEL/PACS bolometer arrays. *Nuc. Inst. Methods (A)*, **567**, 137–139.

Craig, S.C., McGregor, H.M., Atad-Ettedgui, E. *et al.* (2010). SCUBA-2: engineering and commissioning challenges of the world's largest sub-mm instrument at the JCMT. *SPIE*, **7741**, 43.

Groppi, C., Walker, C., Kulesa, C. *et al.* (2008). SuperCam: a 64 pixel heterodyne imaging spectrometer. *SPIE*, **7020**, 26.

Holland, W.S., MacIntosh, M., Fairley, A. *et al.* (2006). SCUBA-2: a 10,000-pixel submillimeter camera for the James Clerk Maxwell Telescope. *SPIE*, **6275**, 45.

Mather, J.C. (1982). Bolometer noise: nonequilibrium theory. *Appl. Opt.*, **21**, 1125–1129.

Mazin, B.A. (2009). Microwave kinetic inductance detectors: The first decade. In *AIP Conf. Proc.*, **1185**, 135–142.

Oosterloo, T., Verheijen, M., and van Cappellen, W. (2010). The latest on Apertif, in ISKAF2010 Science Meeting: http://pos.sissa.it/cgi-bin/reader/conf.cgi?confid = 112

Purcell, E.M. (1985). *Electricity and Magnetism*, 2nd edn. New York: McGraw-Hill.

Rieke, G.H. (2003). *Detection of Light from the Ultraviolet to the Submillimeter*, 2nd edn. Cambridge: Cambridge University Press.

Sayers, J., Golwala, S.R., Ade, P.A. *et al.* (2008). Studies of atmospheric noise on Mauna Kea at 143 GHz with Bolocam. *SPIE*, **7020**, 43.

Van der Ziel, A. (1976). *Noise in Measurements*. New York: Wiley.

Walton A.J., Parkes W., Terry J.G., *et al.* (2004). Design and fabrication of the detector technology for SCUBA-2. *IEE. Proc. – Sci. Meas. Technol.*, **151**, 110–120.

Wilson, T.L., Rohlfs, K., and Hüttemeister, S. (2009). *Tools of Radio Astronomy*, 5th edn. Berlin, New York: Springer.

Chapter 9

Bryan, R.K. and Skilling, J. (1980). Deconvolution by maximum entropy, as illustrated by application to the jet of M87. *MNRAS*, **191**, 69.

Condon, J.J., and Ransom, S.M. (2010). Essential radio astronomy: http://www.cv.nrao.edu/course/astr534/ERA.shtml

Cornwell, T. (2004). SKA and EVLA computing costs for wide field imaging. *SKA Memo 49*: http://www.skatelescope.org/publications/

Fomalont, E. (2006). Error recognition and image analysis. Tenth Synthesis Imaging Summer Workshop: http://www.phys.unm.edu/~kdyer/2006/lectures/

Högbom, J.A. (1974). Aperture synthesis with a non-regular distribution of interferometer baselines. *A&AS*, **15**, 417–426.

Mioduszewski, A. (2010). Array configuration. In *Synthesis Imaging in Radio Astronomy*, Hartebeesthoek Radio Observatory: http://192.96.5.2/synthesis_school/Miod_Array_Design.pdf

Tabatabaei, F.S., Krause, M., and Beck, R. (2007). High-resolution radio continuum survey of M 33. I. The radio maps. *A&A*, **472**, 785–796.

Thompson, A.R., Moran, J.M., and Swenson, G.W. (2001). *Interferometry and Synthesis in Radio Astronomy*, ed. A.R. Thompson *et al.*, 2nd edn. New York: Wiley.

Viallefond, F., Goss, W.M., van der Hulst, J.M., and Crane, P.C. (1986). HII regions in M33. II – Radio continuum survey. *A&AS*, **64**, 237–246.

Chapter 10

Ables, J.G. (1968). Fourier transform photography: a new method for X-ray astronomy. *Proc. Astron. Soc. Australia*, **1**, 172–173.

Atwood, W.B., Abdo, A.A., Ackermann, M. *et al.* (2009). The Large Area Telescope on the Fermi Gamma-Ray Space Telescope Mission. *ApJ*, **697**, 1071–1102.

Canizares, C.R., Davis, J.E., Dewey, D. *et al.* (2005). The Chandra High-Energy Transmission Grating: Design, fabrication, ground calibration, and 5 years in flight. *PASP*, **117**, 1144–1171.

Caroli, E., Stephen, J.B., Di Cocco, G., Natalucci, L., and Spizzichino, A. (1987). Coded aperture imaging in X- and gamma-ray astronomy. *SSRv*, **45**, 349–403.

Den Herder, J.W., Brinkman, A.C., Kahn, S.M. *et al.* (2001). The Reflection Grating Spectrometer on board XMM-Newton. *A&A*, **365**, 7.

Dicke, R.H. (1968). Scatter-hole cameras for X-rays and gamma rays. *ApJL*, **153**, L101–106.

Elmegreen, B.G. and Scalo, J. (2004). Interstellar turbulence I: Observations and processes. *ARAA*, **42**, 211–273.

Fenimore, E.E. and Cannon, T.M. (1978). Coded aperture imaging with uniformly redundant arrays. *App. Opt.*, **17**, 337–347.

Goldwurm, A., David, P., Foschini, L. *et al.* (2003). The INTEGRAL/IBIS scientific data analysis. *A&A*, **411**, L223–L229.

Gunson, J. and Polychronopulos, B. (1976). Optimum design of a coded mask X-ray telescope for rocket applications. *MNRAS*, **177**, 485–497.

Hammersley, A.P. (1986). The reconstruction of coded-mask data under conditions realistic to X-ray astronomy observations. Ph.D. thesis, University of Birmingham.

Hartmann, R., Buttler, W., Gorke, H., *et al.* (2006). A high-speed pnCCD detector system for optical applications. *Nuc. Inst. & Methods* (*A*), **568**, 118–123.

In't Zand, J. (1996) (with updates). Coded aperture camera imaging concept: http://www.sron.nl/~jeanz/cai/coded_intr.html

Kelley, R.L., Mitsuda, K., Allen, C.A. *et al.* (2007). The Suzaku High Resolution X-Ray Spectrometer. *PASJ*, **59S**, 77–112.

Kim, Y., Rieke, G.H., Krause, O., Misselt, K., Indebetouw, R., and Johnson, K.E. (2008). Structure of the interstellar medium around Cas A. *ApJ*, **678**, 287–296.

Kilbourne, C.A., Doriese, W.B., Bandler, S.P. *et al.* (2008). Multiplexed readout of uniform arrays of TES x-ray microcalorimeters suitable for Constellation-X. *SPIE*, **7011**, 4.

McCammon, D., Almy, R., Apodaca, E. *et al.* (2002). A High Spectral Resolution Observation of the Soft X-Ray Diffuse Background with Thermal Detectors. *ApJ*, **576**, 188–203.

Miville-Deschênes, M.-A., Joncas, G., Falgarone, E., and Boulanger, F. (2003). High resolution 21 cm mapping of the Ursa Major Galactic cirrus: Power spectra of the high-latitude H I gas. *A&A*, **411**, 109–121.

Morselli, A., Barbiellini, G., Budini, G. *et al.* (2000). The space gamma-ray observatory AGILE. *Nuc. Phys. B. Proc. Suppl.*, **85**, 22–27.

Paerels, F.B.S., and Kahn, S.M. (2003). High-resolution X-ray spectroscopy with CHANDRA and XMM-NEWTON. *ARAA*, **41**, 291–342.

Porter, F.S., Adams, J.S., Brown, G.V. *et al.* (2010). The detector subsystem for the SXS instrument on the ASTRO-H Observatory. *SPIE*, **7732**, 112.

Rana, V.R., Cook, W.R., Harrision, F.A., Mao, P.H., and Miyasaka, J. (2009). Development of focal plane detectors for the Nuclear Spectroscopic Telescope Array (NuSTAR) mission. *SPIE*, **7435**, 2.

Rybicki, G.B. and Lightman, A.P. (1979). *Radiative Processes in Astrophysics*. New York: Wiley.

Stahle, C.K., Audley, M.D., Boyce, K.R. *et al.* (1999). Design and performance of the ASTRO-E/XRS microcalorimeter array and anticoincidence detector. *SPIE*, **3765**, 128–136.

Strüder, L. and Meidinger, N. (2008). CCD detectors. In *The Universe in X-rays*, ed. J.E. Trümper and G. Hasinger. Berlin, New York: Springer, pp. 51–72.

Tsujimoto, M., Guaninazzi, M., Plucinsky, P.P. *et al.* (2011). Cross-calibration of the X-ray instruments onboard the Chandra, INTEGRAL, RXTE, Suzaku, Swift, and XMM-Newton observatories using G21.5–0.9. *A&A*, **525**, 25–39.

Willingale, R., Sims, M.R., and Turner, M.J.L. (1984). Advanced deconvolution techniques for coded aperture imaging. *Nuc. Inst. Methods (A)*, **221**, 60–66.

Chapter 11

Ahmad, Q.R., Allen, R.C., Andersen, T.C. *et al.* (2001). Measurement of the rate of $ve + d \rightarrow p + p + e^-$ interactions produced by ^8B solar neutrinos at the Sudbury Neutrino Observatory. *Phys. Rev. Lett.*, **87**, 071301.

Arkhipov, A.A. (2006). The GZK puzzle and fundamental dynamics. Eprint arXiv: hep-ph/0607265.

Bionta, R.M., Blewitt, G., Bratton, C.B., Casper, D., and Ciocio, A. (1987). Observation of a neutrino burst in coincidence with supernova 1987A in the Large Magellanic Cloud. *Phys. Rev. Lett.*, **58**, 1494–1496.

Bradt, H. (2004). *Astronomy Methods*. Cambridge: Cambridge University Press.

Davis, R. Jr., Mann, A.K., and Wolfenstein, L. (1989). Solar neutrinos. *Ann. Rev. Nuc. Part. Sci.*, **39**, 467–506.

Halzen, F. and Klein, S.R. (2010). IceCube: An instrument for neutrino astronomy. *Rev. Sci. Inst.*, **81**, 081101.

Hirata, K., Kajita, T., Koshiba, M. *et al.* (1987). Observation of a neutrino burst from the supernova SN1987A. *Phys. Rev. Lett.*, **58**, 1490–1493.

Jarrell, R. (2005). "Radio astronomy, whatever that may be." The marginalization of early radio astronomy. In *The New Astronomy*, ed. W. Orchiston. Astrophysics and Space Science Library, **334**. Dordrecht: Springer, pp. 191–202.

Klepser, S., Kislat, F., Kolanoski, H. *et al.* (2008). Lateral distribution of air shower signals and initial energy spectrum above 1 PeV from IceTop. In *Proceeedings of the 30th International Cosmic Ray Conference*, Vol. **4**, ed. R. Caballero *et al.* Mexico City: Universidad Nacional Autónoma de México, pp. 35–38.

Low, F.J., Rieke, G.H., and Gehrz, R.D. (2007). The beginning of modern infrared astronomy. *ARAA*, **45**, 43–75.

Pierre Auger Collaboration (2004). Properties and performance of the prototype instrument for the Pierre Auger Observatory. *Nucl. Inst. & Methods* (*A*), **523**, 50–95.

Pierre Auger Collaboration (2009). The fluorescence detector of the Pierre Auger Observatory. *Nucl. Inst. & Methods* (*A*), **620**, 227–251.

Stone, E.C., Cohen, C.M.S., Cook, W.R. *et al.* (1998). The Cosmic-Ray Isotope Spectrometer for the Advanced Composition Explorer. *SSRv.*, **86**, 285–356.

Index

Printed in the United States
By Bookmasters